Hazards, Risks, and Disasters in Society

Hazards and Disasters Series

Hazards, Risks, and Disasters in Society

Volume Editors

Andrew E. Collins
Department of Geography/Disaster and Development Network (DDN), Northumbria University, Newcastle, UK

Samantha Jones
Department of Geography/Disaster and Development Network (DDN), Northumbria University, Newcastle, UK

Bernard Manyena
Humanitarian and Conflict Response Institute (HCRI), Manchester University, Manchester UK and Disaster and Development Network (DDN), Newcastle, UK

Janaka Jayawickrama
Post-war Reconstruction and Development Unit (PRDU), University of York, York, UK and Disaster and Development Network (DDN), Newcastle, UK

Series Editor

John F. Shroder
Emeritus Professor of Geography and Geology, Department of Geography and Geology, University of Nebraska at Omaha, Omaha, NE 68182

ELSEVIER

AMSTERDAM • BOSTON • HEIDELBERG • LONDON • NEW YORK • OXFORD
PARIS • SAN DIEGO • SAN FRANCISCO • SINGAPORE • SYDNEY • TOKYO

Elsevier
Radarweg 29, PO Box 211, 1000 AE Amsterdam, Netherlands
The Boulevard, Langford Lane, Kidlington, Oxford OX5 1GB, UK
225 Wyman Street, Waltham, MA 02451, USA

Notices
Knowledge and best practice in this field are constantly changing. As new research and experience broaden our understanding, changes in research methods, professional practices, or medical treatment may become necessary.

Practitioners and researchers must always rely on their own experience and knowledge in evaluating and using any information, methods, compounds, or experiments described herein. In using such information or methods they should be mindful of their own safety and the safety of others, including parties for whom they have a professional responsibility.

To the fullest extent of the law, neither the Publisher nor the authors, contributors, or editors, assume any liability for any injury and/or damage to persons or property as a matter of products liability, negligence or otherwise, or from any use or operation of any methods, products, instructions, or ideas contained in the material herein.

Library of Congress Cataloging-in-Publication Data
Application submitted

British Library Cataloguing in Publication Data
A catalogue record for this book is available from the British Library

ISBN: 978-0-12-396451-9

Contents

Contributors

Supriya Akerkar, Centre for Development and Emergency Practice (CENDEP), Oxford Brookes University, Oxford, UK

David Alexander, Institute for Risk and Disaster Reduction (IRDR), University College London, UK

Tomohide Atsumi, Graduate School of Human Sciences, Osaka University, Osaka, Japan

Per Becker, Centre for Societal Resilience and LUCRAM, Lund University, Stockholm, Sweden

Sarah Bradshaw, Department of Law and Politics, Middlesex University, London, UK

Andrew E. Collins, Department of Geography/Disaster and Development Network (DDN), Northumbria University, Newcastle, UK

John Devavaram, Resource Centre for Participatory Development Studies, Madurai, India

Michael Eburn, ANU College of Law, The Australian National University, Canberra, ACT, Australia

Ross Edgeworth, Disaster and Development Network (DDN), Northumbria University, Newcastle, UK

Maureen Fordham, Department of Geography/Gender and Disaster Network (GDN), Northumbria University, Newcastle, UK

Mohsen Ghafory-Ashtiany, International Institute of Earthquake Engineering and Seismology (IIEES), Tehran, Islamic Republic of Iran

Susanna M. Hoffman, Hoffman Consulting, Telluride, Colorado, USA

Maitland Hyslop, Disaster and Development Network (DDN), Northumbria University, Newcastle, UK

Peter Jackson, University of Leicester, Leicester, UK

Janaka Jayawickrama, Post-war Reconstruction and Development Unit, University of York, York, UK; Disaster and Development Network (DDN), Northumbria University, Newcastle, UK

Samantha Jones, Department of Geography/Disaster and Development Network (DDN), Northumbria University, Newcastle, UK

Elisabeth King, Department of Humanities and Social Sciences, New York University, New York, USA

Julie Koppel Maldonado, American University, Washington DC, USA

Fuad H. Mallick, Postgraduate Programs in Disaster Management (PPDM), BRAC University, Dhaka, Bangladesh

Bernard Manyena, Humanitarian and Conflict Response Institute, Manchester University, Manchester, UK; Disaster and Development Network (DDN), Northumbria University, Newcastle, UK

Anthony Masys, Centre for Security Science (CSS), Defence R&D Canada, Ottawa, ON, Canada

M. Shahjahan Mondal, Institute of Water and Flood Management (IWFM), Bangladesh University of Engineering and Technology (BUET), Dhaka, Bangladesh

Ryo Morimoto, Brandeis University, Waltham, MA, USA; International Research Institute of Disaster Science at Tohoku University, Sendai, Japan; Institute of Comparative Culture, Sophia University, Tokyo, Japan

John C. Mutter, Department of Earth and Environmental Sciences and School of International and Public Affairs, Columbia University, New York, USA

Geoff O'Brien, Department of Geography, Northumbria University, Newcastle, UK

Phil O'Keefe, Department of Geography, Northumbria University, Newcastle, UK

Douglas Paton, School of Psychology, University of Tasmania, Launceston, TAS, Australia

Brenda D. Phillips, Ohio University–Chillicothe, OH, USA

M. Aminur Rahman, Postgraduate Programs in Disaster Management (PPDM), BRAC University, Dhaka, Bangladesh

Mohammad Rezaur Rahman, Institute of Water and Flood Management (IWFM), Bangladesh University of Engineering and Technology (BUET), Dhaka, Bangladesh

Nibedita S. Ray-Bennett, Civil Safety and Security Unit, University of Leicester, Leicester, UK

Hideyuki Shiroshita, Kansai University, Japan

Kees van der Geest, Institute for Environment and Human Security, United Nations University, Bonn, Germany

Dewald van Niekerk, African Centre for Disaster Studies, North-West University, Potchefstroom, South Africa

Sara Walsh, Department of Geography/Disaster and Development Network (DDN), Northumbria University, Newcastle, UK

Koko Warner, Institute for Environment and Human Security, United Nations University, Bonn, Germany

Editorial Foreword

General Hazards, Risks, and Disasters: Hazards are processes that produce danger to human life and infrastructure. Risks are the potential or possibilities that something bad will happen because of the hazards. Disasters are the quite unpleasant result of the hazard occurrence that caused destruction of lives and infrastructure. Hazards, risks, and disasters have been coming under increasing strong scientific scrutiny in recent decades as a result of a combination of numerous unfortunate factors, many of which are quite out of control as a result of human actions. At the top of the list of exacerbating factors to any hazard, of course, is the tragic exponential population growth that is clearly not possible to maintain indefinitely on a finite Earth. As our planet is covered ever more with humans, any natural or human-caused (unnatural?) hazardous process is increasingly likely to adversely impact life and construction systems. The volumes on hazards, risks, and disasters that we present here are thus an attempt to increase understandings about how to best deal with these problems, even while we all recognize the inherent difficulties of even slowing down the rates of such processes as other compounding situations spiral on out of control, such as exploding population growth and rampant environmental degradation.

Some natural hazardous processes such as volcanos and earthquakes that emanate from deep within the Earth's interior are in no way affected by human actions, but a number of others are closely related to factors affected or controlled by humanity, even if however unwitting. Chief amongst these, of course, are climate-controlling factors, and no small measure of these can be exacerbated by the now-obvious ongoing climate change at hand (Hay, 2013). Pervasive range fires and forest fires caused by human-enhanced or-induced droughts and fuel loadings, mega-flooding into sprawling urban complexes on floodplains and coastal cities, biological threats from locust plagues and other ecological disasters gone awry, all of these and many others are but a small part of the potentials for catastrophic risk that loom at many different scales, from the local to planet girdling.

In fact, the denial of possible planet-wide catastrophic risk (Rees, 2013) as exaggerated jeremiads in media landscapes saturated with sensational science stories and end-of-the-world Hollywood productions is perhaps quite understandable, even if simplistically short sighted. The "end-of-days" tropes promoted by the shaggy-minded prophets of doom have been with us for centuries, mainly because of Biblical verse written in the early Iron Age during remarkably pacific times of only limited environmental change. Nowadays however, the Armageddon enthusiasts appear to want the worst to validate their death desires and prove their holy books. Unfortunately we are all entering times when just a few individuals could actually trigger societal breakdown by error or terror, if Mother Nature does not do it for us first. Thus we enter contemporaneous times of considerable peril that the present needs for close attention.

These volumes we address here about hazards, risks, and disasters are not exhaustive dissertations about all the dangerous possibilities faced by the ever-burgeoning human populations, but they do address the more common natural perils that people face, even while we leave aside (for now) the thinking about higher level existential threats from such things as bio- or cybertechnologies, artificial intelligence gone awry, ecological collapse, or runaway climate catastrophes.

In contemplating existential risk (Rossbacher, 2013), we have lately come to realize that the new existentialist philosophy is no longer the old sense of disorientation or confusion at the apparently meaninglessness or hopelessly absurd worlds of the past, but instead an increasing realization that serious changes by humans appear to be afoot that even threaten all life on the planet (Kolbert, 2014; Newitz, 2013). In the geological times of the Late Cretaceous an asteroid collision with Earth wiped out the dinosaurs and much other life; at the present time by contrast, humanity itself appears to be the asteroid.

Misanthropic viewpoints aside, however, an increased understanding of all levels and types of the more common natural hazards would seem a useful endeavor to enhance knowledge accessibility, even while we attempt to figure out how to extract ourselves and other life from the perils produced by the strong climate change that is so obviously underway. Our intent in these volumes is to show the latest good thinking about the more common endogenetic and exogenetic processes and their roles as threats to everyday human existence. In this fashion, the chapter authors and volume editors have undertaken to show you overviews and more focused assessments of many of the chief obvious threats at hand that have been repeatedly shown on screen and print media in recent years. As this century develops, we may come to wish that these examples of hazards, risks, and disasters are not somehow eclipsed by truly existential threats of a more pervasive nature. The future always hangs in the balance of opposing forces; the ever-lurking, but mindless threats from an implacable nature, or heedless bureaucracies countered only sometimes in small ways by the clumsy and often feeble attempts by individual humans to improve our little lots in life. Only through improved education and understanding will any of us have a chance against such strong odds; perhaps these volumes will add some small measure of assistance in this regard.

Social and Cultural Aspects of Hazards, Risks, and Disasters: The chapters presented in this volume represent some new and eclectic thinking that follows a path along a number of philosophical, social, cultural trends that may enlighten or irritate various predispositions; but whatever intellectual journeys one may make with them in pursuit of knowledge, the route can be fruitful. The breadth of topics is impressive and is divided into three parts: from viewpoints on natural hazards in society, to those concerning hazards of technological, social, and economic change, to cross-disciplinary dealings with disasters. Being better able to understand how to take in the frightful death tolls in corrupt societies when shoddy construction allowed through corruption that results in the deaths of so many young children in their schools does get one's attention. That and varied relief measures viewed through the eyes of different religions are equally enlightening. But those viewpoints are just two among many others in this volume; a polemic on capitalism and chapters on climate change and humanitarianism are food for thought. Also included are various other chapters on such things as rights approaches in disasters, perspectives on gender in risks and disasters, simple human dignity, vulnerability and coping behaviors, relations to violent conflict, cyber-security hazards, cultural aspects, and political responses. A great many other aspects of hazards, risks, and disasters in society are presented in this volume as well.

The diversity of these many chapters present expositions on a variety of social topics that are rarely expressed in many treatments of natural hazards. The resulting volume that Editor Andrew Collins has sought to put forward will allow some readers to think about issues that are quite new to them, with the result that some good would have been done in this world of increasing complexity and many forms of danger that we must all face. Perhaps some useful changes in behavior will occur, to the benefit of people who would not otherwise be so rewarded.

John (Jack) Shroder
Editor-in-Chief
July 14, 2014

REFERENCES

Hay, W.W., 2013. Experimenting on a Small Planet: A Scholarly Entertainment. Springer-Verlag, Berlin, 983 pp.

Kolbert, E., 2014. The Sixth Extinction: An Unnatural History. Henry Holt & Company, NY, 319 pp.

Newitz, A., 2013. Scatter, Adapt, and Remember. Doubleday, NY, 305 pp.

Rees, M., 2013. Denial of catastrophic risks. Science 339 (6124), 1123.

Rossbacher, L.A., October 2013. Contemplating existential risk. Earth, Geologic Column 58 (10), 64.

Chapter 1

Introduction: Hazards, Risks, and Disasters in Society

Andrew E. Collins[1], Bernard Manyena[2,4], Janaka Jayawickrama[3,4] and Samantha Jones[1]

[1] *Department of Geography/Disaster and Development Network (DDN), Northumbria University, Newcastle, UK,* [2] *Humanitarian and Conflict Response Institute, Manchester University, Manchester, UK,* [3] *Post-war Reconstruction and Development Unit, University of York, York, UK,* [4] *Disaster and Development Network (DDN), Northumbria University, Newcastle, UK*

ABSTRACT

This introductory chapter outlines why it is important to explore in more depth the relationships between environmental hazards, risks, and disasters in society. It presents an introduction to the challenges presented by mainstream approaches to the human side of disaster studies, whereby perspectives on environmental hazards and human development meet policy and practice. This is informed by analyzing the influences of extreme environmental events on society, exposure factors, and the nature of emergent systems of response. In this field, people are considered as vulnerable and resilient to disaster impacts, suffering, or prospering in times of climate change, development, societal instability, and governance scenarios that can be unpredictable and out of control. This is in part balanced by hope in the emergence of new-found awareness and capacity, to be able to live with hazards and risks, cope with disaster, and prosper socially and economically. A challenge presented by hazards, risks, and disasters is to achieve the capacity to both anticipate the unexpected and act on the known. A wealth of well-grounded emergent knowledge and experience exists to facilitate this, some of the most enlightening and innovative of which is revealed in the selection of contributions to this volume.

1.1 OPENING

Hazards, Risks, and Disasters in Society presents an exploration of how people interrelate with environmental changes and shocks that are variously within or beyond their ability to alter. It includes accounts based on disaster prevention and response approaches with reference to threats that have become increasingly more prevalent. Risk of a disaster at individual or community level is dependent on exposure to emergent and resurgent hazards and the capacity to avoid, adapt, absorb, or control them. Furthermore, despite a rich debate as to

Hazards, Risks, and Disasters in Society. http://dx.doi.org/10.1016/B978-0-12-396451-9.00001-9

what a disaster is (Quarentelli, 1998; Perry and Quarentelli, 2005), definition and experience remain relative to the varied interpretations of heterogeneous people. Consequently, the collection of contributions herein provides critical comment from necessarily cross-disciplinary discussions about how best to deal with hazards, risks, and disasters in society through the societal perspective. We continue by outlining some relatively straightforward principles of hazards, risks, and disasters in society, followed by an overview of the chapter contributions that comprise the volume. We then return to overall emergent points in a concluding chapter.

Hazards, risks, and disasters in society include environmentally related catastrophes within concentrations of human development that can be interpreted in terms of historical, political, and economic contexts. As such, one rationale underlying this volume is that development largely determines the way in which hazards impact on people, whereas disasters alter the scope of development. A summative overview of the more recent state of this perspective is already outlined in some detail in *Disaster and Development* (Collins, 2009), among other sources. The collection of chapter contributions in *Hazards, Risks, and Disasters in Society* reflects how personal and corporate exposure factors, short-term reactions, and longer term responses mediate the manner in which people get understood as vulnerable, resilient, or otherwise. Societies that strengthen themselves or are strengthened mitigate decline and resultant further exposure to what are largely human-induced cycles of environmental, social, and economic change. In a simple delineation, this change may be experienced as improvement by billions of people in economically advantaged societies who become more able to protect against environmental hazards, but as deterioration for billions of other people who are exposed to increased risks. The delineation between those more or less at risk of disaster has been the focus of a long tradition of studies of disaster vulnerability (O'Keefe et al., 1976; Blaikie et al., 1994; Cannon, 1994; Lewis, 1999; Wisner et al., 2004; Bankhoff et al., 2004; Gaillard, 2010; Lewis, 2014). Moreover, in addition to the above, multiple origins occur in the emergence of vulnerability studies through specific hazard and risk categories. For example, it has been central to studies of health risk reduction to view human susceptibility as interacting with socioeconomic vulnerability in relation to both disease hazard and context (Doyal, 1987; Honari and Boleyn, 1999; Collins, 1993, 1996, 1998, 2001, 2003).

Although the rationale of the vulnerability approaches to environmental hazards has become more mainstream discourse in recent years, it remains evident that exposure to risk and disaster remains far from being addressed in practice. An implicit argument is that it is not inevitable that major disasters will occur so much as it is possible for governments, communities, individuals, and industries to choose to affect change toward safer and more resilient societies. For a background to the use of resilience discourse and conceptualization in disaster studies, see, for example, Pelling (2003), IFRC (2004), UNISDR (2005), Paton and Johnston (2006), Manyena (2006), and

Sudmeier-Rieux (2014). A concern for the field of vulnerability analysis and disaster risk is that it is notoriously subject to cultural leanings and critique, whereby derivative notions of resilience become worryingly minimalist, overly accepting of crises, and are not much help for those vulnerable or extremely poor. It is therefore not surprising that following a decade of progress in the Strategy to Build the Resilience of Nations and Communities to Disasters (UNISDR, 2005), the aspiration remains to achieve awareness and capacity for what lies beyond resilience. This volume aims to not repeat the well-trodden ground of vulnerability and resilience studies in relation to hazard and risk, but instead compiles chapter contributions that tend to bring to the fore new detail and innovative ideas, also exposing some of the associated future challenges and opportunities.

Although such a volume is necessarily inductive and eclectic, an introductory and indicative analytical framework for the volume can nonetheless be represented through reference to individual environmental hazards. For example, here we refer to the case of flood risk that can be considered a function of (1) environmental change, (2) people's exposure, and (3) prevention and response systems. These are core drivers of flood risk analysis that lie behind identifying improved flood risk management. Some of the consequent analytical and practical challenges in approaching integrated flood risk management would therefore be as outlined in Table 1.1.

The analytical challenges presented in Table 1.1 suggest inherent complexity to flood risk management, requiring individual analyses at the level

TABLE 1.1 Analytical and Practical Challenges in Integrated Flood Risk Management

Domain of Flood Risk	Analytical Challenge
Environmental change	Predictability/uncertainty, opportunity for precautionary actions, "natural" versus built approaches—hard and soft catchment, river and coastal management, methods of long-term maintenance
People's exposure to flood risk	Perception, socioeconomic enablement, information, communication, expectation, risk culture, age, gender, and other forms of social differentiation
Prevention and response systems	Political will, market forces, capacity, connectedness with proximate and underlying causes, learning cycles, adaptive capacity, role management, centralization/decentralization, hard/soft catchment management (upstream/downstream), maintenance responsibility, insurance

of any one flood event, the weighting of importance of varying components a function of its nature and context. Therefore, rather than emphasizing development of flood risk models with limited applicability to varying environments, societies, and systems of development, progress would be through improving ongoing monitoring, evaluation, and learning for prevention and response. This informs what really is appropriate to addressing complex risks specific in time and relative to knowledge about particular places that are defined by intersecting environmental, social, and economic processes. The approach is key in both the applied and theoretical sense. It can assist in informing smart solutions that balance environmental, social, and economic policy drivers. Though partly idealistic, the aspiration of integrated analyses and action lies at the core of improving flood risk management. It has not yet become "main stream" but is aspirational of what it means to more fully engage hazards risks and disasters in society. The principles here can be extended to other categories of environmental hazards while also being considered as relevant to a multihazard framework.

1.2 CRITICAL PROCESSES AND OUTCOMES OF HAZARDS, RISKS, AND DISASTERS IN SOCIETY

By hazard categories we therefore emphasize critical processes and outcomes that significantly disrupt human well-being over both brief and long time frames. Although hazards, risks, and disasters impact society, individuals, groups, institutions, and organizations offset the effects by becoming strong, organized, healthy, resilient, caring, and creative. Alternatively, political processes and societies become corrupt, inept, and dangerous, exacerbating the impacts of environmental changes on people who are forced to become more vulnerable. Corruption in particular is increasingly recognized as a cause of disaster (Leoni et al., 2011). The situation is dynamic such that disruptive innovations can arise from social organization that is challenged during times of crisis, as well as during times of relative calm. The process of learning an innovation in disaster management is part of the 'development' in *Disaster and Development Studies*.[1] A role of developing prevention and response activities is to "get development out of disaster," otherwise expressed by the United Nations in the following:

1. The editors of this volume have contributed many years to the Disaster Management and Sustainable Development Program at Northumbria University, United Kingdom, the first globally to dedicate itself since 2000 to integrating disaster and development studies. While some understanding of disaster in development and vice versa is as old as development studies itself, there had been a hitherto remarkable lack of research, practice, and policy invested into this fundamental relationship. The approach has since become an aspect of many programs around the world and increasingly is considered central to United Nations strategies.

A disaster with all its negative consequences offers a good opportunity to formulate forward-looking policy concepts pertaining to social development and equity, economic growth, environmental quality and justice, i.e. sustainability.

Living with Risk, UNISDR, 2002 p.21

This aspect of disaster and development studies resembles disaster resilience as forward moving through processes that build back better, go upward, forward, and bring change (i.e., Monday, 2002; O'Brien et al., 2010; Manyena et al., 2011; UNICEF, 2012; Aldunce et al., 2014; Sudmeier-Rieux (2014), provide adaptive capacity (Folke et al., 2002), offset risk by shifts from vulnerability to well-being (Collins, 2009), and so on. This is key, though this concept needs to be perpetually balanced in realism presented by dual emphases of both development-induced disaster and conversely, disasters that prevent development. There is ultimately a need to know how one person's development opportunity can be safe from becoming someone else's disaster threat (Collins, 2009, p. 262).

Pertinent evidence of the challenges ahead are the hundreds of millions of people around the world exposed to impoverishment by development activities, living with polluted environments upon which they depend or who are entrapped by poverty into more risk-exposed locations. The problems of development as disastrous are vast and beyond the scope of this volume, but in summary are to do with "over," "under," "uneven," "sustainable," "appropriate," and more personal aspects of change that define human well-being, being both forward and backward moving. Here we focus more specifically on how predominantly environmental hazards impact societal exposure to disaster through an array of changing risks, and conversely how response strategies variously involving societies may interpret and interact with the hazards and risks that are created.

This volume therefore includes much of use to practitioners and policy makers. Notably, as people better engage prevalent hazards and risks they exercise a process that has become known as disaster risk reduction (DRR), which is frequently referenced throughout the volume. In a context of climatic risks this is also indicative of climate change adaptation (CCA), a further recurrent topic. Ultimately, both DRR and CCA represent quests for development of sustainable environmental and societal futures. Throughout the book, case studies provide the more detailed context and interpretations of hazards risks and disasters in society that shed light on what actions can be built upon and applied within the next generation of DRR, CCA, and development policies.

For hazards, it is important to note that in the context of this volume, no claim is made to "naturalness" as in "natural hazard," instead addressing people's interaction with pertinent threats. By "society" it is acknowledged that a day-to-day positioning and interaction exists of people one with another through a variety of lived interpretations of homogenous and heterogeneous social categories. Variance within society in its simplest demographic form includes differentiation on account of age, gender, wealth, culture, education, skills, rights, location, and health, to name some.

Across the range of contributions, it is accepted that multiple approaches exist to the quantification and qualification of "risk." Although "disaster discourses" are acknowledged as involving oft-referred-to subjective conjectures, the context here is where the likelihoods and impacts of disruptive events are real, wide ranging, and large scale. Away from the challenge of definition, it is known that people around the world are exposed to hazards by force, coercion, or choice, being more or less vulnerable or resilient to the effects of environmental hazards on those terms. For example, mass migrations of people around the Sahel in the face of famine have often not been due to drought and famine directly but to the political gerrymandering of totalitarian governance (Ohta and Gebre, 2007). Meanwhile, millions forced to move out of the way of industrial level developments, such as dam projects, for biofuel development, or through urban redesign did not invite such developments. Many of those displaced people then became impoverished, being also more exposed to environmental hazard. This is charted in detail, from dam projects in nearly all continents, to the impact of coastal industries (such as in South Asia), removals to make way for farms (such as in Brazil), displacements from mine pollution, and power supply accidents (such as in areas Eastern Europe and Russia) to name some.

However, coercion can involve a more subtle process of adjustment whereby human hopes and aspirations are driven by mobility for a better life with greater security, the so-called pull factors in migration studies. In relation to environmental hazards, migration in effect becomes a form of adaptation (Black et al., 2011a). However, for those too poor, infirm, or controlled by political boundaries, migration ceases to be an option—people become entrapped in hazardous locations (Black et al., 2011b; Foresight: Migration and Global Environmental Change, 2011; Collins, 2013a). Although many cannot move out of the way of hazard and risk, recent figures from the United Nations High Commission for Refugees UNHCR) show that the number of refugees and displaced people in the world has exceeded 50 million in 2014, the first time since the Second World War (UNHCR, 2014).

Hazards can be considered environmental, economic, social, psychological, or otherwise. However, consistent with this series as a whole, this volume remains more significantly oriented by driving ecological and climatic concerns of our times. It nonetheless reflects Earth-related hazards and risks that generate disasters or catastrophes within society dependent on historical, political, and economic contexts. Moreover, culture mediates how people may interpret, become more vulnerable or resilient to, hazard exposure. The cycles of environmental, social, and economic degradation to which we have referred are mediated by cultures of practice. The latter includes forms of land management, beliefs about environmental quality, spirituality, and ultimately the maintenance of ecosystem services (Renaud et al., 2013). Of growing importance across the myriad of interpretations of environmental value and sustainability, however, is the observable outcome that societies that strengthen themselves, or are strengthened, mitigate decline and resultant further exposure to these human-induced circumstances.

The volume avoids solely pessimistic accounts of humanity, also showing how a myriad of innovations arise from social organization in times of crisis. Beyond the surge of academic-, policy-, and practice-related definitions of resilience, a wealth of practical actions have been going on around the world aiming to strengthen communities in the face of both disaster risk and climate change. These arguably move beyond resilience. Actions representative of locality- and community-based resilience are regularly brought to the attention of the United Nations through its series of Global Platforms for an International Strategy for Disaster Reduction (UNISDR various, though see United Nations Chief Executive Board for Coordination, 2013). A drive has also occurred to draw together the hitherto disparate policy communities of DRR that underpin the Hyogo Framework for Action of UNISDR (2005) with the findings and aims of the Intergovernmental Panel on Climate Change (IPCC, 2007, 2014), and through the SREX report of 2012 (IPCC, 2012). While increasingly reflective of the demand to address both DRR and CCA together, these have at times been remarkably separate arenas of institutional structure within national and local governments, and also at the United Nations itself.

Some clear examples of documented approaches to locality and community resilience to disaster reduction that are in turn synonymous with actions for protecting against the impacts of climate change, can be read about in the annual *World Disasters Report* of the International Federation of Red Cross and Red Crescent Societies. For example, Chapter 2 of *World Disasters Report* (IFRC, 2009) details a series of projects facilitated by the Disaster and Development Network (DDN) of Northumbria University demonstrating how working alongside local authorities and civil societal partners can be achieved, including the case of establishing (and subsequent analysis of) risk and resilience committees in South Asia and Southern Africa.

Many interesting cases exist of reaction to hazards, risks, and disasters in society. For example, the rapid reactions of the United Kingdom local government and civil society, or longer term, planned responses to flooding, can be compared and contrasted with other experiences around the world. It is notable through this example that at the core of emergency planning and longer term prevention strategies, a commonly identified tension exists between the rights and responsibilities that occur within human agency and institutional structures. The former is dependent on varying levels of community organization, but is made complex by culture. For example, experience from our projects in Nepal showed how community organization with the involvement of municipal authorities allowed for a more effective locality-based risk governance than for self-administered community groups (Jones et al., 2013). However, it is reasonable to argue that the nature and context of hazard, risk, and governance in Nepal is different to the UK—indeed that most countries, subregions, communities, and individuals are different one from the other when considering reactions to risks. It is also pertinent to note examples more associated with specific nation-states.

For the case of the United Kingdom, in the follow-up from the Pitt Review (2008) for flooding, out of 97 or so recommendations, only three had referenced community.[2] The vast bulk of the post-Pitt consultation centered on infrastructure and engineered solutions to flood risk. Although this is in itself a critical infrastructure aspect of protecting people, experience from around the world demonstrates it to be woefully insufficient for dealing with the totality of disaster risk that is by necessity society-centered. This is further emphasized by shifting to greater integration of singularly defined disaster risk categories (such as flood, hurricane, disease epidemic) to multihazard approaches (Cutter et al., 2000; DFID, 2012). It also emphasizes the need to be attentive to slower onset hazards and multiple risk brought about by persistent vulnerability in society, accentuated for the more marginalized and economically compromised members of society (Collins, 2009; Ray-Bennett, 2009; Akerkar and Devavaram, 2014 in this volume; Maldonado, 2014 in this volume; Bradshaw and Fordham, 2014 in this volume).

Regarding what can be done moving forward, it is pertinent to gather a balance between the burgeoning wealth of conceptualizing that surrounds the intellectual debate on resilience to disaster risk and climate change. This would be done by considering the ways in which locality and community are evidenced as advancing this agenda through more sustainable development trajectories (Collins, 2013b). Although this can be analyzed from within the context of individual countries and societies, it is pertinent to explore what has been found in economically advanced states (i.e., a wealth of studies look at lessons about community resilience to Hurricane Katrina, USA) and those classified as low income. In Bangladesh, one of our collaborative research initiatives found location-specific details of improved health and well-being at community level brought a critical meaning to disaster resilience (ESRC, 2006, 2010; Ray Bennett et al., 2010; Nahar et al., 2013). This supports the need for appropriate forms of self-care (Edgeworth and Collins, 2006; Edgeworth, 2014 in this volume) in dealing with a complex of vulnerability issues before, during, and after climatic-related crises in that region (Alam and Collins, 2010). Further work examines the manner in which risk reduction or climate adaptation discourses are operational at multiple levels of governance, and work better as internally or externally driven processes (Mohammad and Collins, 2013).

2. "Citizens and Flooding," Presentation and Panel Discussion by Collins, A.E., for Centre for Public Policy Westminster Seminar on How do we Prevent the Likelihood of Future Flooding: Taking Forward the Select Committee and Pitt Reports—Achieving Proper Infrastructure for the 21st Century, Royal Commonwealth Society, London 14th July, 2008.

1.3 COMPONENTS OF THIS BOOK

The volume responds to three key aspects of hazards, risks, and disasters in society through the following core ingredients.

1. Prevention of and response to hazards, risks, and disasters in society—this provides analysis and case studies indicative of current action on hazards, risks, and disasters in society.
2. Hazards in social, technological, and political-economic change—this details some of the case-driven priorities relevant to current and future challenges of hazards, risks, and disasters in society.
3. Cross-disciplinary and nonmainstream dealing with hazards, risks, and disasters in society—this engages forward-looking theoretical, policy, or practice developments for addressing hazards, risks, and disasters in society.

Overall, the contributing chapters provide detail and synthesis on ways of addressing the nature and context of hazard, risks, and disaster reduction in society as part of this wider book series for which major environmental hazards (such as earthquake, drought, and flood) are addressed in great detail across the eight other volumes. This ninth volume presents societal prevention and responses to hazards, risks, and disasters extending also to insights on the human interrelationships with hazards of technological, social, and economic change that are often not addressed in the context of environmental hazards work. Contributions to the volume demonstrate innovation that can occur through cross-disciplinary working in this field, informing about the processes of human endeavors that can help deal with disaster.

1.4 SUMMARY

It is possible to arrange the 20 chapters that follow in many different ways to reflect the main title of this volume. Each chapter can be read individually as accounts and ideas independent of those presented in the other chapters. However, the contributions were solicited in particular response to three aspects of the subject, namely, prevention and response, contemporary hazards of change, and varied cross-disciplinary perspectives for dealing with disasters in society. Three book sections therefore comprise groups of chapters sourced along these lines, albeit each chapter also contributes in part to all sections. The approach sought is not to consolidate an existing status quo but to seek out new (or more in-depth) perspectives. Chapters are on the whole backed up by selective case-study material that reinforce particular perspectives on hazards, risks, and society demanded by this field that inevitably need to be put under further scrutiny with future research.

Section One introduces pertinent perspectives on people centered prevention and response to natural hazards through a series of six chapters reflecting challenges to current intellectual, policy, and practice. The chapter by Atsumi provides a rich example from Japan of how people volunteering to be involved in disaster reduction avoid institutionalization of such a role, the manner of involvement being reflective of societal values and relationships one to the other. The chapter by O'Keefe et al. delivers what its authors present as a polemical and radical challenge to the field brought about by a political economic perspective on disasters of market forces, unmet climate change responses, and Western humanitarianism. Both chapters present a counterbalance to institutionalization in recognizing that disaster response comes first and foremost from the community affected by disaster. The issue of how disaster risk governance evolves is tackled by Jones et al. who find three dimensions exist to the changing distribution of influences and responsibilities—upward, outward, and downward. The latter represents decentralization characteristic of the more people-centered, preventative approach flagged variously in this volume. However, lack of resourcing for the overburdened societal level reflects tensions that exist in gaining the means to effective DRR investment.

A number of points made so far are then illustrated by the regionally orientated chapter of Becker and van Niekerk, who present their Southern Africa case study by way of flagging the manner in which capacity for disaster risk reduction can be progressed. The point is made that progress in building skilled capacity for disaster risk reduction is already evident through regionally based initiatives, but that these are only sustainable if driven from within the region. Understanding of the rights-based approach, beyond framings of dominant groups, is provided in a detailed account by Akerkar and Devavaram based on the case of marginalized poor women of Tamil Nadu, India, who were widowed in the Tsunami of 2003. The chapter shows how genuinely rights-based approaches demand listening, encouraging agency, attention to dignity, and long-term commitments in disaster recovery. This section of the book then completes with a call from Ray-Bennett et al. for "reflective response" to counter hyperrisks and develop organizational resilience. This is indicated to be the challenge presented by naturally triggered technical (NATECH) disasters, an imperative being that disaster risk reduction needs to operationalize for an interconnected world involving communities of practices and practitioners.

Section Two can be considered to either extend or illustrate some of the opening themes by providing analysis and case material around key areas of current and future hazards in social, technological, and political-economic change. First, the chapter by van der Geest and Warner provides quantifiable analysis of variable vulnerability, coping, and loss and damage in relation to climate events across a selection of countries. Use of the multidimensional vulnerability index and variety of coping measures of varyingly vulnerable households indicate that, even among less vulnerable households of developing-world countries, a majority are unlikely to be able to cope with loss and damage impacts of climate-related events. For the case of Bangladesh,

Rahman et al. reveals the example of coping going on through use of varying forms of flood shelter. However, the needs and aspirations of people living across these floodplains are shown to vary, requiring multiple versions of this infrastructural response sensitive to the nature of problems that people face during flood crises.

Taking the recently intensified case of cyber security hazards, the chapter by Hyslop then considers the contrasting environments and difficulties created in the world of cyberspace. This highly relevant chapter to contemporary crisis management impinges on life through military, organizational, critical infrastructure, and criminal and moral domains, reluctantly leading the author to conclude that some form of regulation is required to protect generally unresilient societies from cyber security hazard. In their chapter, King and Mutter show how the wider threats evident through interrelated fields of natural disasters and violent conflicts create common consequences in society, and that this is therefore most likely an artificial divide. Bridging the divide between disaster and conflict studies is particularly relevant in prevention whereby both peacebuilding and DRR can save lives and improve human security. A further core case is addressed in the chapter by Maldonado who, focusing on the tribal communities of coastal Louisiana, shows how in the face of transformational changes, society may engage everyday forms of resistance as part of the adaptation process. More equitable distribution of knowledge sharing and integration, public participation, local input, and science with multiple perspectives are part of this process of creating greater justice in times of global environmental change.

The following two chapters that complete the second section look at two primary drivers of thematic investigations into the human side of hazard, risk, and disaster in society, being political response (Alexander) and disaster through the lens of gender (Bradshaw and Fordham). Alexander provides definitions of politics in relation to DRR and resilience. Having considered political organization of society in the context of architecture and effectiveness of emergency management systems it is concluded that the ethical framework within which politics are conducted has a strong bearing on the ability to manage disaster risk. This has been exemplified by lack of concern for women's issues and also underpins the rising interest in governance and participation. Bradshaw and Fordham then provide detail of the impact of disasters on women and girls, particularly in the developing world. This highlights how disasters are gendered events whereby women and girls experience them differently from men. The exclusion of gender from much policy in disaster risk reduction, and concerns about forms of women's inclusion, leads the authors to highlight gendered risk as demanding reconceptualization of "disaster," and disaster as a development issue.

Section Three provides chapters on topics currently considered to be nonmainstream aspects of dealing with hazards, risks, and disasters in society. Though aspects of this quest may be found in the other sections of the volume as well, these chapters represent some of the less regularly cited cross-disciplinary analyses that may progress ways in which theory, policy, and practice touched upon in this book as a whole could further develop. Firstly, the chapter by Eburn advances how disaster risk reduction must advance in the "shadow of the law" whereby responsibility must be negotiated with stakeholders. In the interests of a safer community, Eburn, who is also a barrister (lawyer), identifies some of the legal boundaries that help define the limits of shared responsibility, albeit here based on a sample of nations with a shared legal history.

The chapter by Edgeworth focuses on the case of local-level resilience and risk reduction in relation to infectious diseases in Bangladesh demonstrating the role there can be for self-care. Though it cannot be a panacea for all forms of disaster reduction, it highlights how people are not passive observers of events that unfold around them. Adjustment to environmental hazards is here exemplified as being through self-care continuity of local strategies utilized in the home. The chapter by Hoffman, from an anthropological perspective, focuses more specifically on how culture matters as the critical factor in hazard, risk, and disaster recovery, to understand society and working with human behavior. Paton, in his chapter then contributes insights into the ingredients for conceptualizing and operationalizing readiness planning. This unveils some of the societal processes critical to the development of resilience and adaptive capacity within DRR strategies. Although indicated as needing of further work, the approach could progress at individual, family, community, and societal levels providing options for change within an all-hazards perspective. The chapter by Morimoto then returns in part to one of the mainstay tools for hazard and risk assessment, namely, the use of mapping. However, here it is used to demonstrate forward-looking engagement with the complex interpretational processes by which the combined earthquake, tsunami, and nuclear disaster in Japan in 2010 could be viewed. As such we discover explicitly here the imperative for disaster researchers to move forward aware of potential fallacies of using singular material frameworks in interpreting disaster risk.

In a chapter with some resonance to Chapter Two on disaster volunteerism, Phillips extends the discussion around healing that can be experienced post-disaster, citing findings and ideas about the role of therapeutic communities. Though remaining with many questions and the need for further work on this front, it is suggested that, if it were better understood, practitioners might be able to generate and better engage therapeutic communities of the future. The chapter by Ghafory-Ashtiany reawakens a religious perspective, the majority of the world's population residing within individual or societally organized faith and practice. For the case of Abrahamic religions, this can bring an opportunity for an effective risk communication and education approach to DRR, not compromising science, though creating a revived and reformed religious incentive to engage risk reduction.

REFERENCES

Foresight: Migration and Global Environmental Change, 2011 Foresight: Migration and Global Environmental Change., 2011. Final Project Report. Government Office for Science, London.
Alam, E., Collins, A.E., 2010. Cyclone disaster vulnerability and response experiences in coastal Bangladesh. Disasters 34 (4), 931–953.
Aldunce, P., Beilin, R., Handmer, J., Howden, M., 2014. Framing disaster resilience the implications of the diverse conceptualisations of "bouncing back". Disast. Prev. Manag. 23 (3), 252–270.
Akerkar, S., Devavaram, J., 2014. Understanding rights based approach in disasters: a case for affirming human dignity. In: Collins, A.E., et al. (Eds.), Hazards, Risks and Disasters in Society. Elsevier.
Bankhoff, G., Frerks, G., Hilhorst, D. (Eds.), 2004. Mapping Vulnerability: Disasters, Development and People. Earthscan, London.
Black, R., Bennett, S.R.G., Thomas, S.M., Beddington, J.R., 2011a. Climate change: migration as adaptation. Nature 478, 447–449.
Black, R., Adger, W.N., Arnell, N.W., Dercon, S., Geddes, A., Thomas, D., 2011b. The effect of environmental change on human migration. Global Environ. Chang. 21s, 3–11.
Blaikie, P., Cannon, T., Davis, I., Wisner, B., 1994. At Risk: Natural Hazards, People's Vulnerability and Disasters, first ed. Routledge, London.
Bradshaw, S., Fordham, M., 2014. Double disaster: disaster through a gender lens. In: Collins, A.E., et al. (Eds.), Hazards, Risks and Disasters in Society. Elsevier.
Cannon, T., 1994. Vulnerability analysis and the explanation of 'natural' disasters. In: Varley, A. (Ed.), Disasters, Development and the Environment. John Wiley, Chichester, pp. 13–30.
Collins, A.E., 1993. Environmental influences on the distribution of incidence of cholera: a case study in Quelimane, Mozambique. Disasters 17 (4), 321–340.
Collins, A.E., 1996. The geography of cholera. In: Drasar, B.S., Forrest, B.D. (Eds.), Cholera and the Ecology of *Vibrio cholerae*. Chapman and Hall, London, pp. 255–294.
Collins, A.E., 1998. Environment, Health and Population Displacement: Development and Change in Mozambique's Diarrhoeal Disease Ecology. Making of Modern Africa Series. Ashgate, Aldershot.
Collins, A.E., 2001. Land degradation, health ecology and development. Land Degrad. Dev. 12 (3), 237–250.
Collins, A.E., 2003. Vulnerability to coastal cholera ecology. Soc. Sci. Med. 57, 1397–1407.
Collins, A.E., 2009. Disaster and Development. Routledge, London.
Collins, A.E., 2013a. Applications of disaster risk reduction to migration influenced by environmental change. J. Environ. Sci. Policy 27 (S1), S112–S125.
Collins, A.E., 2013b. Editorial - Linking disaster and development: further challenges and opportunities. Special Edition Environ. Hazards 12 (1), 1–4.
Cutter, S.L., Mitchel, J.T., Scott, M.S., 2000. Revealing the vulnerability of people and places: a case study of Georgetown County, South Carolina. Ann. Assoc. Am. Geogr. 90 (40), 713–737.
DFID, 2012. Department for International Development, Multi-Hazard Disaster Risk Assessment (V2). CHASE, DFID, London.
Doyal, L., 1987. The Political Economy of Health. Pluto Press, London.
Edgeworth, R., Collins, A.E., 2006. Self-care as a response to diarrhoea in rural Bangladesh: empowered choice or enforced adoption? Soc. Sci. Med. 63, 2686–2697.
Edgeworth, R., 2014. Self-care in Bangladesh: local level resilience and risk reduction. In: Collins, A.E., et al. (Eds.), Hazards, Risks and Disasters in Society. Elsevier.

ESRC., 2006 and 2010. Project Overview and Final Report documents of Collins, A.E. et al. - ESRC Research Grant RES-167-25-0241, The Meaning of Health Security for Disaster Resilience in Bangladesh: A Health Security Framework for Disaster Risk Reduction. Available at Economic and Social Research Council, UK. www.esrc.ac.uk.

Folke, C., Carpenter, S., Elmqvist, Gunderson, L., Holling, C.S., Walker, B., 2002. Resilience and sustainable development: building adaptive capacity in a world of transformations. Ambio 31 (5), 437–440.

Gaillard, J.C., 2010. Vulnerability, capacity and resilience: perspectives for climate and development policy. J. Int. Dev. 22 (2), 218–232.

Honari, M., Boleyn, T. (Eds.), 1999. Health Ecology: Health, Culture and Human-Environment Interaction. Routledge, London.

IFRC, 2004. Focus on Community Resilience. World Disaster Report 2004. IFRC, Geneva.

IFRC, Collins, A.E., 2009. The people centred approach to early warning systems and the 'Last Mile'. In: International Federation of the Red Cross and Red Crescent Societies (IFRC), Focus on Early Warning, Early Action, pp. 39–68. World Disaster Report, (Chapter 2).

IPCC, 2007. Climate Change 2007: The Physical Science Basis, Summary for Policymakers. Contribution of Working Group 1 to the Fourth Assessment Report of the Intergovernmental Panel on Climate Change. WMO and UNEP, Geneva (updated for 2014).

IPCC, 2012. Managing the Risks of Extreme Events and Disasters to Advance Climate Change Adaptation. A Special Report of Working Groups I and II of the Intergovernmental Panel on Climate Change. Cambridge University Press, Cambridge, UK, and New York, NY, USA.

IPCC, 2014. Climate Change: Impacts, Adaptation and Vulnerability. WMO and UNEP, Geneva.

Jones, S., Aryal, K., Collins, A.E., 2013. Local-level governance of risk and resilience in Nepal. Disasters 37 (3), 442–467.

Leoni, B., Radford, T., Schulman, M., 2011. Disaster through a Different Lens: Behind Every Effect, There Is a Cause. United Nations Office for Disaster Risk Reduction (UNISDR), Geneva.

Lewis, J., 1999. Development in Disaster-prone Places: Studies of Vulnerability. IT Publications (Practical Action), London.

Lewis, J., 2014. The susceptibility of the vulnerable: some realities reassessed. Disast. Prev. Manag. 23 (1), 2–11.

Maldonado, J.K., 2014. Everyday practices and symbolic forms of resistance: adapting to environmental change in coastal Louisiana. In: Collins, A.E., et al. (Eds.), Hazards, Risks and Disasters in Society. Elsevier.

Manyena, B., 2006. The concept of resilience revisited. Disasters 30 (4), pp. 433-350.

Manyena, S.B., O'Brien, G., O'Keefe, P., Rose, J., 2011. Disaster resilience: a bounce back or bounce forward ability? Local Environ. 16 (5), 417–424.

Mohammad, S.M., Collins, A.E., 2013. Disaster risk reduction in Bangladesh as an adaptation strategy for climate change. In: Proceedings of ANDROID Conference and Doctoral School, Limassol, Cyprus, 23–24th October, 2013, pp. 93–101.

Monday, J.L., 2002. Building back better: creating a sustainable community after disaster. Nat. Hazards Inf. 3, 1–11. Available at. www.colorado.edu/hazards/publications/informer/informer.html (Last visited: 19/06/2013).

Nahar, P., Collins, A.E., Bhuiya, A., Alamgir, F., Ray Bennett, N., Edgeworth, R., 2013. Indigenous indicators of health security in relation to climatic disasters in Bangladesh. Environ. Hazards 12 (1), 32–46. Special Edition - Linking Disaster and Development: further challenges and opportunities.

O'Brien, G., O'Keefe, P., Gadema, Z., Swords, J., 2010. Approaching disaster management through social learning. Disast. Prev. Manag. 19 (4), 498–508.

Ohta, I., Gebre, Y.S. (Eds.), 2007. Displacement Risks in Africa: Refugees, Resettlers and Their Host Population. Kyoto University Press.

O'Keefe, P., Westgate, K., Wisner, B., 1976. Taking the naturalness out of natural disasters. Nature vol. 260, 566–567.

Paton, D., Johnston, D. (Eds.), 2006. Disaster Resilience: An Integrated Approach. Charles Thomas Publishers, USA.

Pelling, M., 2003. The Vulnerability of Cities: Natural Disasters and Social Resilience. Earthscan, London.

Perry, R.W., Quarantelli, E.L., 2005. What is a Disaster? New Answers to Old Questions. International Research Committee on Disasters, USA.

Quarantelli, E.L., 1998. What is a Disaster? Perspectives on the Question. Routledge, London.

Ray-Bennett, N.S., 2009. The influence of caste, class and gender in surviving multiple disasters in Orissa, India. Environ. Hazards 8, 5–22.

Ray-Bennett, N., Collins, A.E., Bhuiya, A., Edgeworth, R., Nahar, P., Alamgir, F., 2010. Exploring the meaning of health security for disaster resilience through people's perspectives in Bangladesh. Health Place 16, 581–589.

Renaud, F.G., Sudmeier-Rieux, K., Estrella, E. (Eds.), 2013. The Role of Ecosystems in Disaster Risk Reduction. United Nations University Press.

Sudmeier-Rieux, K.I., 2014. Resilience – an emerging paradigm of danger or of hope? Disast. Prev. Manag. 23 (1), 67–80.

UNICEF, 2012. Build Back Better: A Twelve-month Update on UNICEF's Work to Rebuild Children's Lives and Restore Hope since the Tsunami. UNICEF, New York.

United Nations Chief Executive Board for Coordination, 2013. United Nations Plan of Action on Disaster Risk Reduction for Resilience. United Nations, New York.

UNHCR, 2014. United Nations High Commission for Refugees – Global Trends 2014. UNHCR, Geneva.

UNISDR, 2002. Living with Risk: A Global Review of Disaster Reduction Initiatives. UNISDR, Geneva.

UNISDR, 2005. Hyogo framework for action 2005-2015: building the resilience of nations and communities to disasters. In: World Conference on Disaster Reduction. United Nations International Strategy for Disaster Reduction, Geneva. http://www.unisdr.org/we/coordinate/hfa (Last visited: 19/06/2013).

Wisner, B., Blaikie, P., Cannon, T., Davis, I., 2004. At Risk: Natural Hazards, People's Vulnerability and Disasters, second ed. Routledge, London.

Section 1

Perspectives on People-Centred Prevention and Response to Natural Hazard

Chapter 2

Against the Drive for Institutionalization: Two Decades of Disaster Volunteers in Japan

Tomohide Atsumi
Graduate School of Human Sciences, Osaka University, Osaka, Japan

ABSTRACT
The present chapter describes two decades of disaster volunteers in Japan and draws out implications for society. According to the author's own long-term fieldwork and participation in a nonprofit organization, an ethnographic history of disaster volunteers is described. It is divided into four stages. The first stage is the dawn of disaster volunteers after the 1995 Great Hanshin-Awaji Earthquake. The powerful drive of institutionalization over the expansion of disaster volunteerism occurred at the second stage. In the third stage, an extension of activities occurred. Finally, at the fourth stage, currently a networking of volunteers is moving against the drive for institutionalization. This discussion is followed by an example of action research for keeping a disaster volunteers network embedded in Japanese society. The author attempted to motivate previous survivors to help survivors of the Tohoku region in East Japan in return for the support they had received in the past. It is concluded that the almost two decades of disaster volunteerism give us a glimpse of the future of Japanese society with the network of disaster volunteers, which would work against the drive for institutionalization, at least, intermittently.

2.1 INTRODUCTION

Volunteer activities in disaster fields are common on the global context. For instance, a number of volunteers of disaster nonprofit organizations (NPO) consisting of National Voluntary Organizations Active in Disaster have served various domestic disasters in the US, whereas many volunteers of nongovernmental organizations have been active in worldwide disasters. For another

Hazards, Risks, and Disasters in Society. http://dx.doi.org/10.1016/B978-0-12-396451-9.00002-0

instance, many religious groups (e.g., Christian charity) provide disaster assistance, whereas nonreligious organizations (e.g., Red Cross and Red Crescent) play a role of disaster relief in many countries.

Disaster volunteers, in general, can offer emotional support toward the survivors, which the state is not good at providing. The role of disaster volunteering in the neoliberal climate is getting more important than before because the lower the service delivered by the state, the more active the citizens are to be. Disaster volunteers in Japan are not exceptional. However, the history of Japanese disaster volunteers is unique and short.

The present chapter describes the two decades of disaster volunteers in Japan and draws out the implications for society. The 20-year history can be divided into four stages. The first stage is the dawn of volunteer activities at disaster sites, which did not emerge before the 1995 Great Hanshin-Awaji Earthquake. A powerful drive of institutionalization over the expansion of disaster volunteerism occurred at the second stage as well as over the extension of activities at the third stage. Finally, a movement of networking of volunteers against the drive for institutionalization is the current, fourth stage.

According to my own long-term fieldwork and participation in an NPO, the Nippon Volunteer Network Active in Disaster (NVNAD), I describe an ethnographic history of disaster volunteers. I then provide an example of action research for keeping the disaster volunteers network embedded in Japanese society. I assess the contributions by disaster volunteers to Japanese society over the almost two decades and those likely to be made in the future.

2.1.1 An Ethnographic Overview of the Two Decades of Disaster Volunteers in Japan

At 5:46 AM on Tuesday, January 17, 1995, the Great Hanshin-Awaji Earthquake (henceforth called the Kobe Earthquake) struck the regions of Hanshin and Awaji Island in Japan, claiming more than 6,400 lives and injuring more than 15,000 people in a densely populated urban coastal corridor whose central city is Kobe. In the impacted region, more than 300,000 survivors were displaced from their homes and faced the hardship of finding shelter, food, and water in cold winter weather. Since 1995, Japan has experienced several earthquakes and other disasters culminating in the Great Tohoku Earthquake and Tsunami of March 11, 2011, which claimed the lives of nearly 20,000 people and may prove to be the costliest natural disaster in world history.

Following the Kobe Earthquake, volunteers came from all parts of the country to assist survivors and their communities logging more than 1 million person-days of activity. Although volunteerism in general has a long history in Japan (Rausch, 1998) and a long tradition of community-based assistance exists via neighborhood associations (Taniguchi, 2010), the overwhelming number of convergent volunteers in disaster response had never before been experienced. Hence, the year 1995 was named the "renaissance of volunteerism" as well as "first year of the (disaster) volunteer" or "volunteer revolution" (Atsumi, 2001; Homma and Deguchi, 1996; Yatsuzuka, 1999).

2.1.2 First Stage: The Context for Volunteer Convergence in Japan

Beginning on the afternoon of January 17, 1995, volunteers and donated goods converged in the Nishinomiya City Hall and other locations in the area. Nishinomiya city administrators and officers like others in the earthquake-impacted region were struggling to cope with the damage to their homes and communities, and the ordinary functions of the city hall had come to a halt. No procedure or protocol existed for coordinating volunteers because disaster volunteers were not expected, but a hastily organized reception process was set up at city hall largely by volunteers. Volunteers, therefore, coordinated themselves according to their own daily experiences. Such a state of emergent struggle for order and situational definition prevailed in many places in the city, but survivors, neighbors, and volunteers helped each other in whatever ways they were able. A week later, these emergent groups and the city administrators realized that one of the major problems to avoid was the duplication of efforts between government and emergent volunteer groups. Hence, the dominant emergent group at the Nishinomiya City Hall assumed leadership among 13 other emergent groups and established the Nishinomiya Volunteer Network on February 1, 1995. A year later, it was reorganized and crystallized as a new organization; the NVNAD was established on the first anniversary of Kobe Earthquake.

Several disaster research scholars have pointed out that the Japanese disaster prevention system depends more on governmental agencies compared to systems in the United States (Britton, 2006; Suzuki et al., 2003; Tierney and Goltz, 1997). Prior to the 1995 Kobe Earthquake, people in Japan had generally recognized national and local governmental agencies as the main response agents. In fact, government-centered disaster management policies and practices appear to have achieved a dramatic decrease in the number of victims of natural disaster. This dramatic decrease was sufficient for most Japanese people to trust the efficiency of governmental management of disaster. As a result of the dominant role of the government, citizen participation in disaster response was quite low. Accordingly, few studies of disaster volunteers have been made in Japan, focusing on their activities during the response stage of a major emergency (Suzuki et al., 2003).

2.1.3 Second Stage: Expansion, Institutionalization, and Evolving Mission

During the 5-year period between 1995 and 2000, the NVNAD leadership and volunteers gained important experience in responding to domestic as well as international disasters of various kinds. They gained legitimacy through national legislation that recognized the contributions of volunteers and their organizations and incorporated them into official disaster management operations. In addition, this was a time in which volunteer organizations recognized the need for coordinated action among themselves and in which the first efforts to address coordination were made. It was also a time in which the NVNAD recognized that institutionalization brought with it a tendency to adopt bureaucratic forms of organization in which the affective needs of survivors were frequently overlooked. For instance, according to a manual for disaster volunteer, more attention is paid on how to organize and operate a disaster volunteer center rather than on how to focus on each and every survivor.

Consecutive disasters during this period, including floods, earthquakes, and volcanic eruptions, provided opportunities for volunteers to gain valuable experience in response to natural hazards. Human-induced disasters occurred as well. On January 2, 1997, more than 3,700 kl of heavy oil was spilled from a Russian tanker that ran aground in the Sea of Japan. The largest number of disaster volunteers since the Kobe Earthquake participated in the response to this accident. The NVNAD was quickly activated and sent a staff member to one of the affected towns. At its office in Nishinomiya, it played the role of an information clearinghouse on the spill and coordinated numerous spontaneous volunteers as well as donated goods. It also helped the affected area to open a reception center for volunteers. In other activations during this period, the NVNAD sent its members to domestic disasters including major floods in 1998, an earthquake in Tottori, and a volcanic eruption in Hokkaido in 2000 and to participate in international earthquake response in Turkey and Taiwan in 1999.

As for the broader social context during this time period, first the national government and subsequently local governments began to recognize the significance of disaster volunteers and revised their master plans to include volunteers in disaster management plans and procedures. The broader society also recognized the significance of volunteers not only for disaster response but also for social welfare, environmental issues, and other critical social issues. As a result, the Act on Promotion of Specified Non-profit Activities was enacted in 1998, including disaster volunteers as part of its mandate. The NVNAD became incorporated under this law as the first NPO in Hyogo Prefecture in 1999.

Cultures within organizations of disaster volunteers in Japan were evolving and differentiating from one another. For some, the primary objective was to develop procedures to allocate volunteers to disaster-related tasks in an efficient and coordinated manner to provide victims and survivors with the basic necessities of shelter, food, water, sanitation, and other material needs. Other organizations became concerned that the affective and emotive needs of survivors were of paramount importance and oriented their organizations to meeting these needs. This latter organizational culture took hold in the NVNAD.

2.1.4 Third Stage: Extension to Recovery from Two Earthquakes in Niigata

Consistent with this emerging orientation toward the affective and emotional support for disaster survivors, the NVNAD also recognized the importance of postdisaster revitalization of communities. The Niigata Chuetsu Earthquake occurred at 5:56 PM on October 23, 2004, a few months before the 10th anniversary of the Kobe Earthquake. It was not only a major earthquake with more than 60 fatalities, but also a new phenomenon to those volunteers familiar with disaster in a metropolitan area, as the earthquake destroyed many small villages scattered throughout the mountains. These villages had been facing depopulation and aging populations before the earthquake. Most of the young had left for jobs in urban areas and mainly the elderly remained cultivating small terraced rice fields and maintaining traditional culture.

Hence, the disaster response efforts here were different from those in Kobe. In fact, disaster NPOs tried to apply the lessons from Kobe (e.g., how to settle in shelters), but the major attention soon shifted from the short-term response period to the long-term recovery of each village because each mountainous village had experienced depopulation and aging population before the earthquake in culturally and traditionally unique context from one village by another. Although recovery as conceptualized in Japan connotes physical reconstruction, the volunteer organizations, including NVNAD prefer to call this phase "revitalization," which emphasizes cultural and social, as well as physical, recovery.

The NVNAD dispatched members to Niigata Prefecture the day after the earthquake and has continued to be involved in its revitalization process. The NVNAD opened its own local office in cooperation with other NPOs in the nationwide network called J-Net. Some of the student volunteers began to visit the mountainous villages and joined in local events and traditional festivals. Through these activities, the student volunteers learned about societal issues of depopulation, modernization, industrialization, and aging demographic structure, as well as traditional Japanese mountain cultures and human relationships. These activities were part of the new orientation toward revitalization adopted by the NVNAD and contributed to long-term support of the survivors.

Niigata Prefecture was once again hit by an earthquake on July 16, 2007, an M6.8 event that caused 10 deaths, 1,339 injuries, and $5 billion in damage. Experienced volunteers and leaders of disaster NPOs met in Kariwa Village at the local headquarters of the Council of Social Welfare to teach effective methods of disaster response management to local volunteers. These local volunteers were expected to manage the volunteer center, though they themselves were victims of the disaster and little provision had been made for addressing their emotional needs or those of their families. The NVNAD once again confronted the struggle between the coordination of response and focusing on the individual survivors. Internally, the NVNAD leadership agreed that volunteer activities themselves should not be a goal of disaster response and recovery, but a means for helping survivors. However, establishing and operating the disaster volunteer center led by drive for institutionalization became the purpose of this response effort at the expense of meaningful contact with individual survivors.

To address this anomaly regarding the desire of volunteers to assist victims in a meaningful way, including victims who were also local volunteers, the idea of providing a footbath emerged. Student volunteers from Kobe, Osaka, and Niigata formed a team called the Chuetsu-Kobe Footbath Team and provided a footbath at shelters and temporary housing in villages affected by the 2007 Chuetsu Earthquake. A footbath does not just relax people physically by putting their feet in warm water for 15–20 min, but is also a tool for communication between survivors and volunteers. A volunteer may ask an elderly person how she evacuated from the village while providing a footbath, whereas the elderly survivor may talk about his or her grandchildren to the volunteer. Without the footbath, it would be very strange and difficult for student volunteers to start such communications because they are from outside the community. Especially after major life events such as a disaster, people would like to have someone to listen to their stories. Of course, if volunteers detect any serious problems (e.g., mental health issues) through discussions during the footbath, they are expected to report them to local leaders. Otherwise, conversation records are shared among volunteers and used for training and future response and recovery activities.

The NVNAD aggregated their experiences in the two Chuetsu Earthquakes and other disasters and concluded that every individual survivor should receive intensive personal attention. By 2009, the organization had abandoned the single-minded focus on effective coordination of volunteers and based on the experience of disaster response in the Chuetsu Earthquakes, decisively led the organization to focus on the survivors instead of organizational efficiency. The NVNAD now returned to the original survivor-centered approach against the drive for institutionalization (i.e., a tendency to take precedence of bureaucratic forms and procedure of organizations over focusing on each and every survivor) and the long-term perspective for reconstruction that characterized response to the Kobe Earthquake.

2.1.5 Fourth Stage: Toward New Era of Disaster Volunteerism after the 3.11 Earthquake

When the Great East Japan Earthquake occurred on March 11, 2011, many nonprofit disaster organizations in Japan were activated promptly, one of them being the NVNAD. As the current president of the NVNAD and a researcher at Osaka University, the author held an emergency meeting of the board members of the NVNAD to discuss what it should do, how, and where. Members at the meeting consensually agreed to pursue the following three projects. First, the organization dispatched an advance party to the Tohoku region to identify a community waiting for volunteer work. Second, it communicated with key persons in Niigata Prefecture to confirm whether they were willing to accept refugees from the neighboring prefecture, Fukushima, in which many residents were forced to evacuate from their nuclear-contaminated homes. The NVNAD had kept a close relationship with former survivors in Ojiya City who were affected by the 2004 Niigata Chuetsu Earthquake and those in Kariwa village affected by the 2007 Niigata Chuetsu-Oki Earthquake. Third, it started fundraising at both the local and national levels to make long-term assistance possible.

The advance team, consisting of four researchers including two from NVNAD, flew into Aomori airport located in the north end of the disaster region. They selected this airport because many disaster NPOs and volunteers gathered in Tokyo, which was the south end of the disaster region and started to move up north along the tsunami area. In other words, the team expected that relatively fewer number of volunteers would be in the northern area and, if so, more volunteers were needed in the northern area. They started to collect information from the north down to the south and finally identified a small village, Noda, as the NVNAD's focal point of service due to the most damage with the least number of volunteers.

Noda's downtown was completely swept away; 28 of the 4632 residents were killed by the tsunami, and more than 300 houses completely collapsed. Since it is located far to the north of the disaster region without easy access from the prefectural capital city, relatively few volunteers came in mid-March. The NVNAD, based on the report of its advance team, decided to support this village intensively.

The activities for the first years are summarized as follows: Following the initial phase, the office of NVNAD located in Nishinomiya began to send volunteers by bus, which took 18 h to reach Noda village. Mostly university students responded to the NVNAD's appeal and joined its volunteer programs. It sent the first volunteer bus on March 29. It has sent a bus every month (i.e., 12 times for the first year, 6 times for the second, and twice for the third year). Each bus brought about 20 volunteers from Nishinomiya city. Staff members, including the author, made efforts to establish a local network of volunteers working in Noda village. Concurrently, Hirosaki University, Hachinohe College of Engineering and Technology in Aomori Prefecture, and universities in the Kansai region (i.e., Osaka University and Kwansei Gakuin University) were all sending student volunteers to Noda Village. The NVNAD attempted to link all these schools and established a network of volunteers for Noda Village, called "Team North Rias" (TNR) in May, 2011. During the first summer and autumn, disaster volunteers used TNR's local office in Noda Village and repeatedly visited people living in temporary housing. The NVNAD is currently a core member of TNR and still sends a volunteer bus almost monthly.

As for refugees from Fukushima, the NVNAD offered indirect support through former survivors in Niigata. A staff member of NVNAD called Kariwa Village a few days after the 3.11 earthquake. Its local Social Welfare Council was about to receive a large number of refugees from Fukushima Prefecture. However, the local staff members were not sure how many refugees came to their village, what they brought with them, and how much they could spend for refugee care. Therefore, the NVNAD donated 500,000 Yen (about US$ 5,000) to the council and prepared mattresses and other materials for refugees. The village government opened five shelters scattered around the village and received 307 refugees as of April 21. The staff members of the local Social Welfare Council and local residents voluntarily took care of the refugees. When the author asked why local social welfare personnel and volunteers worked so hard for the refugees, they unanimously told him that they wanted to return what they had received from Nishinomiya and other places all over Japan during their own disaster in 2007.

2.2 ACTION RESEARCH: AN EXAMPLE

As ethnographic description above indicates, nowadays in Japan, when a natural disaster occurs, many survivors help each other, and professional supporters and volunteers outside of the affected area join in helping victims and survivors there. Disaster researchers have focused on this high level of mutual aid, whether including disaster volunteers or not, and labeled it in various ways: "postdisaster utopia," "altruistic community," "heroic and honeymoon phases," and "stage of euphoria" in classic studies such as Wolfenstein (1957) and Fritz (1961). In 2009, journalist Rebecca Solnit published a book titled *A Paradise Built in Hell*, and these phrases became well known to the public in Japan because this book was translated into Japanese and happened to be published only a few months before the Great East Japan Earthquake and Tsunami. One of the phrases above, "disaster utopia," which was the Japanese title of the book, has become popular in Japan. Obviously, many survivors after disasters live, not in paradise, but in hellish situation, and care should be taken in using this word, even if it has been used in previous studies. Hence, the term paradise is used judiciously in the following section to indicate the situation where people in the disaster field behave collectively in altruistic ways toward each other without much emphasis on ordinal norms in a society.

2.2.1 Research Questions

As Solnit (2009) noted, "the real question is not why this brief paradise of mutual aid and altruism appears but rather why it is ordinarily overwhelmed by another world order" (p. 97). The unique feature of her perspective is to show the ordinary world against that pursued by an anarchist such as Kropotkin, namely, the world of mutual aid. In other words, the paradise of mutual aid arises, but realistically, it eventually diminishes. To look at the matter from a different angle, this up-and-down trend is not limited to a recent academic finding. It had been described already in one of the classic literary masterpieces of Japan, *Hojoki*, which was written in 1212. Therefore, it was already common knowledge that people help each other immediately after disasters but that this paradise faded.

Demonstrably, such a paradise as defined above (i.e., people in the disaster field behave collectively in altruistic ways toward each other without much emphasis on ordinal norms in a society) was built and overwhelmed after the Great East Japan Earthquake. For instance, the number of disaster volunteers registered at disaster volunteer centers in the Tohoku region, which was affected by the March 11 earthquake and tsunami, increased until July but steeply decreased after several months (Japan National Council of Social Welfare, 2012). It is true that people helped each other and enjoyed being in paradise when Japan, even though a modern, industrialized society, was smothered in a sense of helplessness. However, this paradise was overwhelmed within a matter of months. Therefore, it is reasonable to assume that, if we could find a way to open a permanent door to the paradise, a breakthrough could come in Japanese society. The most real and practical problem is, hence, how to maintain or resume this paradise in society.

The research question for this action research is as follows: How can we maintain the disaster paradise or resume the paradise in society if it diminishes immediately? As described above, previous survivors from the 2004 and 2007 earthquakes in Niigata Prefecture remembered that they had been helped by disaster volunteers and wanted to give something back to them in return. That is why they supported refugees from Fukushima Prefecture with great hospitality. According to the interviews, they unanimously said that it was a good chance for them to pay back their debt. In other words, past survivors in Niigata experienced the paradise built in their own disaster and looked for an opportunity to rebuild it for the current survivors from the Great East Japan Earthquake in a unique way of indirect reciprocity. Perhaps here we can detect a clue for the answer to the research question. Namely, it can be hypothesized that the paradise can be resumed, even after a while, if previous survivors find an opportunity to help the current survivors in a reciprocal way not of pay-it-back, but of pay-it-forward. The present action research attempts to examine this hypothesis.

2.2.2 Action Research

The author contacted staff members of the Kariwa Social Welfare Council in summer 2011 and asked if they could visit Noda Village in Iwate and support local survivors there. He expected that they would hesitate to do so because they were busy with the refugees in their own villages. However, they thanked him for motivating them to visit the disaster-affected area and indicated that they would definitely like to help survivors there. One of the staff members of the local social welfare council said, "It is like the drop of a hat! It is actually a great opportunity for us to give something to the survivors in return for what we received from Nishinomiya and other places in 2007" (field note, September 6, 2011). Accordingly, the author coordinated their schedules with that of the NVNAD's volunteer bus. He also made arrangements with TNR in Noda Village for receiving volunteers and coordinating them with local survivors.

Early on the cold morning of December 10, 2011, two buses arrived at Noda Village, one from Kariwa and the other from Nishinomiya. The former survivors of Kariwa could not wait to set up their activities (e.g., pounding steamed rice into cakes). When they visited families at temporary housing one by one, they delivered rice cakes and talked with each other. I helped their activities, and at the same time, I took notes if I heard any impressive conversation. After handing a souvenir from Niigata to a resident in temporary housing in Noda, a volunteer from Kariwa said to the author with tears in his eyes, "Now I was able to return my debt, and I realized that it is extraordinarily important to help people in trouble" (field note, December 10, 2011). His wife, another volunteer, also said, after her activities at other temporary housing, "I became friends with an old lady living in temporary housing because we had the same hobby. I would like to visit her again. It is so good to help each other" (field note, December 11, 2011). Volunteers from Kariwa and Nishinomiya held a party in the evening at the community center of the temporary housing. The former leader of the Kariwa Social Welfare Council, who was participating as a volunteer this time, suggested, "Let's compare the taste of Sake from Iwate with that from Niigata" (field note, December 10, 2011). This phrase demonstrates that they were not merely offering support with frowning faces, but that they truly enjoyed helping each other.

Note that these events took place in December of 2011, 9 months after the earthquake. It is obvious that the disaster paradise had already been lost. However, volunteers from a former disaster area, Kariwa, helped the current survivors and resumed the disaster paradise in Noda Village. A few months later, the current leader of the Social Welfare Council told the author that he had budgeted a visit to Noda Village four more times in the next fiscal year.

2.2.3 Discussion

In this action research, we found that the disaster paradise can be maintained, at least intermittently against the drive for institutionalization. In a paradise, people do not examine how bureaucratically effectively people help each other, but focus on how heartwarmingly people help each and every individual. Hence, building the disaster paradise intermittently is against the drive for institutionalization. That is to say, even if the paradise diminishes in time after a disaster, it is possible for previous survivors to revive it when the initial paradise is gone. This case is also an example of relay of support among disaster-affected areas. The NVNAD supported Kariwa Village in 2007, whereas volunteers of Kariwa helped survivors in Noda Village, Iwate. All of the interviews indicated that rebuilding the paradise was a good opportunity to help the current survivors in the sense that they paid their debts not backward but forward. Thus, this network can be called the Pay-it-forward network (PFN). At least two issues are to be examined here in the nature of the PFN, psychologically as well as sociologically.

Psychological implications: Motives and Shared Emotions of Unshared Pain. Why can a paradise be resumed intermittently? According to the interviews, the motive behind the PFN was a feeling of indebtedness. In fact, previous survivors in both Kariwa and Ojiya thought about how to return what they had received, and that is why they seemed to be relieved after they had volunteered for current survivors. Hence, it is implied that the feeling of debt is a key motive of the PFN.

One might say that returning one's debt to current survivors is a very simple matter psychologically. In other words, the former survivors, because they shared the emotions of surviving disaster with the current survivors, gave the latter the same kindness they had received in the past. However, this way of thinking about it is not entirely true simply because each experience of disaster is unique and different from another, and it is simply impossible to understand others' experiences fully. For instance, a person who has experienced an earthquake does not necessarily understand the emotional reactions of another person who lost everything to a tsunami. Therefore, previous and current survivors do not necessarily have common emotions with each other.

What, then, is shared between survivors of the past and the present? I suggest that they actually share the emotion that nobody can understand their experiences. In other words, they share the emotion of unshared pain. It is possible to refer to a clinical finding here. Jensen et al. (2011) indicated that a sufferer's pain is more likely to be relieved when another person offers support without attempting to relate to the pain than when someone supports as if he or she understands the pain. That is to say, it is suggested that the shared emotion of unshared pain may facilitate relief activities. It is psychologically implied, therefore, that the paradise can be rebuilt intermittently because it provides a good opportunity for the former survivors to pay their debt not back but forward to current survivors with the unshared emotion of pain. In other words, even if the former survivors do not share the emotion of pain with the current survivors, they try to pay their debt forward by helping the current survivors.

Sociological implications: Expansion of PFN and its Impact on Japanese Society. What about the expansion of the PFN? It is obvious that the rebuilt paradise will also diminish eventually. Is there any social impact if the paradise is intermittently rebuilt? Let us consider the PFN from a network analysis perspective. A classic work in social psychology gives us a crucial insight for this examination. According to Milgram (1967), a person is only six steps away from any stranger through shared acquaintances so that it is a small world. This notion was elaborated on mathematically by Watts and Strogatz (1998). Assuming that there are many points on a circle, if one point can connect only with its neighbor, then it takes many steps to reach a point across the circle. However, surprisingly, they found that a random connection of any two points facilitates the connecting ties of the whole points on the circle. The PFN is a set of random connections in this paradigm. It is implied that the ties of a PFN may lead society to a new era of human bonds. Modern developed societies are too institutionalized and have no room to move, like nomads, to contact other "human beings" directly. PFN may lead to a new society of human relationships.

For future research: If the nature of PFN is more thoroughly examined, it will be not only a network of disaster relief, but also a new type of human ties, based on psychological motivation and emotion, which has the potential to expand over the whole society of Japan experiencing many disasters. At least two issues should be considered for future research.

First, systematic collection of data is required. The present study is based on long-term fieldwork, but provided some anecdotal excerpts from the author's field notes. It surely indicates the existence of PFN and explores its theoretical and practical potential; however, it is suitable for the future research to provide more detailed evidence based on various methods. It is also interesting to explore cultural and political factors. Let us focus on East Asia, for instance. Japan, Taiwan, and China were hit by devastating earthquakes in 1995, 1999, and 2008. They have supported each other; however, it is not clear if PFN exists beyond cultural and political obstacles. It is recommended for future research not only to take on systematic data gathering, but also to keep conducting fieldwork to determine a policy of international and cross-cultural PFN.

Second, the PFN leads us to examine reciprocity in general. Besides a tradition of anthropology and sociology of giving (e.g., Mauss, 1925), reciprocity has been examined in evolutionary biology and psychology (e.g., Nowak and Sigmund, 2005). Nowak and Sigmund (2005) classified reciprocity first into two categories, direct and indirect reciprocity, and then divided the latter into upstream and downstream. Direct reciprocity means that person A helps person B and B helps A in turn, whereas indirect reciprocity includes the third person C in two different ways. One is upstream reciprocity (person B, who has just received help from person A, goes on to help person C), and the other is downstream reciprocity (person A has helped person B and therefore receives help from person C). The PFN corresponds to upstream reciprocity, although it puts more emphasis on collective activities rather than individual relationships as evolutionary biology. Further theoretical investigation, as well as data analyses by simulation, would be a fruitful source of understanding human reciprocity.

For future practice: At least two aspects should be considered for the potential of the PFN for future practice. First, because, as indicated, nonprofit disaster organizations have come to play an extremely important role in Japan since the 1995 Kobe Earthquake (e.g., Atsumi and Goltz, in press), these are the organizations that should make former survivors collectively remember the significance of a PFN. In fact, it was obvious that the PFN here was motivated by a disaster NPO, NVNAD. Hence, it is disaster NPOs that can make it possible to resume the paradise by forming the PFN.

Second, it might be a good idea to motivate "future survivors" in Japan. Japan will always have to worry about earthquakes, typhoons, volcanoes, and many other natural disasters. For instance, a set of huge earthquakes are expected in the metropolitan region, including Tokyo. If the Japanese can be motivated to help survivors of the Great East Japan Earthquake for example, then they could join the PFN in advance without receiving any support immediately. This network could be called the "Pay-it-Beforehand" network. This network is a good candidate for expanding the PFN in that it is likely to leave a feeling of indebtedness and, accordingly, expand the PFN.

It may seem as if a disaster sends us to hell. However, people can build and rebuild the paradise of mutual aid, with the assistance of disaster NPOs, through PFN. When we become able to support disaster survivors through a PFN, that will be the time our society will change.

2.3 GENERAL DISCUSSION AND CONCLUSIONS

The renaissance of volunteerism in Japan was not a temporary boom, as it has remained stable since 1995 and continues today. Ethnographic research indicated that disaster volunteers in Japan have been established through the stages of convergence, expansion, extension, and network. It also clarified that some disaster volunteers and organizations were settled in the society by drive for institutionalization, while others such as NVNAD against the drive. Action research conducted here indicated that PFN can be against drive for institutionalization in a sense that, in disaster paradise, survivors are more important than bureaucratic procedure.

In conclusion, disaster voluntarism since the 1995 Kobe Earthquake and the almost two decades of its development described in this study have perhaps given us a glimpse of the future of Japanese society, one characterized by empathy and community regarding values to counteract the individualism, institutionalization, and alienation fostered by urban life of the post-World War II era. The most obvious expression of this emerging ethos is the footbath, which establishes an emotional, affective, and physical connection between one individual to another. It implies that the next era may be characterized by a more direct and engaging relationship with others. While social networking (e.g., Facebook) may connect us electronically, connecting directly with each other in everyday life in Japan is more challenging.

Although this view is rather optimistic, we now have a hopeful strategy, the PFN, which would work against the drive for institutionalization, at least, intermittently.

REFERENCES

Atsumi, T., 2001. Wisdom of Volunteers [Borantia no Chi]. Osaka University Press, Osaka (in Japanese).

Atsumi, T., Goltz, J.D., 2014. Fifteen years of disaster volunteers in Japan: a longitudinal fieldwork assessment of a disaster non-profit organization. Int. J. Mass Emerg. Disasters 32 (1), 220–240.

Britton, N., 2006. National planning and response: national systems. In: Rodriguez, H., Quarantelli, E.L., Dynes, R.R. (Eds.), Handbook of Disaster Research. Springer, New York, pp. 347–367.

Fritz, C.E., 1961. Disaster. In: Merton, R.K., Nisbet, R.A. (Eds.), Contemporary Social Problems: An Introduction to the Sociology of Deviant Behavior and Social Disorganization. Harcourt, New York, pp. 651–694.

Homma, M., Deguchi, M., 1996. Volunteer Revolution: Applying Experience of the Earthquake to a Citizen's Movement [Borantia kakumei: Daishinsai deno keikenn o shimin katsudoue]. Toyokeizai Shimposha, Tokyo (in Japanese).

Japan National Council of Social Welfare, 2012. Statistics of Disaster Volunteers in Tohoku. http://www.saigaivc.com (in Japanese) (accessed 01.08.12.).

Jensen, M.P., Moore, M.R., Bockow, T.B., Ehde, D.M., Engel, J.M., 2011. Psychosocial factors and adjustment to chronic pain in persons with physical disabilities: a systematic review. Arch. Phys. Med. Rehabil. 92 (1), 146–160.

Mauss, M., 1925. 'Essais sur le don' L'Annee sociologiques. In: Nouvelle Serie, vol. 1.

Milgram, S., 1967. The small-world problem. Psychol. Today 2, 60–67.

Nowak, M.A., Sigmund, K., 2005. Evolution of indirect reciprocity. Nature 437 (7063), 1291–1298. http://dx.doi.org/10.1038/nature04131.

Rausch, A.S., 1998. The emerging consciousness of japanese voluntarism. Int. J. Jpn. Sociol. 7, 1–21.

Solnit, R.A., 2009. Paradise Built in Hell: The Extraordinary Communities that Arise in Disasters. Viking, New York.

Suzuki, I., Suga, M., Atsumi, T., 2003. Disaster volunteers in Japan: an historical review and the current movement since the 1995 Kobe earthquake ["Nihon niokeru saigai borantia no doukou: Hansin-Awaji daisinsai o keikitosite"] Jpn. J. Exp. Soc. Psychol. 42, 166–186 (in Japanese with English abstract).

Taniguchi, H., 2010. Who are volunteers in Japan? Nonprof. Volunt. Sect. Q. 39 (1), 161–179.

Tierney, K.J., Goltz, J.D., 1997. Emergency Response: Lessons Learned from the Kobe Earthquake. See http://www.udel.edu/DRC/preliminary/260.pdf.

Watts, D.J., Strogatz, S.H., 1998. Collective dynamics of "Small-world" networks. Nature 393, 440–442.

Wolfenstein, M., 1957. Disaster: A Psychological Essay. Free Press, Glencore.

Yatsuzuka, I., 1999. The activity of disaster relief volunteers from the viewpoint of social presentations: social construction of "Borantia" (voluntarism) as a new social reality after the 1995 Great Hanshin earthquake in Japan. Prog. Asian Soc. Psychol. 2, 275–290.

Chapter 3

Disastrous Disasters: A Polemic on Capitalism, Climate Change, and Humanitarianism

Phil O'Keefe [1], Geoff O'Brien [1] and Janaka Jayawickrama [2,3]

[1] *Department of Geography, Northumbria University, Newcastle, UK,* [2] *Post-war Reconstruction and Development Unit, University of York, York, UK,* [3] *Disaster and Development Network (DDN), Northumbria University, Newcastle, UK*

ABSTRACT

The way people have thought about disasters has changed in almost immeasurable ways. They are no longer considered as "Acts of God" but caused either by natural hazards or by human actions. This shift in thinking has led to different ways of dealing with disasters. What is becoming apparent is that more and more disasters are being produced by human actions. We cause them but seem unable to find the political will to stop doing so. Further, the role of humanitarian agencies has changed. At one time bodies such as the Red Cross would be associated with helping the injured in the battlefield. The nature of present-day warfare or complex emergencies and terrorism means that humanitarian bodies are dealing with many vulnerable refugees and internally displaced people. The following polemic reflects on these changes.

We had better start somewhere to trace the polemic. And that somewhere is probably with the authors.

Phil O'Keefe completed his doctoral fieldwork in Central Province, Kenya, a place of ridge valley landscapes with steep slopes. In making observations on soil erosion, against the United Nations (UN) Forestry and Agricultural Organization recommendations that 1 ha with a 15° slope would require more than 800 days of annual labor to build and maintain soil terraces, he noted the absence of labor on family farms. Men had gone to the city, Nairobi, for work; women worked piecemeal for the large monocultural agribusinesses at the very time labor was needed on their own family farms. Soil erosion, he concluded was caused by urbanization. Later he would see the same process writ large under the apartheid system of Bantustans (O'Keefe, 1983). Later, at the Bradford University Disaster Research Unit, he would lead the analysis of vulnerability that produced a radical paradigm shift in the study of disasters (O'Keefe, 1976). Nature was, on observation, never neutral; risk was constructed by people.

Hazards, Risks, and Disasters in Society. http://dx.doi.org/10.1016/B978-0-12-396451-9.00003-2

Geoff O'Brien came to a similar conclusion but by a different route. After graduating as an electrical engineer, he led global environmental oversight for a number of multinational oil companies around the world. Changing his focus midcareer, he focused on greening small and medium-sized enterprises. Initially focused on energy and environment, these studies broadened to risk in developed economies. Producing a comparative analysis of the United States' and the United Kingdom's risk response systems, he argued that what had been made resilient were the emergency services, the "First Responders", not the communities at risk (O'Brien, 2008).

Janaka Jayawickrama came to academia as a practitioner of disaster management and humanitarian aid delivery. He was acutely concerned that, at best, Western interventions were couched in terms of "Do No Harm" when the reality was that many interventions were inappropriate and culturally insensitive. For him, the delivery of psychosocial interventions, based on individual diagnosis of posttraumatic stress disorder was far from the services disaster-hit communities required (Jayawickrama et al., 2014). He continues to build new models, and ethics, for disaster intervention.

Not surprisingly, given these varied but converging backgrounds, our shared view of disasters and humanitarian assistance is somewhat radical. It is braced in a critique of the geographical paradigm of natural hazards research (Kates, 1997). It emerges from an understanding of people–land relationships within a tradition of political economy of nature (Cline-Cole and O'Keefe, 2006). This leads to a number of theses that are now discussed.

3.1 THESIS 1: WITH THE RISE OF CAPITALISM, WE HAVE GONE FROM THE HUSBANDRY OF NATURE TO THE PRODUCTION OF NATURE: THAT CHANGE IN RELATIONSHIP TO NATURE PRODUCES NEW RISKS

Under globalized capitalism, nature is produced (Smith and O'Keefe, 1980). In the production of nature, people change their relationship to nature from that of husbandry, an organic relationship with the environment, to one of mastery, even if that mastery is never fully achieved. People also change their relationship to each other as individuals, communities, and classes. Importantly, these changes also change the nature of risk and thus disaster type (Beck, 1992, 2009).

It is relatively easy to illustrate these arguments. Fracking increases earthquake risk; flooding is increased by land-use change; climate change heralds increasing extremes of temperature and precipitation beyond what people accept as "normal" weather. It is not simply the physical world but the biological one where fundamental changes are occurring, for example, human immunodeficiency virus/acquired immunodeficiency syndrome, severe acute respiratory syndrome epidemics, and bovine spongiform encephalopathy outbreaks. Technology itself generates new risks, not least with nuclear engineering and biotechnologies.

Science can only bring a limited solution to these problems not least because the modern positivist science project comes encumbered with a dual vision of nature. This dual vision holds that nature is "external" to humans, while simultaneously nature is "universalist" incorporating humans. But a global capitalism necessarily produces a humanly nature in the realm of exchange, or market, values. Talk of a postnormal science to address complex problems simply reinforces the contradictory duality of the scientific approach to nature (O'Brien and O'Keefe, 2014).

Under globalized capitalism, new risk is produced that science cannot control.

3.2 THESIS 2: THE UNMET CHALLENGE OF CLIMATE CHANGE

Accelerated climate change and increasing variability is the greatest universal environmental threat we face. This threat can be viewed in two ways. First, the immediate threat exists from increasingly severe weather events and second, the long-term threat to global ecosystems occurs as the planet warms. The international community has recognized this threat and through the United Nations Framework Convention on Climate Change (UNFCCC) has tried to develop solutions. Progress, however, has been agonizingly slow. Initially the main thrust of the work of the Convention was to deal with the long-term threat by focusing on reducing greenhouse gas emissions. However, the vehicle devised to reduce emission rates, the Kyoto Protocol, has failed due, in part, to its architecture and effective campaigning and lobbying by climate skeptics funded through the energy sector. It is hoped that a new agreement to reduce greenhouse gas emissions will be ready for implementation in 2020. Whether the negotiators can really address the scale of change needed in the global energy system is an open question (O'Keefe et al., 2010). In the meantime, we are emitting our way into an unknown future. No indication exists as to what greenhouse gas concentration and what timetable will be agreed upon. Little chance seems to exist of us keeping the average global temperature rise below 2 °C. We are producing a dangerous and unknown future.

Interestingly, dealing with the consequences was given a low priority in the early days of the Convention. Adjusting to consequences, or adaptation, was thought only to be an issue for the Less Economically Developed Countries, such as Small Island Developing States. However, the litany of extreme weather events across the globe has rapidly raised adaptation up the agenda. A recent report by the Intergovernmental Panel on Climate Change (IPCC, 2014) paints a very bleak picture of what the world we are creating could look like: increased floods and droughts and more severe weather events that will threaten food security. Adaptation is now an issue for all nations. The problem is that if we do exceed the 2 °C temperature rise, then we enter an era that is termed dangerous climate change. Stay below that temperature rise and we should be able to adapt. Go above that temperature, and we could be exposed to irreversible changes and the collapse of ecosystems to which we would be unable to adapt (O'Brien and O'Keefe, 2014).

In short, we need to move to a low-carbon energy economy as soon as possible in order to avoid a future catastrophe. In the interim, we need to use our ingenuity to adapt to the consequences. The UNFCCC has been in place for some 30 years. To date, little real progress has been made and this does not augur well for the negotiations up to 2020. Reducing greenhouse gas emission appears, on the face of it, relatively straightforward. Much of the technology needed to both use energy efficiently and produce it cleanly is available. The problem lies in the lack of political will to opt for solutions rather than being market based. In many cases, we are "locked into" a number of technologies, for example, the United States seems to have opted strongly for fracking. Though gas is cleaner than coal, the investment decisions made today are likely to be in place for 40–50 years. We are getting to grips with renewable technologies and their use, in the main, of low-density, variable primary energy sources, such as wind and solar. But a gargantuan effort on production and deployment will be needed in order to produce significant energy from renewable resources (O'Keefe et al., 2010). Ultimately, it is up to our political leaders to agree to a stabilization target and timetable. Energy planners can then get to work on planning the replacement of high carbon-emitting technologies with low ones. This may mean scrapping some systems before their due date. We must be brave enough to do that!

Adapting to the changes locked into the climate system is also a massive challenge. But, as Stern (2007) pointed out, it is more cost-effective to address the problems now rather than deal with the consequences in the future. Progress on adaptation is being made and despite the huge uncertainties it is possible that we can develop sustainable approaches to this multifaceted challenge (O'Brien and O'Keefe, 2014). Recognition has been occurring between the climate and disaster management communities of the overlap of disaster risk reduction and climate change adaptation techniques. This does raise the question of whether or not we have the right institutional arrangements. Adaptation is about people and their livelihoods and protecting them from climate risks. This is the role of the United Nations Development Programme (UNDP). Reducing risk is the role of the United Nations International Strategy for Disaster Reduction. Promoting ecological well-being is the role of the United Nations Environment Programme. We could frame all of this with a Convention on Climate Adaptation that could still be linked with the IPCC. We could then streamline the institutional structures and have lines directly to the national, subnational, and local levels. This new body would have sovereign institutional responsibility for developing adaptation within a framework of human and ecological well-being (O'Brien et al., 2012).

The final issue to address is funding. Changing the energy system and developing adaptation measures will not come cheap. The UNFCCC has agreed a Green Climate Fund that will have funding of US$ 100 billion per year from 2020 for both mitigation and adaptation. But as O'Brien and O'Keefe (2014) have pointed out, the International Energy Agency estimate that moving to low-carbon energy could require some US$ 10.5 trillion by 2030 and that adaptation costs by various organizations such as the World Bank, UNFCCC, Oxfam, and UNDP range between US$50 billion and US$171 billion per year. The Green Climate Fund is unlikely to meet all needs. Many of the wealthier nations will need to dig deeper. Perhaps we could put some of the US$1.7 trillion per year we spend on the military to better use. All funding mechanisms involve privatizing the commons and commodifying the global atmospheric system through trading emission levels. Is that really the way to go?

3.3 THESIS 3: HUMANITARIAN ASSISTANCE IS A CORE TOOL OF WESTERN COUNTRIES' FOREIGN POLICY

Sadako Ogata, a former High Commissioner for Refugees, argued that "There are no humanitarian solutions to humanitarian crises (since) mass displacement, hunger disease et cetera, are the humanitarian fallout of political failures" (Weir, 2006, p. 4). Humanitarian assistance includes local, national, and international efforts provided by individuals, communities, and national and international organizations addressing people in need where the prime motivation is to prevent and alleviate human suffering. The four fundamental principles underlying humanitarian interventions are humanity, impartiality, neutrality, and independence of the humanitarian actors (United Nations, 1991). These principles are not usually observed in humanitarian interventions, not least in those tied to military intervention such as Kosovo, Iraq, and Afghanistan.[1]

Humanitarian aid addresses the needs of a population after natural disasters or complex emergencies. Natural disasters are dominated by meteorological phenomena (drought, flood, and hurricane) and, to a much lesser extent, geological ones (earthquakes, volcanoes, and tsunamis). Intervention in the aftermath of natural disaster is dwarfed by intervention in complex emergencies. Complex emergencies, characterized by high levels of civilian casualties, deliberate destruction of livelihoods, and the collapse

1. It was the politics of war in Biafra that led to the founding of MSF and a political commitment to advocacy rather than neutrality.

of the rule of law, have increased the number of displaced people either as refugees or internally displaced people (IDP), particularly since the end of the Cold War (Middleton and O'Keefe, 1997). The current crisis in Syria has generated 2.6 million refugees and 6.5 million IDP. In general, the humanitarian system was not designed to deal with such massive problems.

The adoption of the four "Geneva Conventions" underpinned what has been called the "Dunantist" position, named after the founder of the Red Cross, where the delivery of humanitarian aid was seen as independent and long term. This contrasts sharply with the "Wilsonian" position where humanitarian actors are used as an extension of state, often military, intervention. A third type of organization is the religion-based one, which frequently seeks to use humanitarian action to proselytize (Macrae and Harmer, 2003). As of 2014, reviewing humanitarian action in 2013, a total of US$ 17.9 billion was in international response, of which US$ 12.9 billion came from governments, specifically US $ 11.6 billion from Organisation for Economic Co-operation and Development countries (GHA, 2013). Afghanistan, Pakistan, Somalia, and the West Bank dominated these actions accounting for more than US$ 5 billion.

O'Dempsey and Munslow (2009) highlighted the extent to which humanitarianism has been instrumentalized. "Objective humanitarian need, on the basis of human suffering, is never the criteria used to determine whether or not the international community will act. Power and money determine the working of international politics, tempered to a variable degree, by the humanitarian impulse, which is influenced by the media and electoral sentiment" (O'Dempsey and Munslow, 2009, p. 446).

Humanitarian aid is predominantly delivered in Muslim countries, in situations frequently overseen by Western military forces in wars they themselves have started or condoned. Not neutral, impartial, or independent but as an extension of Western foreign policy. Do you sincerely wish to be a humanitarian worker?

3.4 THESIS 4: THE GROWTH INDUSTRY OF HUMANITARIANISM AND ACCOUNTABILITY

Since 1989, when humanitarian aid on a global scale was US$ 0.5 billion, it has grown by a factor of 40 to some US$ 20 billion a year. This significant rise, largely fueled by an increase in public expenditure under neoliberal regimes that were generally committed to the withering of state expenditures, is unparalleled in any other sector.

Some 50 percent of the expenditures are through the UN system (chiefly the World Food Programme (WFP), United Nations High Commissioner for Refugees, and United Nations International Children's Emergency Fund), although much of the final distribution of UN resources is delivered by international non-governmental organizations (iNGOs). iNGOs themselves deliver 25 percent of all humanitarian aid, with government contributions dominating individual iNGO budgets. The Médecins Sans Frontières (MSF) family is an exception to this statement. Figure 3.1 shows the broad routings on humanitarian monies. The important routes on the diagram are from donor governments to the UN system and iNGOs and from the UN system to iNGOs. Those routes command the most flow. What the affected population makes of

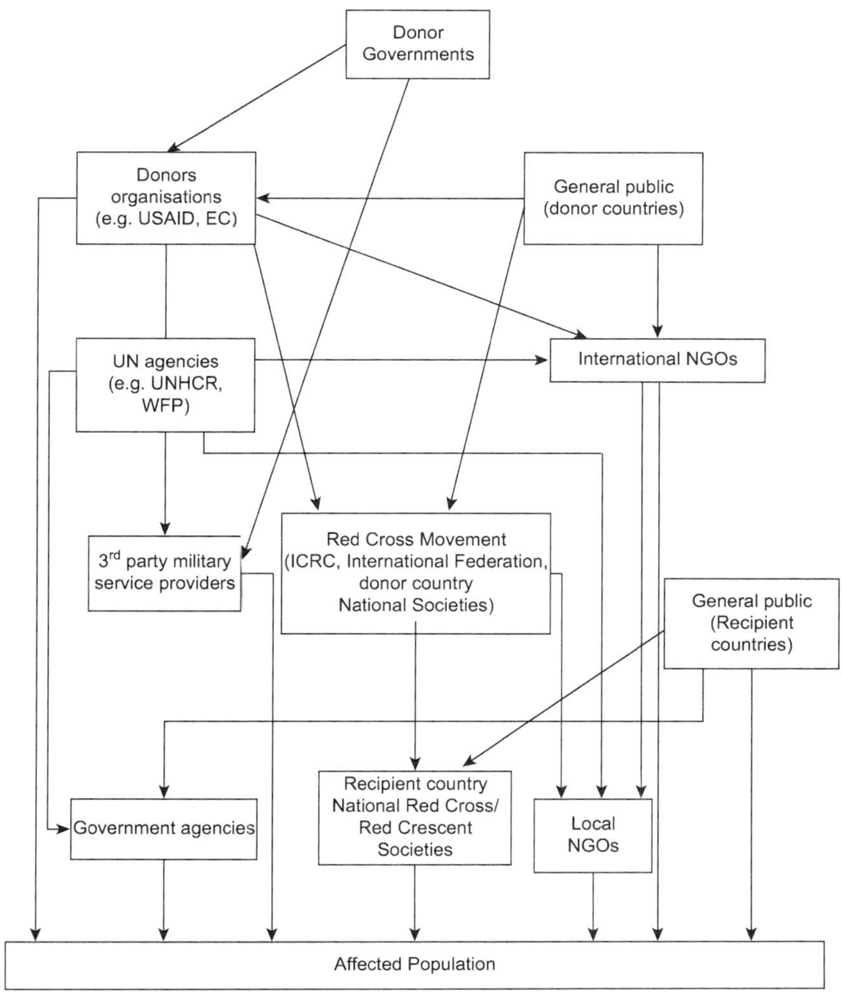

FIGURE 3.1 Resource flows in humanitarian aid. *Source: Hallam (1998).*

all the players on the ground, frequently with different rules of engagement, remains largely unanswered.

The iNGOs essentially form an oligopoly. This oligopoly is distinguished by its members signing "Heads of Agreement" with the major UN agencies, especially UN WFP, to act as distributors in UN programs of humanitarian aid. In addition, they form joint fund-raising appeals where membership is restricted, excluding smaller iNGOs. Perhaps the best example of this is the operation of the Disaster Emergency Committee (DEC) in the United Kingdom currently. The DEC brings 13 leading UK aid charities together in times of crisis: Action Aid, Age International, British Red Cross, Catholic Fund for Overseas Development (CAFOD), Care International, Christian Aid, Concern Worldwide, Islamic Relief, Oxfam, Plan UK, Save the Children, Tearfund, and World Vision. Most are global families, frequently headquartering those institutions in London or Washington. They are essentially businesses with high corporate remuneration and high rates of growth.

This growth largely came after 1990, financed usually through development-aid budgets. As development aid itself was increasingly evaluated from the 1970s onward, a demand occurred from major donors for similar evaluations of humanitarian spending, not least because humanitarian expenditure was, in most cases, more than 5 percent of the overall development budgets. Humanitarians widely resisted such invasion and questioning of purpose. After all, they were doing "good" (Borton et al., 2005).

3.5 THESIS 5: CURRENT HUMANITARIAN AID IS DOMINATED BY A GROWTH IN LOCAL WARS

Although it is true that the number of natural disasters is rising, the growth of humanitarianism has been dominated by the response to local wars. The euphemism for local war is "complex emergencies," implying more than a single cause but associated with violence. But it is people who count.

The Guardian, 20th of June, 2014, reported that official figures for refugees, including IDPs, had risen to over 50 million people making a population base of "the 24th largest country in the world." In the accompanying graphic of places that receive refugees, it is obvious that most refugees, i.e., those crossing international borders, go to neighboring states, namely, Pakistan, Iran, Lebanon, Turkey, Kenya, Egypt, and Jordan. The United Kingdom is noticeable by its absence, although France, Germany, and the United States each take in over 200,000 refugees (Sherwood, 2014).

A body of international law exists that governs refugees, although it is rarely justiciable. No such thing exists for IDPs. An attempt was made to codify practice for improved delivery to IDPs but it exists only as exhortation (Borton et al., 2005). Instead of "Do No Harm," the humanitarian equivalent of the Hippocratic Oath, a tendency occurs to do nothing about IDPs.

Essentially this is a foreign policy of disdain frequently in wars generated by Western interests. As MSF noted in a speech to North Atlantic Treaty Organization:

"This lack of trust is all the stronger when UN and/or western militaries claim to use military force in furtherance of "humanitarian" goals, thereby unilaterally pronouncing a unity of purpose between humanitarian non-governmental organizations (NGOs) and intervention/occupation forces. In 2001 no less than Colin Powell proclaimed 'NGOs are such a force multiplier for us, such an important part of our combat team.' Even more unhelpful, humanitarians have been labeled as sources of information. It should be obvious to you in the military that if we are part of your team, if we are on your side, if we are providing you with information, if we are advancing towards the same goals as you, then we fall directly into the crosshairs of the other side. It's nothing personal, but we can't afford this sort of unity."[2] While there may be a market for universities in training the military in humanitarian action, is it really where humanitarian experts wish to go?

3.6 THESIS 6: THE NGOS AS AN OLIGOPOLY

The UN system, or the Red Cross and Red Crescent in some cases, is the framework for the delivery of humanitarian assistance. On the ground, however, much of the delivery is undertaken by iNGOs. Leading NGOs usually have "Heads of Agreement" with UN agencies, i.e., contracts to deliver to targeted beneficiaries. Each iNGO has its own mission and thus coordination, a buzzword for leadership, becomes difficult. From observation, the less money there is for a disaster, the more there is likely to be coordination: competition for funds seems to generate control of humanitarian action.

The best overview of who makes it to the oligopoly is probably provided by the United Kingdom's DEC. This self-selected membership raises monies in unified appeals after large-scale global disasters accessing free media and banking services. Monies are divided between the INGOs largely based on a formula of how much they had in previous disasters. By 2014, the DEC had launched some 64 appeals, which raised more than UK£ 1.1 billion.

The DEC, like the iNGOs themselves, is largely self-policing using evaluation for learning purposes, although selling those evaluations to donors as proof of accountability. Most iNGOs now have internal evaluation personnel directly linked to institutional policy making. The beneficiary voice is quite weak in evaluation production. This self-selection of membership and self-policing of performance is not surprising because charities are essentially multimillion, multimarket businesses in competition with each other. The oligopoly preserves market share.

2. Full speech is available at: http://www.msf-me.org/en/specialreports/news-media/special-reports/nato-speech-rheindalen-germany.html.

3.7 THESIS 7: A MISTAKEN BELIEF EXISTS THAT FIRST RESPONDERS, BE THEY EITHER EMERGENCY SERVICES OR HUMANITARIAN AGENCIES, PROMOTE COMMUNITY WELL-BEING

In any disastrous situation, the first responders are those that are alive and able to function. Many examples occur of people helping others. Many instances occur of people who have worked together as they realized that it would be many days before any form of official help would arrive (Ferdinand et al., 2012). It is the resilience of people that is an important factor in their ability to respond to disasters and contributes to them being first responders. We do not in any way decry the efforts made by emergency responders or humanitarian agencies but argue that they are mission focused and do not necessarily have the wider community well-being agenda in mind. Successful predisaster planning must focus on building both personal and community resilience and enhancing social capital.

Reich (2006) discussed three social organizing principles for resilience to disasters. These are control, coherence, and connectedness. In the case of control, external help can create a state of dependency. Short-term help is often needed but as soon as possible people should be given resources to rebuild their lives. In terms of coherence people need to understand what has happened. For connectedness people need to part of any recovery process—communities that recover together will stay together.

Many responder agencies may address perhaps one or two of these social organizing factors, but it is unlikely that they will address all three. An integrated approach does need to occur that recognizes the majority of people are inherently resilient, and by being sensitive to this, agencies can work with affected communities in ways that enhances the recovery process.

3.8 TOWARD A CONCLUSION

None of what we have written above essentially questions the need for humanitarian assistance or that most humanitarian workers have good intentions. What we ask for is recognition that response to disasters comes first and foremost from the community affected by disaster. What we also are arguing is that the nature of disaster itself will change, to be dominated by greater extremes in climate events, although, at the moment, wars dominate humanitarian intervention. We need to recognize that cities themselves will be a locus for disasters as populations become urbanized and subject to new hazards. The institutions that deliver in disaster, especially the INGOs, need to question their own industry practices and standards and above all reconsider whether a nineteenth-century model of charity under benevolent capitalism is appropriate for humanitarian action in the twenty-first century. Above all, there must be recognition that risk is socially produced in the production of nature.

REFERENCES

Beck, U., 1992. Risk Society: Towards a New Modernity (London).

Beck, U., 2009. World at Risk. Polity Press, Cambridge, UK.

Borton, J., et al., 2005. Support to Internally Displaced Persons: Learning from Evaluations. SIDA, Stockholm.

Cline-Cole, R., O'Keefe, P., 2006. Mainstreaming the African environment in development. Rev. Afr. Political Econ. 33 (109), 391–406.

Ferdinand, I., O'Brien, G., O'Keefe, P., Jayawickrama, J., 2012. The double bind of poverty and community disaster risk reduction: a case study from the Caribbean. International Journal of Disaster Risk Reduction 2, 84–94. ISSN:2212-4209. http://nrl.northumbria.ac.uk/10147/. Available at: http://dx.doi.org/10.1016/j.ijdrr.2012.09.003.

Global Humanitarian Assistance (GHA). Report 2013, 2013. Global Humanitarian Assistance. Online at: http://www.globalhumanitarianassistance.org/wp-content/uploads/2013/07/GHA-Report-2013.pdf.

Hallam, A., 1998. Evaluating humanitarian assistance programmes in complex emergencies. Good Practice Rev. 7 (102). Retrieved January 2013, from. http://www.odihpn.org/hpn-resources/good-practicereviews.

IPCC, 2014. Summary for policymakers. In: Edenhofer, O., Pichs-Madruga, R., Sokona, Y., Farahani, E., Kadner, S., Seyboth, K., Adler, A., Baum, I., Brunner, S., Eickemeier, P., Kriemann, B., Savolainen, J., Schlomer, S., von Stechow, C., Zwickel, T., Minx, J.C. (Eds.), Climate Change 2014, Mitigation of Climate Change. Contribution of Working Group III to the Fifth Assessment Report of the Intergovernmental Panel on Climate Change. Cambridge University Press, Cambridge, United Kingdom and New York, NY, USA.

Jayawickrama, Janaka, O'Keefe, Phil, O'Brien, Geoff, 2014. The dilemmas of psychosocial interventions in humanitarian assistance. In: Negotiating Relief: The Dialectics of Humanitarian Space. Hurst, New York, USA, pp. 1–9.

Kates, R.W., 1997. Human adjustment. In: Hanson, S. (Ed.), 10 Geographic Ideas that Changed the World. Rutgers University Press, New Brunswick.

Macrae, A., Harmer, A. (Eds.), 2003. Humanitarian Action and the "Global War on Terror": A Review of Trends and Issues, HPG Report 14. ODI, London.

Middleton, N., O'Keefe, P., 1997. Disaster and Development: The Politics of Humanitarian Aid. Pluto Press, London.

O'Brien, G., O'Keefe, P., 2014. Managing Adaptation to Climate Risk: Beyond Fragmented Responses. Routledge, London and New York.

O'Brien, G., O'Keefe, P., Jayawickrama, J., 2012. Institutions or solutions. Letter to Editor Int. J. Environ. Stud., 1–2.

O'Brien, G., 2008. Disaster Management, Climate Change and Social Resilience (Ph.D.). Northumbria University.

O'Dempsey, T.J., Munslow, B., 2009. Complex political emergencies in the war on terror era. In: Desai, V., Potter, R.B. (Eds.), The Companion to Development Studies. Hodder Education, London, p. 466.

O'Keefe, P., O'Brien, G., Pearsall, N., 2010. The Future of Energy Use. Earthscan, UK.

O'Keefe, P., 1983. The causes, consequences and remedies of soil erosion in Kenya. Ambio xii (6).

O'Keefe, P., 1976. Taking the naturalness out of natural disasters. Nature 260.

Reich, J.W., 2006. Three psychological principles of resilience in natural disasters. Disaster Prev. Manage. 15 (5), 793–798.

Sherwood, H., 2014. Global Refugee Figure Passes 50 m for First Time Since Second World War. The Guardian. Friday 20 June 2014, Online at: http://www.theguardian.com/world/2014/jun/20/global-refugee-figure-passes-50-million-unhcr-report.

Smith, N., O'Keefe, P., 1980. Geography, marx and the concept of nature. Antipode 12 (2), 30–39.

Stern, N., 2007. The Stern Review: The Economics of Climate Change. Cambridge University Press.

United Nations, 1991. Strengthening of the Coordination of Humanitarian Assistance in the United Nations. General Assembly. A/RES/46/182.

Weir, E.A., 2006. Conflict and Compromise: UN Integrated Missions and the Humanitarian Imperative. Kofi Annan International Peacekeeping Training Centre.

Chapter 4

Disaster Risk Governance: Evolution and Influences

Samantha Jones[1], Bernard Manyena[2,3] and Sara Walsh[1]
[1] *Department of Geography/Disaster and Development Network (DDN), Northumbria University, Newcastle, UK,* [2] *Humanitarian and Conflict Response Institute, Manchester University, Manchester, UK,* [3] *Disaster and Development Network (DDN), Northumbria University, Newcastle, UK*

ABSTRACT
Disaster risk reduction (DRR) represents a shift in the paradigm of disaster management from "response and recovery" to "prevention and preparedness." International organizations have been key players in advancing this agenda. This chapter seeks to explore the challenging nature of contemporary endeavors in disaster risk governance, which are intricately linked to the neoliberal agenda of "hollowing out" state functions. Under this agenda a reduced role for the state has emerged and an opening of the governing arena to a wider multitude of nonstate actors. This chapter discusses three dimensions to the changing distribution of influence and responsibility in disaster risk governance. The first is the "upward" dimension, wherein governments are becoming more accountable to global institutions. Second, the "outward" or mainstreaming the agenda of disaster risk reduction requires sectors to integrate disaster and development into their activities to develop better prevention and preparedness. Third, the "downward" or decentralization of disaster risk governance arguably enables local communities to formulate realistic and implementable prevention, preparedness, response, and recovery plans. In this complex and changing governance landscape of disaster risk reduction, as the neoliberal state is hollowed out and responsibilities are reoriented upward, outward, and downward, the question arises: "who really governs DRR?"

4.1 INTRODUCTION

Changes in politics and economics over the last 20 years under the neoliberal agenda that have led to a "hollowing out" of the nation-state can also be framed in terms of a shift from "government" to "governance." The term "governance" is often used to denote the "multitude of actors and processes that lead to collectively binding decisions" (van Asselt and Renn, 2011, p. 431) or "the structures and processes by which communities share power" (Lebel et al., 2006). This suggests a reduced role for the state and an opening of the governing arena to a wider multitude of nonstate actors. Concepts of disaster risk

Hazards, Risks, and Disasters in Society. http://dx.doi.org/10.1016/B978-0-12-396451-9.00004-4

governance also convey a similar idea. For example, the United Nations (UN) Development Programme (UNDP, 2013, p. 1) views disaster risk governance as "the way in which the public authorities, civil servants, media, private sector, and civil society coordinate at community, national and regional levels in order to manage and reduce disaster and climate related risks." Bang (2013, following OECD 2010) noted that governance influences the way that national and subnational actors are willing and able to coordinate their actions to manage and reduce disaster-related risk. A consequence of this broadening governance, as Bulkeley and Jordan (2012, citing Agnew and Corbridge, 1995 and other authors) explained, is that the functions of the nation-state are being redistributed "upward" to international institutions and transnational networks, "downward" to regional and local tiers of authority, and "outward" to a range of nonstate actors. This not only has created a diffusion of power (cited by Batterbury and Fernando, 2006), but also has given rise to a policy world comprising diverse, overlapping, and integrated networks often operating beyond effective control of formal governmental structures (Goodwin, 1998).

This chapter explores this wider governance landscape in the context of disaster risk reduction (DRR). DRR represents a shift in the paradigm of disaster management from "response and recovery" to "prevention and preparedness," which have been discussed extensively in the literature (for example, Wisner et al., 2004; Bankoff, 2004). International organizations have been key players in advancing this agenda. In particular, the International Decade for Natural Disaster Reduction (IDNDR) and the Hyogo Framework for Action 2005–2015 (HFA) appear to have promoted an "upward," "outward," and "downward" disaster risk governance, which broadly tends to resonate with the neoliberal agenda. For that reason, this chapter adopts the "upward," "outward," and "downward" structure to elaborate on the meaning of disaster risk governance.

4.2 EVOLUTION OF DISASTER RISK GOVERNANCE

While disaster risk governance has been in place for a relatively short time, the evolution of the thinking and practices behind it have a long history. Evidence of practices in disaster risk management (DRM) as early as 3200 BC can be found in modern-day Iraq, where a social group, Asipu, would advise community members of the action to take to deal with disasters (Coppola, 2011, p. 4). Coppola (2011) noted that disaster management was then limited to single acts and addressed individual hazards such as floods, drought, and fire. Surprisingly, many of the accomplishments were organized, comprehensive, and effective in reducing disaster impacts. Notwithstanding the single hazard focus that persisted over centuries, there was growing disaster awareness among governing bodies. For example, the management of drought impacts became centralized by setting up drought planning committees during the nineteenth century by the British and India governments (Coppola, 2011). The structures that were created were response focused, and therefore much influenced by the notion of disasters as being Acts of God and Acts of Nature, rather than being Acts of Men and Women (Furedi, 2007). From this vantage point, disasters were unpreventable as they were caused by supernatural forces: God and Nature (Wijkman and Timberlake, 1984; Furedi, 2007). Hazards were disasters per se. Disaster governance was informed by technical solutions such as the construction of levees and flood defenses to reduce the impact of hazards.

The hazard paradigm was bolstered in some ways by the civil defense of the Cold War (1948–1989) where contingency plans, based on vulnerability maps and scenarios, were made to relocate large civilian populations in the event of an unpredictable nuclear attack. Most nations developed centralized civil defense organizations that were supported by legal frameworks and financial and human resources (Coppola, 2011). The civil defense's command-and-control model derived from a militaristic system, which is a top-down approach of disaster response, informed disaster governance. When disasters of natural origin occurred, the civil-defense legal, institutional, and organizational frameworks were employed, using, in the main, military rather than civilian assets. Meanwhile, as disasters continued to increase in frequency, intensity, and impact, questions were raised on the efficacy of the hazard-focused, civil-defense model. The questions went deeper into the debate about disaster causation. In their seminal paper, "Taking the naturalness out of natural disasters," O'Keefe et al. (1976) argued that disasters were more a consequence of socioeconomic vulnerability than natural factors. This reasoning shifted disaster causation from "Acts of God" and environmental determinism to "vulnerability," which is underpinned by not least physical, social, and economic factors. Inevitably, this called for a new form of disaster risk governance.

4.3 UPWARD DISASTER RISK GOVERNANCE

The redistribution of state functions upward to the global level in disaster governance involves a redistribution of power and influence upward to institutions such as the UN and World Bank. The early steps in this shift can be traced back to the establishment of United Nations Disaster Relief Office (UNDRO) in the 1970s. UNDRO became the global focal point for disaster relief matters. UNDRO, as its name suggests, was a response-focused organization, and therefore much influenced by the hazard paradigm. The influence of international organizations became more pronounced in 1987 when the UN adopted a resolution (42/169) declaring the 1990s the IDNDR (Alexander, 1991; Lechat, 1990). Resolution 42/169, paragraph 8 (United Nations, 1987), introduced global governance of DRR as governments were called upon:

... to keep the Secretary-General informed of their countries' plans and of assistance that can be provided so that the United Nations may become an international centre for the exchange of information, the storing of documents and the coordination of international efforts concerning the activities in support of the objective and goals ... thus enabling each Member State to benefit from the experience of other countries.

Thus, IDNDR initiated the upward shift in disaster risk governance where Member States would report to the Secretary-General on country progress on DRR. This process was further strengthened by the creation of the United Nations International Strategy for Disaster Risk Reduction (UNISDR) in 1999 (which replaced UNDRO and IDNDR). UNISDR is the designated focal point in the UN system for the coordination of disaster reduction activities of governments, UN agencies, regional organizations, and civil society organizations. In 2005, it was bolstered further through the HFA, which was adopted at the World Conference on Disaster Reduction in Japan. The expected outcome of HFA over the 2005–2015 period is "substantial reduction of disaster losses, in lives and in the social, economic, and environmental assets of communities and countries." It articulates three strategic goals, five priorities for action, and four cross-cutting issues. To reduce disaster risks, the HFA encourages the integration of disaster prevention, mitigation, preparedness, and vulnerability reduction into development programs. It is also assumed that strong disaster institutions, mechanisms, and capacities at all levels, in particular at the community level, can systematically contribute to building resilience to hazards.

UNISDR guides, monitors, and reports on the progress of the implementation of the HFA. To fulfill its mandate, UNISDR has created its own structures, or "platforms" at global, regional, and national levels. The platforms are mechanisms for coordination and policy guidance on DRR that are multisectoral and interdisciplinary in nature, with public, private, and civil society participation involving all concerned entities locally, within a country, region, and globally (UNISDR, 2005). The assumption here is that DRR requires the knowledge, capacities, and inputs of a wide range of sectors and organizations, including UN agencies at local, national, regional, and global levels (UNISDR, 2005). The platforms tend to be embedded in legal frameworks or are loose assemblage of agencies and individuals. Whatever form the platforms take, they are "technical platforms" for sharing information and not necessarily to make policy decisions.

Of all the platforms, the national platforms tend to be more critical to UNISDR, as other platforms (local, subregional, and regional) do not complete the HFA monitor that became accessible in 2008. The subregional and regional platforms are by-passed by the UNISDR system and may not be aware of what their Member States report to UNISDR. Nonetheless, every 2 years since 2007, the national platforms are required to report on their progress on HFA implementation, using the HFA monitor ratings (Table 4.1).

The country reports are compiled by central government agencies, commonly referred to as National Disaster Management Organisations (NDMOs), which are mandated to coordinate DRR. The reports, at least in theory, should be compiled with the participation of national platform members. The midterm review of the HFA noted a lack of evidence on the

TABLE 4.1 HFA Self-Assessment Ratings

Level	Generic Description of Level of Progress
5	Comprehensive achievement has been attained, with the commitment and capacities to sustain efforts at all levels.
4	Substantial achievement has been attained, but with some recognized deficiencies in commitment, financial resources, or operational capacities.
3	There is some institutional commitment and capacities to achieving DRR but progress is not comprehensive or substantial.
2	Achievements have been made but are incomplete, and while improvements are planned, the commitment and capacities are limited.
1	Achievements are minor and there are few signs of planning or forward action to improve the situation.

Source: UNISDR (2013a).

effectiveness of different governance systems for DRR and weaknesses in tackling issues of accountability, transparency, and participation in administrative and decision-making processes at the local and national levels. However, the 2011 Global Assessment Report on DRR states that about half the countries and territories (42 of 82) reported substantial or comprehensive achievement in HFA Priority 1, which focuses on governance. There seems to be recognition of the importance of reporting to UNISDR, which has gradually become the institution of practice and a constitutive element of disaster governance.

Several mechanisms occur at the national level that support the upward shift of disaster risk governance. Chief among them is the UN system, particularly through the UN Country Team, which produces the United Nations Development Assistance Framework (UNDAF). The UNDAF document contains actionable and fundable country development priorities. As DRR tends to be associated with development, the bulk of the disaster activities fall under the UNDP. The UNDP, mainly through its Crisis Prevention and Recovery Thematic Trust Fund, in turn funds the NDMOs to support DRR governance at the national and regional levels. In this way, the UN system has put in place mechanisms to accelerate the upward shift of disaster risk governance, which has been coated with resource mobilization incentives, even if the sustainability of some of these incentives maybe questionable.

4.4 OUTWARD DISASTER RISK GOVERNANCE: MAINSTREAMING

4.4.1 NDMOs and National Platforms

At the national level, disaster management has historically been the responsibility of NDMOs (discussed above). These have tended to comprise a single unit or department within a central, provincial/state, and local government. These disaster management organizations are often underresourced, relief and response focused (rooted in the earlier paradigm discussed in Section 2), have little to no authority to compel other ministries or jurisdictions (whether they are private or public) to prepare or mitigate against hazards, or have not worked in a cross-sectoral environment. Thus, the changes proposed by the UN system highlighted above pose a significant governance challenge for nation-states and a substantial change in the way governments have traditionally administrated disaster management. They are now being called upon to coordinate and compel DRR policy integration throughout complex bureaucracies, without adequate supportive legislation and resources.

Despite the "hollowing out" of the state that has occurred in the last 20 years, the state is still called upon to integrate DRR internally throughout sectors and administrative levels and externally to private partners (e.g., insurance, utilities) and other development partners such as nongovernmental organizations (NGOs). The UNISDR (2005) guides nation-states to "integrate risk reduction, as appropriate, into development policies and planning at all levels of government, including in poverty-reduction strategies and sectors and multi-sector policies and plans." To achieve this, national platforms are recommended to be "the coordination mechanism for mainstreaming DRR into development policies, planning and programmes in line with the implementation of the HFA" (UNISDR, 2007a, p. 4). National platforms serve two functions: they facilitate an "upward" interface between the state and the international DRR governance bodies, as discussed above, and also provide an "outward" mechanism for implementing DRR integration and mainstreaming throughout all politicoadministrative levels.

Nunan et al. (2012) argue that the need to mainstream and integrate environment policy is recognition that a single government department on its own cannot address the complexity of issues associated with environmental policy; the same can be said for DRR policy. In Pelling and Holloway's (2006, p. 16) discussion on the importance of legislation to support the DRR mainstreaming agenda, they offer this definition of DRR mainstreaming:

"Mainstreaming" derives from the metaphor of a small, isolated flow of water being drained into the mainstream of a river where it will expand to flow smoothly without loss or diversion. Therefore "mainstreaming risk reduction" describes a process to fully incorporate disaster risk reduction into relief and development policy and practice. It means radically expanding and enhancing disaster risk reduction so that it becomes normal practice, fully institutionalized within an agency's relief and development agenda.

In order to "institutionalize" DRR, it needs to be integrated throughout a state's organizational structures and policy processes. Lafferty and Hovden (2003) suggested that policy integration needs to occur both horizontally across sectors and vertically within sectors. For government, this means expanding the disaster risk reduction agenda so that all planning in distinct departments such as health, transportation, economic development, and natural resource management internalize DRR into policy and practice. A central coordinating body that can compel ministries and departments to enact policy integration is an important component of mainstreaming (Lafferty and Hovden, 2003; Nunan et al., 2012). Many national platforms are not adequately resourced (e.g., lacking funding, human resources, and authority) to fulfill this function.

UNISDR (2013b) undertook a review of national platforms, surveying 50 in order to gauge how they function and the challenges they faced, and to provide recommendations to increase national platform adoption in all HFA signatory countries. Among many issues, the review found that national platforms vary between nation-states in how they are administered and where they are located within the political hierarchy. A majority of national platforms surveyed ($n = 36$) were integrated within the government administration, and only a few ($n = 3$) organized through an NGO structure (UNISDR, 2013b, p. 4). The national platforms identified many gaps and challenges in the report. Relevant to this discussion, national platforms reported a "lack of expertise and capacity in mainstreaming disaster risk reduction into sectoral activities and networks" (UNISDR, 2013b, p. 9); "low potential to influence the ministerial and state bodies regarding prevention and crisis management"; and "insignificant implementation of laws on disaster prevention resulting in failure to eliminate disaster risks" (UNISDR, 2013b, p. 10).

Despite these significant weaknesses and limitations, the Global Assessment Report (2011, p. 77) indicated that "[T]he institutional arrangements for DRM in many countries have certainly evolved, from traditional single-agency 'civil protection of defence' structures to multi-sector systems and platforms." However, despite the progress being made in HFA Priority Area One at the national level, it is not stimulating DRR action at the local level. The Global Network for Disaster Reduction Views from the Frontlines (2009, p. 36) report finds that "Nationally formulated DRR policies and plans are not generating widespread systemic change in local practices, and importantly, they are not engaging vulnerable and marginalized people as vital and active partners in building disaster-resilient communities." The failure of government to implement DRR has created a void that is largely being filled by development partners particularly from the NGO sector.

4.4.2 Nongovernmental Organizations

In addition to mainstreaming DRR "outward" within the state apparatus, the governance agenda, as noted above, has entailed the inclusion of a wider range of stakeholder groups in disaster management. Rajan (2002) noted that disaster management in the developing world is characterized by missing expertise in the state. He noted that with respect to environmental governance, a variety of actors have in the past two decades begun to attempt to strengthen capacity in sectors where it has traditionally been absent, catalyze systems of transparency and accountability, and promote the erection of democratic and devolved governance frameworks. A parallel situation exists with disaster management, particularly in weaker states, whereupon (international) NGOs have been playing a key role in capacity building.

In DRR specifically, it has been noted that over the past years, some NGOs have committed themselves to advocate for policy changes; others have been active in mainstreaming DRR, whereas many have engaged in capacity building, knowledge transfer, and awareness-raising in communities at risk (UNISDR, 2006). This suggests that NGOs work at multiple levels of governance, from the national (e.g., advocacy roles to influence policy making) to the local level (e.g., supporting community-based disaster risk reduction (CBDRR)). In poorer countries receiving official development assistance, this activity has been facilitated through an increased proportion of aid being channeled to and through NGOs (Suleiman, 2013; citing Edwards and Hulme, 1996) as NGOs have found a new place within the neoliberal global order (Frewer, 2013).

This role for NGOs in DRR has received widespread support. UNISDR (2006), for example, note that the HFA, although placing primary responsibility for achieving disaster resilience on national governments, highlights the importance of involving civil society, NGOs, community organizations, and voluntary groups in DRR. The UNISDR secretariat believes that "building the resilience of nations and communities to disasters cannot be done without the active participation of NGOs" (UNISDR, 2006, p. 1). Benson et al. (2001, p. 200) also suggested that it is recognized that "governments alone cannot achieve significant, sustainable hazard risk reduction" and greater emphasis needs to be placed on local level and community approaches "supported by NGOs." International NGOs are often seen as being more effective and efficient than governmental actors because of: (1) their more grassroots approaches, (2) their tendency to work with poorer and more marginalized groups in society and pursue the "participatory approach to development," and (3) their principled motivations, clear mission and objectives, adaptability, progressive development agendas, and lower overhead costs (Suleiman, 2013; Davis and Murdie, 2012; Benson et al., 2001; Hulme and Edwards, 1997; Lewis and Kanji, 2009). These advantages are also regarded to apply to NGOs engaged in DRR. UNISDR (2006), for example, sees the advantages of NGOs as being able to operate at the grassroots level, enjoy higher organizational flexibility, and often work on behalf of the most "needy" groups.

However, the enthusiasm toward NGOs as a vehicle for development has long been under scrutiny within academia (Frewer, 2013) and a number of papers in the literature critique the role of NGOs in disaster response. NGOs tend to reflect and reproduce both the hierarchies of the societies in which they are embedded, such as neopatrimonial networks, clientelism, corruption, and inequality (Frewer, 2013; Suleiman, 2013; O'Reilly, 2010). NGOs have been criticized for (1) facilitating neoliberal policy change (or at least being co-opted to serve hegemonic development agendas), (2) being more accountable to donors (upward accountability) than "civil society," (3) being self-interested, (4) functioning in a similar way to businesses, (5) privileging those who have the education and skills to compete for donor funding, (6) lacking coordination and generating fragmentation and duplication, and (7) being economically dependent on external funding, which has sustainability implications (Booth, 2011; Ebrahim, 2003; Frewer, 2013; Lewis and Kanji, 2009; Suleiman, 2013; Townsend et al., 2004).

4.5 DOWNWARD DISASTER RISK GOVERNANCE: DECENTRALIZATION

Downward disaster risk governance could be viewed as being synonymous with decentralization. Definitions of decentralization are inconsistent in the literature. Pomeroy and Berkes (1997, p. 469) provided a broad definition: "the systematic and rational dispersal of power, authority and responsibility from central government to lower level institutions." According to Rodden (2004), decentralization is most often viewed as a shift of authority toward local government and away from central government. Others, however, are less specific about decentralization involving state bodies. For example, Ribot et al. (2010), define decentralization as the transfer of meaningful discretionary powers to local representative authorities. Ribot (1999, cited by Lane, 2003) explicitly included nonstate bodies, as decentralization is seen to be the devolution of state powers or assets to local decision-making bodies "including nonstate associations." Below, both are considered with respect to DRR.

4.5.1 Decentralization within the State

Decentralization, as part of neoliberal reforms, has emerged as a major strategy for many nation-states to achieve development goals, provide public services, and undertake environmental conservation (Agrawal and Ostrom, 2001; Perreault, 2005). It promises improvements in efficiency, responsiveness to spatially variable conditions, and greater opportunities for citizen participation and empowerment (Poteete and Ribot, 2011). Scott and Tarazona (2011) suggested that decentralization entails the restructuring of authority according to the principles of subsidiarity (i.e., responsibilities and resources should be decentralized down to the lowest level that can effectively perform necessary tasks).

A distinction can be made between DRR taking place in the context of decentralization and the specific decentralization of DRR functions. Decentralization processes are well underway throughout most of the developing world (Scott and Tarazona, 2011), with some countries having decentralized government actions in response to pressure from international donors (Escaleras and Register, 2012). The underlying premise is that decentralization may improve the provision of public goods and services by "bringing decision makers closer to those affected by their decisions." This means that in most countries it is very likely that DRR will be carried out in the context of decentralized governance. Some quantitative research has been conducted to explore whether government structure is important in limiting disaster losses by exploring the relationship between decentralization and disaster casualties. Skidmore and Toya (2013, p. 102) remarked that "generally our findings suggest that decentralized governments are more effective in disaster preparations and/or responses relative to more centralized government systems." Escaleras and Register (2012) similarly found a consistent and negative statistically significant relationship between a country's degree of fiscal decentralization (proportion of government expenditure controlled by subnational governments) and its death rate from natural disasters as well as the number of persons affected by disasters. They argued then, that their results are supportive of international donor agencies trend in coupling their development assistance to increasing decentralization, at least with respect to hazard risk management.

The literature exploring the extent to which specific DRR functions should be carried out at the local level is perhaps more contested. It has been pointed out that natural disasters, although capable of striking entire nations, more typically affect a local or regional part of a country (Escaleras and Register, 2012). Scott and Tarazona (2011) pointed out that disaster risk is always experienced locally. Bollin et al. (2003, cited by Escaleras and Register, 2012) called for an elevated involvement for local actors in risk reduction for this reason. The UNISDR recognized the need to decentralize and institutionalize risk mitigation at the community and local levels (Bang, 2013) and the HFA clearly states that local authorities should be empowered to manage and reduce disaster risk by having access to information, resources, and the authority to implement solutions (Scott and Tarazona, 2011).

The advantages of decentralized DRR include that (1) members of local government should be the first to respond when disaster occurs (Bang, 2013); (2) local knowledge, combined with delegated spending authority may prove useful in managing disaster risk (Escaleras and Register, 2012); and (3) a host of localized public services may play a role in reducing loss of life, including building code and zoning enforcement, police and fire services, transportation, health services, disaster preparation and response, and critical infrastructure investment (Skidmore and Toya, 2013). On the other hand, a broad consensus exists in the literature that governments need to play a role in designing and implementing comprehensive DRM systems (Scott and Tarazona, 2011) and some experts argue that the national government must retain DRR responsibility, policy, and programming to ensure it has adequate political profile and resources (for example, GAR, 2009; cited by Scott and Tarazona, 2011). In a decentralized governance system, it is more likely that specific challenges for DRR may be faced such as coordination issues, financing difficulties, and capacity constraints (Scott and Tarazona, 2011).

No easy answer occurs to the question of how to reconcile the "top-down national government-driven" approach and the "bottom-up community-led approach" (Scott and Tarazona, 2011). However, the principle of subsidiarity seems key, that specific functions need to be carried out at the national level (such as intersectoral and cross-jurisdictional coordination and policy integration and command of the military) and also those may be more effective if carried out at the local level to enable locally relevant preparedness measures to be adopted (requiring fiscal as well as political decentralization). The challenge may be to ensure that the most appropriate level is ascertained for each disaster management function, that the appropriate authorities are charged with being accountable for providing those functions, and equipped with adequate resources to do so.

4.5.2 Decentralization beyond the State

Under the banner of decentralization in its broadest sense and the redistribution of authority downward, community-based disaster management (CBDM); community-based disaster preparedness (CBDP), and CBDRR have emerged. The value of community-based approaches and projects and programs, which are often developed or supported by NGOs, is that they may focus on "grassroots" community involvement, empowerment, and ownership; give a voice to local people and their representatives; and be based on identifiable local needs and priorities (Batterbury and Fernando, 2006; Scott and Tarazona, 2011). According to Herbert-Cheshire and Higgins (2004) often as part of the decentralization process, communities become enmeshed in networks of relations that assist them in acquiring the capacities necessary to govern themselves. Support exists in the academic literature for CBDRR and developing "disaster-resilient communities," and specific empirical studies present such initiatives in a positive light (see, for example, Sanderson, 1997; Lebel et al., 2006; IFRC, 2009; Tadele and Manyena, 2009; Allen, 2006; Jones et al., 2013).

It has also been noted that a danger may exist "that CBDP could become an added burden for community institutions and actors, especially when not well matched with increased access to resources or decision-making powers" (Allen, 2006, p. 97). Disaster management roles often have been passed to civil-society actors as state expenditure is reduced (Allen, 2006), and where responsibilities have been devolved, revenue transfer for their fulfillment has seldom followed (Coombes, 2007). It has been reported that households may be reluctant to invest their own resources (including time) in community projects (Allen, 2006), which in part may be due to the priorities and perceptions of risk and disaster relative to other concerns. Tackling disaster risk may be seen as a low priority when compared to meeting basic needs and addressing livelihood matters, not least because DRR rarely produces swift or highly visible results (Paton and Johnston, 2001; DFID, 2006). This point is substantiated by field research (e.g., Bajeck et al., 2008; Chen et al., 2006). In addition, a lack of ownership of projects has been reported to be a problem where initiatives are driven by external institutions (Kamoto et al., 2013) or local participation is not strongly encouraged in government-led projects (Zhang et al., 2013). Where there is local interest in CBDM local elites may be most able to attract funding, reproducing embedded distributions of power and vulnerability (Pelling, 1999, 2007).

4.5.3 Local Government and CBDRR

Potentially, it seems that local government can play an important role in bringing together the benefits of "top-down national government-driven" and "bottom-up community-led" approaches (Scott and Tarazona, 2011). For example, upwards and downwards accountability of local authorities may help to safeguard against the "elite capture" of the agenda and resources at the community level, for example (Lebel et al., 2006; Brown and Lassoie, 2010). Yet, although many communities have a long history of "informally" reducing risk, formal local governments have emerged in developing countries within the last 30 years, and much of the political and institutional framework for supporting DRR is even newer (Scott and Tarazona, 2011). Furthermore, inadequate financial, human, and material resources can present a challenge for local government (Manyena, 2006). Most local governments do not employ staff skilled in DRR and local governments lack experience in managing disaster-affected populations (Bang, 2013). Within local government, corruption, the quest for political (party) support, and minimal local participation can be obstacles to the effective local governance of DRR (Bang, 2013). Echoing King's call, clearly a need exists for, and significant benefit is to be gained from "a much fuller engagement of government at the local level" (King, 2008, p. 488).

A number of authors have suggested that ideally, CBDP initiatives should be founded on preexisting local institutions and work within existing structures rather than create new or parallel structures and empower local people by supporting them in becoming increasingly self-reliant (Maksrey, 1989; Allen, 2006; Tadele and Manyena, 2009). However, where new institutions have been formed and have been successful, it is usually because they are nested within the supportive institutions of the state (Johnson and Nelson cited by Coombes, 2007) or supported by local political authority (Larson, 2003). A number of studies in DRR substantiate this. Pelling (2011) found that the most successful CBDRR projects he reviewed across three countries were built on strong preexisting partnerships with strong buy-in from local and municipal governments. Williams et al. (2010) similarly argued that recognition by and support from the local government are needed regardless of the centrality of community motivation to engage in risk reduction activity. Jones et al. (2013) found that "risk and resilience committees" in Nepal were more effective when they operated independently from political parties but were embedded within local government structures.

A number of authors have pointed to the fact that effective decentralization paradoxically depends on strong centralized government institutions (e.g., Veron et al., 2006, p. 1937). Scott and Tarazona (2011) turned this around with respect to DRR and suggested that good local governance is key to strong national governance. Clearly, this is a mutually reinforcing relationship. If DRR roles and responsibilities were allocated on the basis of a well-informed application of the principle of subsidiarity and accompanied by appropriate level of funding and expertise, an important role can be played at all levels of the governance hierarchy.

4.6 CONCLUSIONS

This chapter has sought to explore the challenging nature of contemporary endeavors in disaster risk governance, which are broadly and intricately linked to the neoliberal agenda of "hollowing out" state functions. Increasingly, disaster risk governance has evolved into at least three dimensions. First, an "upward" shift is apparent and this has taken the form of governments being accountable to global institutions, which have set up nonlegal binding mechanisms and structures in the form of global, regional, national, and local platforms through which they govern. Second, the "outward" or mainstreaming DRR requires sectors to integrate disaster and development into a single framework. One of the biggest challenges is that mainstreaming DRR requires a change of mindset from response to prevention orientation. The NDMOs, which were mostly created during the civil defense era of the Cold War, are still skewed toward the response rather than prevention (Manyena et al., 2013). Moreover, most developing world governments have financial, human, and material resource challenges to effectively provide technical support to sectors to mainstream DRR into their programs. Meanwhile, NGOs fill the gap in service delivery left by a "rolled back" state (Grindle, 2004). Third, the "downward" or decentralization of disaster risk governance loosely refers to any delegation and transfer of powers or functions of government from the national level to any subnational level. Decentralizing DRR, arguably, enables local communities, not only as first responders to disasters, but a fundamental human right to participate and formulate more realistic and more implementable prevention, preparedness, response, and recovery plans. However, without devolution of power and resources, local government and/or communities may only feel this as an added burden and may be reluctant to engage. In this complex and changing governance landscape of DRR, as the neoliberal state is hollowed out and responsibilities are reoriented upward, outward, and downward, the question arises: “who really governs DRR?”

REFERENCES

Agnew, J., Corbridge, S., 1995. Mastering Space: Hegemony, Territory and International Political Economy. Routledge, London.

Agrawal, A., Ostrom, E., 2001. Collective action, property rights and decentralisation in resource use in India and Nepal. Politics Soc. 29 (4), 485–514.

Alexander, D., 1991. Natural disasters: a framework for research and teaching. Disasters 15, 209–226.

Allen, K., 2006. Community-based disaster preparedness and climate adaptation: local capacity-building in the Philippines. Disasters 30 (1), 81–101.

van Asselt, M.B.A., Renn, O., 2011. Risk governance. Journal of Risk Research 14 (4), 431–449.

Bajeck, R., Matsuda, Y., Okada, N., 2008. Japan’s Jishu-bosai-sokshiki community activities: analysis of its role in participatory community disaster risk management. Nat. Hazards 44 (2), 281–292.

Bang, H.N., 2013. Governance of disaster risk reduction in Cameroon: the need to empower local government. Jamba J Disaster Risk Stud. 5 (2), 1–10.

Bankoff, G., 2004. The historical geography of disaster: ’Vulnerability’ and ’local knowledge’ in western discourse. In: Bankoff, G., Frerks, G., Hilhorst, D. (Eds.), Mapping Vulnerbaility. Disasters, Development and People. Earthscan, London.

Batterbury, S.J., Fernando, J.L., 2006. Rescaling governance and the impacts of political and environmental decentralisation: an introduction. World Dev. 34 (11), 1851–1863.

Benson, C., Twigg, J., Myers, M., 2001. NGO initiatives in risk reduction: an overview. Disasters 25 (3), 199–215.

Bollin, C., Cárdenas, C., Hahn, H., Vatsa, K.S., 2003. Natural Disasters Network: Comprehensive Risk Management by Communities and Local Governments. Inter-American Development Bank. Available at: http://www.gtz.de/en/themen/uebergreifende-themen/nothilfe/1817.htm/S.

Booth, D., 2011. Towards a theory of local governance and public goods provision. IDS Bull. 42 (2), 11–21.

Brown, H.C.P., Lassoie, J.P., 2010. Institutional choice and local legitimacy in community based forest management: lessons from Cameroon. Environ. Conserv. 37 (3), 261–269.

Bulkeley, H., Jordan, A., 2012. Transnational environmental governance: new findings and emerging research agendas. Environ. Plann. C: Gov. Policy 30, 556–570.

Chen, L.C., Liu, Y.C., Chan, K.C., 2006. Integrated community-based disaster management program in Taiwan: a case study of Shang-An village. Nat. Hazards 37 (1–2), 209–223.

Coombes, B., 2007. Defending community? Indigeneity, self determination and institutional ambivalence in the restoration of Lake Whakaki. Geoforum 38 (1), 60–72.

Coppola, D.P., 2011. Introduction to International Disaster Management. Butterworth-Heinemann, Boston.

Davis, D.R., Murdie, A., 2012. Looking in the mirror: comparing INGO networks across issue areas. Rev. Int. Organ. 7 (2), 177–202.

DFID (Department for International Development), 2006. Reducing the Risk of Disasters – Helping to Achieve Sustainable Poverty Reduction in a Vulnerable World. A DFID Policy Paper. March. DFID, London.

Ebrahim, A., 2003. Accountability in practice, mechanisms for NGOs. World Dev. 31 (5), 813–829.

Edwards, M., Hulme, D., 1997. Non-Governmental Organisations, States and Donors: Too Close for Comfort? Macmillan, Basingstoke.

Escaleras, M., Register, C., 2012. Fiscal decentralisation and natural hazard risks. Publ. Choice 151, 165–183.

Frewer, T., 2013. Doing NGO Work: the politics of being "civil society" and promoting "good governance" in Cambodia. Aust. Geographer 44 (1), 97–114.

Furedi, F., 2007. The changing meaning of disaster. Area 39, 482–489.

Global Assessment Report on Disaster Risk Reduction, 2009. Risk and Poverty in a Changing Climate. UNISDR [Online] Available at: http://www.preventionweb.net/english/hyogo/gar/report/index.php?id=9413&pid:34&pil:1 (accessed 14.02.14).

Global Network of Civil Society Organisations for DRR, 2009. Clouds but Little Rain: Views from the Frontlines – a Local Perspective of Progress towards Implementation of the Hyogo Framework for Action. [Online] Available at: http://www.globalnetwork-dr.org/resources.html (accessed 13.02.14).

Goodwin, M., 1998. The governance of rural areas: some emerging research issues and agendas. J. Rural Stud. 14 (1), 5–12.

Grindle, M.S., 2004. Good enough governance: poverty reduction and reform in developing countries, governance: an international journal of policy. Adm. Inst. 17 (4), 525–584.

Herbert-Cheshire, L., Higgins, V., 2004. From risky to responsible: expert knowledge and the governing of community-led rural development. J. Rural Stud. 20 (3), 289–302.

Hulme, D., Edwards, M., 1997. NGOs, states and donors: an overview. In: Hulme, D., Edwards, M. (Eds.), NGOs, States and Donors: Too Close for Comfort? Macmillan Press Ltd, Basingstoke.

IFRC, 2009. World Disaster Report 2009: Focus on Early Warning, Early Action. IFRC, Geneva.

Jones, S., Aryal, K., Collins, A., 2013. Local level governance of risk and resilience in Nepal. Disasters 37 (3), 442–467.

Kamoto, J., Clarkson, G., Dorward, P., Shepherd, D., 2013. Doing more harm than good? Community based natural resource management and the neglect of local institutions in policy development. Land Use Policy 35, 293–301.

King, D., 2008. Reducing hazard vulnerability through local government engagement and action. Nat. Hazards 47 (3), 497–508.

Lafferty, W., Hovden, E., 2003. Environmental policy integration: towards an analytical framework. Environ. Politics 12 (3), 1–22.

Lane, M., 2003. Decentralization or privatization of environmental governance? Forest conflict and bioregional assessment in Australia. J. Rural Stud. 19 (3), 283–294.

Larson, A.M., 2003. Decentralisation and forest management in Latin America: towards a working model. Public Admin. Dev. 23 (3), 211–226.

Lebel, L., Anderies, J.M., Campbell, B., et al., 2006. Governance and the capacity to manage resilience in regional socio-ecological systems. Ecol. Soc. 11 (1), 19. http://www.ecologyandsociety.org/issues/view.php?sf=22.

Lechat, M.F., 1990. The international decade for natural disaster reduction: background and objectives. Disasters 14, 1–6.

Lewis, D., Kanji, N., 2009. Non-governmental Organisations and Development. Routledge, London.
Manyena, B., 2006. The concept of resilience revisited. Disasters 30 (4), 433–450.
Manyena, S.B., Mavhura, E., Muzenda, C., Mabaso, E., 2013. Disaster risk reduction legislations: is there a move from events to processes? Global Environ. Change 23 (6), 1786–1794.
Maskrey, A., 1989. Disaster Mitigation: A Community-based Approach. In: Development Guidelines No. 3. Oxfam, Oxford.
Nunan, F., Campbell, A., Foster, E., 2012. Environmental mainstreaming: the organisational challenges of policy integration. Public Adm. Dev. 32, 262–277.
O'Keefe, P., Westgate, K., Wisner, B., 1976. Taking the naturalness out of natural disasters. Nature 260, 566–567.
O'Reilly, K., 2010. The promise of patronage: adopting and adapting neo-liberal development. Antipode 41 (1), 179–200.
Paton, D., Johnston, D., 2001. Disasters and communities: vulnerability, resilience and preparedness. Disaster Prev. Manage. 10 (4), 270–277.
Pelling, M., Holloway, A., 2006. Legislation for Mainstreaming Disaster Risk Reduction. Tearfund [Online] Available at: http://www.preventionweb.net/english/professional/publications/v.php?id=1198 (accessed 27.09.13).
Pelling, M., 1999. The political ecology of flood hazard in urban Guyana. Geoforum 30 (3), 249–261.
Pelling, M., 2007. Learning from others: the scope and challenges for participatory disaster risk assessment. Disasters 31 (4), 373–385.
Pelling, M., 2011. Urban governance and disaster risk reduction in the Caribbean: the experiences of Oxfam GB. Environ. Urbanisation 23 (2), 383–400.
Perreault, T., 2005. State restructuring and the scale politics of rural water governance in Bolivia. Environ. Plann. A 37 (2), 263–284.
Pomeroy, R.S., Berkes, F., 1997. Two to tango: the role of government in fisheries co-management: lessons from Asia. Marine Policy 25, 197–208.
Poteete, A., Ribot, J.C., 2011. Repertoires of domination: decentralisation as process in Botswana and Senegal. World Dev. 39 (3), 439–449.
Rajan, S.R., 2002. Disaster, development and governance: reflections on the lessons of Bhopal. Environ. Values 11, 369–394.
Ribot, J.C., 1999. Decentralisation, participation and accountability in Sahelian Forestry: legal instruments of political administrative control. Africa 69 (1), 23–65.
Ribot, J.C., Lund, J.F., Treue, T., 2010. Democratic decentralisation in sub-Saharan Africa: its contribution to forest management, livelihoods and enfranchisement. Environ. Conserv. 37 (1), 35–44.
Rodden, J., 2004. Comparative federalism and decentralisation: on meaning and measurement. Comp. Politics 36 (4), 482±.
Sanderson, D., 1997. Reducing risk as a tool for urban improvement: the Caqueta ravine, Lima, Peru. Environ. Urbanisation 9 (1), 251–262.
Scott, Z., Tarazona, M., 2011. Studies on Disaster Risk Reduction, Decentralisation and Political Economy. Global Assessment Report on Disaster Risk Reduction. ISDR, UNDP, and Oxford Policy Management.
Skidmore, M., Toya, H., 2013. Natural disaster impacts and fiscal decentralisation. Land Econ. 89 (1), 101–117.
Suleiman, L., 2013. The NGOs and the grand illusions of development and democracy. Voluntas 24, 241–261.

Tadele, F., Manyena, B.S., 2009. Building disaster resilience through capacity building in Ethiopia. Disaster Prev. Manage. 18 (3), 317–326.

Townsend, J.G., Porter, G., Mawdsley, E., 2004. Creating spaces of resistance: development NGOs and their clients in Ghana, India and Mexico. Antipode 36 (5), 871–889.

UNDP, 2013. Issue Brief: Disaster Risk Governance. www.undp.org/content/dam/undp/library/crisis%20prevention/disaster/Issue_brief_disaster_risk_reduction_governance_11012013.pdf.

UNISDR, 2005. Hyogo Framework for Action: Building the Resilience of Nations and Communities to Disasters [Online] Available at: http://www.unisdr.org/we/inform/publications/1037 (accessed 13.02.14).

UNISDR, 2006. NGOs and Disaster Risk Reduction: A Preliminary Review of Initiatives and Progress Made. Background paper for a consultative meeting on 'A global network of NGOs for community resilience to disasters', Geneva 25–26 Oct 2006. United Nations.

UNISDR, 2007a. Guidelines: National Platforms for Disaster Risk Reduction [Online] Available at: http://www.unisdr.org/we/inform/publications/601 (accessed 21.02.14).

UNISDR, 2007b. Building Disaster Resilient Communities: Good Practices and Lessons Learned. UNISDR, Geneva.

UNISDR, 2011. Global Assessment Report on Disaster Risk Reduction. Geneva, Switzerland: United Nations International Strategy for Disaster Reduction [Online] Available at: http://www.preventionweb.net/english/hyogo/gar/2011/en/home/download.html (accessed 29.09.2014).

UNISDR, 2013a. National HFA Monitor 2013–2015.

UNISDR, 2013b. Findings of the Review of National Platforms for Disaster Risk Reduction 2012–2013 [Online] Available at: http://www.unisdr.org/files/35266_nationalplatforms submitted for publication.finalreport.pdf (accessed 21.02.14).

United Nations, 1987. International Decade for Natural Disaster Reduction. A/RES/42/169. New York.

Veron, R., Williams, G., Corbridge, S., Srivastava, M., 2006. Decentralised corruption or corrupt decentralisation? Community monitoring of poverty alleviation schemes in eastern India. World Dev. 34 (11), 1922–1941.

Wijkman, A., Timberlake, L., 1984. Natural Disasters: Acts of God or Acts of Man. Earthscan, London.

Williams, L., Collins, A.E., Bauaze, A., Edgeworth, R., 2010. The role of risk perception in reducing cholera vulnerability. Risk Manage. 12 (3), 163–184.

Wisner, B., Blaikie, P., Cannon, T., Davies, I., 2004. At Risk: People's Vulnerability, Hazards and Disasters. Routledge, London.

Zhang, Y., Yi, L., Zhao, D., 2013. Community based disaster management: a review of progress in China. Nat. Hazards 65, 2215–2239.

Chapter 5

Developing Sustainable Capacity for Disaster Risk Reduction in Southern Africa

Per Becker [1] and Dewald van Niekerk [2]

[1] *Centre for Societal Resilience and LUCRAM, Lund University, Stockholm, Sweden,* [2] *African Centre for Disaster Studies, North-West University, Potchefstroom, South Africa*

ABSTRACT

Southern Africa suffers from an acute lack in skilled capacities for disaster risk reduction and climate change adaptation. The region, with its unique disaster risk profile, is mostly bound by its inability to effect change and transformation for sustainable development and resilience building. Reasons for these limitations are largely due to the lack of capacities in disaster risk reduction with the public, semiprivate, as well as private sectors. However, since the 1990s, some significant work in disaster risk governance in the region has led to some of the most forward-thinking policies and acts in the world. Therefore, the disaster risk reduction communities in the various Southern African states have not been idle. The major constraint now faced is the translation of these policies and plans into action, and this is where the need for more capacities becomes strikingly obvious. Research and academic institutions in the region have outlived all of the regimes and regime changes. These institutions are ideally suited to address the policy/practice gap that exists within the disaster risk reduction domain. In this chapter, we argue for a focus shift in developing sustainable capacities for disaster risk reduction in Southern Africa by first acknowledging and second harnessing the existing skills base in the region. We believe that contemporary capacity development for disaster risk reduction is predominantly ad hoc, short term, project focused, and microsized, and should be changed from within the region, by building on existing institutions and networks that are driven by Southern African institutions.

Hazards, Risks, and Disasters in Society. http://dx.doi.org/10.1016/B978-0-12-396451-9.00005-6

5.1 INTRODUCTION

Southern Africa is a complex and dynamic region. After an era of conflict in the subcontinent, gradual stabilization during the two previous decades (1990–2010) has left active conflict in only one country, namely, the Democratic Republic of the Congo (UCDP, 2012). The 15 member states[1] of the Southern African Development Community (SADC) span countries from low to very high human development (UNDP, 2013). Although nine SADC members are still considered to be least developed countries, only two of these had annual gross domestic product (GDP) growth rates of less than 4 percent in 2012 and five of them experienced staggering growth rates of around 7 percent (World Bank, 2013). Much seems to be in place for rapid human development, but alarming inequalities remain, or even deepen, within the populations.

Part of the challenge for sustainable development in Southern Africa is connected to reoccurring and seemingly increasing disasters (Botha and van Niekerk, 2013; Brauch, 2011; United Nations International Strategy for Disaster Reduction, 2013). Madagascar, Mozambique, and the small island states of the region (e.g., Seychelles and Mauritius) are repeatedly affected by cyclones, which may become more frequent and intense in the light of a changing climate (Webster et al., 2005; Nordhaus, 2006). Most of the region is vulnerable to floods and droughts (Ogallo, 2010; Villholth et al., 2013), and stable weather patterns of the past seem to have become out of order for at least the last decade or so (Young et al., 2010), thus disrupting agricultural production even in the traditional breadbasket of Africa. Hailstorms frequently destroy crops (Chagutah, 2009) and wildfires roam the dry countryside (Jansen Van Vuuren, 2013; Van Wilgen, 2009). The hazard profile of the region also includes earthquakes and volcanoes (Haulle, 2012), predominantly along the Great Rift Valley, as well as landslides and sinkholes (Coetzee and van Niekerk, 2013). Diseases and epidemics continuously take their toll on human lives and suffering, particularly lower respiratory infections, diarrheal diseases, and human immunodeficiency virus/acquired immunodeficiency syndrome (WHO, 2011). Pests and invasive species threaten crops and animals on which so many depend for their food security and livelihood. Traffic accidents, industrial accidents, and pollution are soaring in the urban centers of these developing countries (Lozano et al., 2013; Mills-Knapp et al., 2012).

In addition to the various and often increasing hazards of the region, population growth, migration, and rapid urbanization place increasing numbers of people in harm's way. Globalization and the increasing complexity of modern society increase our interdependence in such a way that "the effects

1. Angola, Botswana, The Democratic Republic of the Congo, Lesotho, Madagascar, Malawi, Mauritius, Mozambique, Namibia, Seychelles, South Africa, Swaziland, Tanzania, Zambia, Zimbabwe.

of distant events can be highly significant elsewhere and specific local developments can come to have considerable global consequences" (Held and McGrew, 2001), as illustrated by the relatively limited volcano eruption in Iceland having a direct impact on the Tanzanian tourism and flower industries. Food prices are no longer determined by what is produced locally, but by an increasingly ravenous global market. Therefore, not only hazards are changing in the region, but also vulnerability to hazards.

The Southern African risk landscape is transforming and further accentuating the importance of effective disaster risk reduction (DRR). While the global Hyogo Framework for Action (2005) has advocated capacity development as the tool to substantially reduce disaster losses for almost a decade now, most appraisals of the practical implications of it are rather discouraging (e.g., GNCSODR, 2009; CADRI, 2011; GNCSODR, 2011; GNCSODR, 2013). Moreover, the lack of appropriate skills and abilities in transdisciplinary DRR hampers effective prevention and mitigation measures (Van Niekerk, 2012). The purpose of this chapter is to elaborate on the challenges for capacity development for DRR in Southern Africa and to suggest a way forward. This will firstly be done by explaining the context in which DRR in Southern Africa functions, secondly the challenges for DRR in Southern Africa are discussed, and lastly a number of good examples for capacity development in Southern Africa will enjoy attention.

5.2 SETTING THE CONTEXT

In December 1987, the United Nations General Assembly adopted Resolution 42/169, which proclaimed the years 1990–1999 as the International Decade for Natural Disaster Reduction (IDNDR) (WMO, 1997; Smith, 2009; UNISDR, 2002; UN, 1987; Lechat, 1990). During this decade, a concerted international effort was attempted to reduce the loss of life, property, livelihoods, and social and economic disruption caused by the violent impact of nature on vulnerable conditions. The aim of this decade was to ensure a shift from the reactive approach toward natural disasters to that of proactive planning and prevention (Housner, 1989; Lechat, 1990; Smith, 2009). Although heavily reliant on technological solutions to natural hazard mitigation (Pelling and Wisner, 2012), the IDNDR succeeded in focusing international attention on issues of disaster reduction (United Nations, 1994) and in doing so paved the way for the follow-up decade, the International Strategy for Disaster Reduction (ISDR), 2000–2010.

Smith (2009) notes that the 2000 onward era is characterized by a continued growth in scale and diversity of research on hazards where many of the ideas provided by the social science community have taken root. In this era, an apparent continuation occurs in the shift in emphasis from hazards to disaster risk and the spotlight on human vulnerability is maintained. The ISDR thus brought the human element into the equation of understanding disaster risk. It is also in the post-2000 era that significant growth in the DRR knowledge base occurred, to the extent that some argue for the formation of a new discipline (Van der Waldt, 2013).

It is in the post-2000 era that many scholars started to argue for an integrated focus on issues of DRR and climate change adaptation (Ireland, 2010; Prabhakar et al., 2009; Schipper and Pelling, 2006), although evidence suggests that these two concepts are still treated as separate entities by governments within the Southern African region (Becker et al., 2013). Notwithstanding this lack of integration, civil society in the Southern African region has been quite active in driving the "capacity development for DRR" agenda. This is evident in the proliferation of formal, nonformal, and informal training interventions in the subregion since 2000. It is worth noting that in many Southern African countries the main drive toward legislative and policy reform in the DRR sector has been the research, training, and education sectors (Pelling and Holloway, 2006).

However, the integration (also called mainstreaming) of DRR into development planning is not a new idea (Blaikie et al., 1994; Cuny, 1983; Schipper and Pelling, 2006; Smith, 2009). From the 1990s onward, multiple disciplines and sectors have been making the linkage between disasters and development (United Nations Development Programme, 2004; United Nations International Strategy for Disaster Reduction, 2004; Wisner et al., 2004). The emphasis in mainstreaming DRR in development is almost logical if one considers the immense impact that disasters can have on development gains. However, such logic is not always observable in practice. Many social scientists became vocal on the need to consider disaster risk issues as social problems and not merely as natural occurrences (Birkmann, 2006; Cardona, 2004; Hewitt, 2013). Comfort et al. (1999) noted that widespread failure had occurred in recognizing and addressing the ways in which changes in land use, settlement policies, population distribution, and the attendant degradation of habitats dramatically increase hazard exposure and vulnerability, and ultimately increase the risk of disasters.

Reducing disaster risks in the most vulnerable African communities therefore requires a radical shift in the thinking of governments from judging disaster risks only in terms of harm to recognizing that disaster risk contains many opportunities. Currently, disaster and risk management in a number of African countries still revolves around planning for disasters and reactive measures to disasters. Much attention is given to preparedness planning and contingency measures in the guise of disaster reduction. DRR is mostly not implemented, or actions for traditional disaster management are merely used as synonyms for disaster risk management. Critical and open evaluation of existing plans, policies, and legislation is limited or totally lacking (Van Niekerk and Coetzee, 2012). Development and humanitarian aid still play a significant role within African development (Kellett and Sweeny, 2011; Moyo, 2009). Vulnerable African states are compelled to play into the hand of donors to ensure external funding for their development as well as disaster reduction needs.

However, it is now widely recognized by the government and civil society actors that the most likely solution to disaster problems is the implementation of successful developmental projects toward vulnerability and risk reduction, environmental management, socioecological system sustainability, and sustainable livelihoods (United Nations, 2005; United Nations Development Programme, 2004; De Graaf, Musters and Chapin, 1998; Bryan et al., 2009). Even if there is verbal agreement by African leaders, one must ask how implementation of such a comprehensive approach to disaster reduction is likely to play out in a continent where human development (measured by movement toward the minimal Millennium Development Goals) is seriously in question (Moyo, 2009; Parnell and Walawege, 2011). In short, how does one translate a comprehensive approach to both disaster risk management and development planning into tangible and sustainable actions?

The above problems are quite pronounced in the Southern African setting. A number of Southern African states are faced with the most basic security and development problems (Hewitt, 2013). Issues of integration of DRR and development are mostly not very high on national agendas. The reasons are shortage of adequate funding, resources, and most importantly challenges in human capacity, skills, and knowledge (Andrade Perez et al., 2010; Coetzee and Fourie, submitted for publication; Hagelsteen and Becker, 2013; Kellett and Sweeney, 2011). However, some very heartening cases also occur where successes in DRR and sustainable development are present. Examples such as DRR in Poverty Reduction Strategy Papers (PRSPs) of Tanzania and Zambia, the emphasis on DRR in the Madagascar Action Plan, and the decentralization and legal incorporation of DRR in local development plans in South Africa and Lesotho can be cited.

5.3 CHALLENGES FOR EFFECTIVE DISASTER RISK REDUCTION IN SOUTHERN AFRICA

The implementation of disaster risk solutions takes place in specific political, economic, environmental, and sociocultural contexts (see United Nations Development Programme, 2004). Thus, the concrete implementation of policy based on this new paradigm cannot be separated from the realities of governance in Southern African countries and the institutional inertia created by decades of implementing "development" policies and projects in certain ways. Therefore, one of the main reasons for the patchy uptake and implementation of a comprehensive approach to DRR and development can be traced to early acceptance by African elites of the Western-imposed notion of export and growth-led development. However, as Moyo (2009) argued, African governments should relook at how development is taking place on the continent. Van Niekerk and Wisner (2014) mentioned the lack of DRR/development integration as one of the major stumbling blocks toward reducing the losses associated with disasters.

The Millennium Report (United Nations, 2000) recommended a fivefold strategy for reducing losses from disasters, which includes

- Strategies to reduce disaster losses need to be mainstreamed in PRSPs;
- Infrastructure investment to incorporate DRR;
- Social safety nets for the vulnerable, particularly through government provisions;
- Early warning capacities and information campaigns supported by governments;
- Precrisis emergency and contingency planning.

However, if the development focus shifts to include social protection and community participation (as the World Bank's PRSPs), one is left with a conundrum: How does one justify the spending of public money and the application of resources to prevent something that might not happen when day-to-day developmental pressures (for health care, roads, schools, etc.) are greater (United Nations Development Programme, 2004)?

Many influential voices of the international community cry out the need for addressing issues of disaster risk, and many of them frame a way forward in terms of capacity development for DRR (e.g., ISDR, 2005; Field et al., 2012; UNCSD, 2012). Therefore, while the Southern African countries must develop their own capacities for DRR, external actors can play important roles in supporting such development. However, when looking at past capacity development for DRR, a rather disheartening picture emerges:

> *While the importance of capacity is widely recognised, how it emerges, how to develop and evaluate it and how to sustain it is for many less clear. There are a number of experiences, tools and resources that are now available in the field of disaster risk reduction and relate to capacity development. Lessons of past experience, for example, point to many inappropriate approaches with short-lived impacts on the part of development cooperation partners. There is however the need for many to better familiarise with the link between capacity, its development and disaster risk reduction. The evidence and knowledge available within the disaster risk reduction community on how to support the development of capacity "in practice" is still not widely systematised and shared, although examples do exist."*
>
> (CADRI, 2011: pp. 7–8)

This statement from the Capacity for Disaster Reduction Initiative (CADRI) provides a sobering perspective of the contemporary position of internationally supported capacity development for DRR. Spahn et al. (2010) state that capacity development brings forward major challenges when implemented in practice, as visible in their study on tsunami preparedness in

Indonesia. Spahn et al. (2010) noted that a need for clear institutional arrangements exists at different levels, making way for a more streamlined and effective development process. They elaborate by stating that due to a lack of information and knowledge on DRR, local actors often have limited influence and ability before or during a disaster.Individuals and organizations, of course, are addressing this challenge, but it is unfortunate so that many others must abandon considerable parts of their habitual mode of thinking concerning capacity development for it to have any lasting effects on DRR in Southern Africa.

Contemporary capacity development for DRR is predominantly ad hoc, short term, project focused, and microsized (Hagelsteen and Becker, 2013). It is mostly equated with training of individuals (Hagelsteen and Becker, 2013), which of course often may be necessary but is rarely sufficient when their organization is not capable of utilizing and maintaining their newly acquired knowledge and skills (Eade, 1997; Schulz et al., 2005). In other words, training is an important instrument for capacity development, but sustainable capacity for DRR can never be developed if it is not properly institutionalized within the country or region in question (Becker, 2009; Hagelsteen and Becker, 2013). For example, South Africa has long been cited as a best practice example of DRR policy and legislative development and implementation. However, research by Botha et al. (2011) and Coetzee and Fourie (submitted for publication) indicate that the institutionalization of skills and capacity development remains a huge stumbling block at local government level. They found that also in South Africa, capacity development is treated as an ad hoc activity without much forward planning or focus on continuity.

Capacity development must instead entail a combination of activities, addressing challenges concerning human and material resources, organization, systems of organizations, and legal and institutional frameworks (Becker et al., 2014). It must be holistic and long term. Instead, contemporary capacity development for DRR is most often fragmented, short-term activities, aimed at quick visible results to serve political purposes (McEntire, 2007). This unfortunately may even undermine the overall capacity for DRR in developing countries. Although quick positive feedback is important for creating commitment in a particular initiative (Kotter and Cohen, 2002), such short-term activities should always only be the first phase of a more long-term agenda.

5.4 ADDRESSING CAPACITY DEVELOPMENT IN SOUTHERN AFRICA

Considering the challenges for DRR, and perhaps even more so for capacity development for DRR, the future for vulnerable people in Southern Africa appears bleak at best and unbearable at worst. Without sufficient human and material resources, it is difficult to reduce disaster risk. However, having the necessary resources does not in any way guarantee success. Effective DRR also requires organization, not only within organizations as such, but also between them in the larger institutional framework. Here legislation, policy, and other formal institutions become important in providing persistent and predictable guidelines for behavior and interaction among individuals and organizations that facilitate coexistence and collective activities by reducing the need for constant negotiation (Handmer and Dovers, 2007). Norms, values, and other informal institutions have similar utility as formal institutions, though lacking the same formal sanctioning. Developing sustainable capacity for DRR in Southern Africa requires in other words comprehensive purposeful initiatives that address all these levels if necessary. It is thus a truly daunting task, but a task that one cannot shy away from or reduce to ad hoc short-term projects just because more adequate approaches are difficult. One initiative by the African Centre for Disaster Studies at North-West University in South Africa, funded by the United States Agency for International Development (USAID), aims to use open access and open knowledge principles as driving force to assist in capacity development in the region (see Box 5.1).

Influential guidelines for capacity development state that it concerns countries' own development processes, is an ongoing process of change that requires time, depends on the countries' own development level and path

BOX 5.1 DRR Knowledge Shop

The DRR Knowledge Shop was developed to provide a unique DRR body of knowledge, free of copyright protection to role-players in the Southern African region. The DRR Knowledge Shop works on the principles of open knowledge. The idea of the DRR Knowledge Shop is that the provision of free, open, and easily accessible training and learning materials (called Knowledge Products) will lead to flexible, multisphere, and transdisciplinary DRR training programs aimed at various practitioners working in related disaster risk management fields. Such training programs can be context and content specific and address a specific target audience. The DRR Knowledge Shop aims to stimulate multiple training interventions within various Southern African training institutions, utilizing the common body of knowledge. These training interventions range from short courses to Masters and PhD programs. Key to the success of the DRR Knowledge Shop is the necessity for adapting the various products to suit a particular environment and circumstances. As these knowledge products are changed and adapted, they are once again included in the DRR Knowledge Shop as a new/alternative iteration from which the next user can learn. Some successes in utilizing and applying the products of the DRR Knowledge Shop are already visible with a number of formal and nonformal training interventions offered throughout the region. Not only will this approach ensure quality and uniformity, but also the SADC region will have a common standard in DRR knowledge.

The Knowledge Shop can be accessed at www.acds.co.za.

without set formulas or blueprints, and relates as much to broader societal challenges and systemic issues as to training, skills development, and technology transfer (UNDP, 2009; CADRI, 2011). In other words, although outside stakeholders have important roles to assist disaster-prone developing countries achieve their own development objectives, there must be clear local ownership of the capacity development process (CADRI, 2011). This is fully in line with the principles of the Paris Declaration on Aid Effectiveness (OECD, 2009), but not always implemented in capacity development initiatives in Southern Africa. A common rationale for such deviations away from fundamental principles among external partners, though tainted by both orientalism and ideas of "white man's burden," is that the internal partner is too weak, not capable or willing to shoulder the responsibility. However, it may be fair to assume that what is experienced as lack of capacity or willingness, could also be a symptom of internal partners being reluctant to have their work co-opted by outsiders, their plans overrun by short-term but well-funded projects, or their agendas filled with activities that they have had little influence on from the start. Regardless of the case in such circumstances, it poses a classic causality dilemma that needs to be resolved. This can be achieved by explicitly involving other types of institutions that are designed for long-term change, i.e., universities and research institutions. Universities, perhaps, are the most stable type of institutions in Southern Africa, as many of them have outlived regime change, armed conflict, and disasters. The universities of subregions have a crucial role to play in developing sustainable capacity for DRR.

This idea of the importance of universities should not be underestimated. For instance, when contemplating the reasons for the success of the DRR programs of countries like Cuba, it is clear that the capacities for research and education within the country play central roles (Thompson and Gaviria, 2004). With in-house ability for scientific knowledge production and dissemination concerning DRR in a disaster-prone developing country itself, it is possible to contextualize knowledge and solutions and to maintain a critical mass of people applying them in practice. We believe the great Thomas Huxley was right when stating that "[t]he great end of life is not knowledge but action" (Huxley, 1882, p. 89). However, "a person who does not have access to information cannot take responsibility," but a "person who has information cannot resist from taking responsibility" (Carlzon, 2008, our translation from Swedish). Contextualized knowledge is thus key to everything. It is the key to designing cost-effective ways to organize DRR that are appropriate to the challenges of each particular country. It may take some time, but a steady stream of well-educated individuals will also eventually spur changes in legislation and policy, and even in norms and values as new ideas take precedence in the minds of the population.

The good news is that functioning universities exist in all countries in Southern Africa and several of them already have departments or centers focusing on DRR. For instance, the Disaster Management Training Centres at

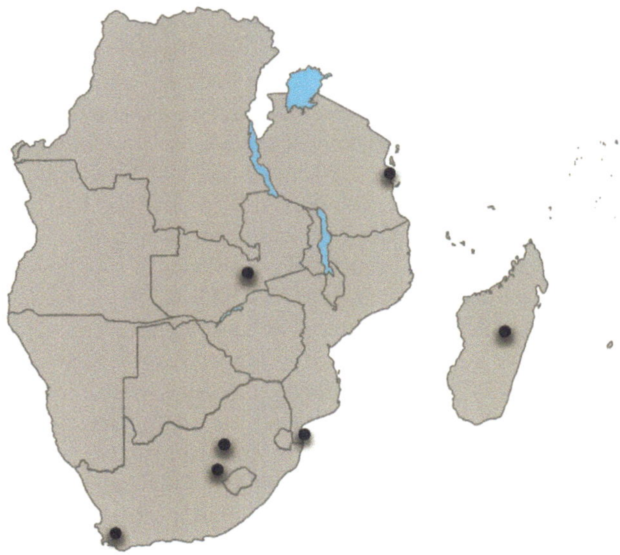

FIGURE 5.1 Examples of locations of universities with departments or centers focusing on DRR.

Ardhi University in Tanzania and Mulungushi University in Zambia, University of Antananarivo in Madagascar, Universidade Técnica de Moçambique in Mozambique, African Centre for Disaster Studies at North-West University in South Africa, Disaster Management Training and Education Centre for Africa at the University of the Free State in South Africa, and Disaster Mitigation for Sustainable Livelihoods Programme at Stellenbosch University in South Africa (listed north to south in Figure 5.1).

It is obvious that not all universities in Southern Africa can pride themselves of excellence in research and education. We have to remember that it is a developing region. However, the institutions exist, and it is fair to assume that they will continue to be there long after any internationally supported initiatives have ended. Supporting universities to develop their capacities for knowledge production and dissemination is therefore a long-term strategy that is likely to have sustainable results. Many donors already acknowledge this and support Southern African universities in a whole range of topics. For example, Sida supports universities in Tanzania and Mozambique; Department for International Development (DFID) supports universities in South Africa, Tanzania, and Zimbabwe; and USAID supports South Africa and its partner organizations. These initiatives are important and we believe the international community should consider scaling up the support to develop capacities for scientific knowledge production and dissemination concerning DRR. One very good example of such initiative is the USAID-supported partnership of African universities called

BOX 5.2 Periperi U

Periperi U is a partnership of African universities that spans the continent and is committed to building local DRR capacity. It started out in 2006 with five members and has since grown to 10 universities in 10 countries including Ethiopia, Uganda, Kenya, Tanzania, Mozambique, Madagascar, South Africa, Ghana, Senegal, and Algeria. In terms of capacity development, the Periperi U program aims to develop active teaching and training, research, and policy advocacy capacity in Africa. The short courses presented by the partners offer subjects like food livelihood security, introduction to DRR in urban areas, food hygiene and safety, emergency planning, and nutrition in emergencies. A number of articles have been published by Periperi U that are freely available for download from their Web site. With a solid foundation and growing demand, partners are continuing to develop their formal academic programs related to disaster risk with the goal of further strengthening skilled human capacity in DRR in Africa. More than 1500 professionals and practitioners have received training in their local languages with more than 200 skilled graduates with over 400 current students.

Periperi U (Box 5.2), which includes four of the universities listed earlier. However, such networks could be made even more effective for capacity development if more explicitly focusing on peer-to-peer learning, not only in the form of south–south cooperation within the region, but also by including universities from outside the region and from more affluent countries, which also have a lot to learn from the experiences and expertise of their Southern African colleagues. Activities could involve sharing curricula, finding

BOX 5.3 The Southern African Society for Disaster Reduction

The Southern African Society for Disaster Reduction (SASDiR) is a community of experts within the field of DRR in the regional context of SADC. It was founded in 2012 with partners in South Africa, Namibia, Swaziland, Sweden, Mozambique, Lesotho, Malawi, and Zambia, ranging from individuals to institutions. The society is open to any individual, organization, or government department that identifies itself with DRR and contributes to lessening the impacts of hazards on vulnerable conditions within SADC communities, which could lead to disasters. The main objectives of SASDiR are to facilitate interaction between DRR role-players in civil society, government, and institutions of higher learning and research; to build on the current body of knowledge of DRR for the SADC region and beyond; to share innovative disaster reduction practices and research among its members; and to contribute to standardization of practices and baseline information on DRR for the region. SASDiR is currently developing a mentoring and internship program aimed at bringing together expertise for in-service training and learning. The society believes that there is a wealth of DRR knowledge in SADC, which is exponentially growing and which can be shared where a need arises.

partnerships, comparing data, joint project, and exchange of expertise. Another good example of a network that has already proven a vibrant platform for such collaboration is the Southern African Society for Disaster Reduction (Box 5.3), through which partners, for instance, have worked on joint research projects supported by USAID, UN Office for the Coordination of Humanitarian Affairs (OCHA), and Food and Agriculture Organization (FAO).

The way forward for capacity development for DRR in Southern Africa requires the international community to not only acknowledge that it is a planned long-term process in rhetoric and policy, but to adjust its actual activities accordingly. Supporting Southern African universities to develop their capacities for research and education in DRR, by supporting networks of peer-to-peer learning and south–south cooperation, is an important strategy for facilitating sustainable development in such a complex and dynamic region.

5.5 CONCLUSION

Disasters continue to pose grave challenges for sustainable development in Southern Africa, and the capacity development efforts of the international community have so far shown little long-term impact. In short, Southern Africa has seen enough ad hoc, short-term, project-focused, and microsized initiatives. Although there may be circumstances that demand immediate attention and action to address a particular urgent issue, such activities can never resolve the challenges of DRR for sustainable development in the region. Therefore, Southern Africa first and foremost needs to strengthen its own capacity for knowledge production and dissemination. Knowledge production through research is needed to explain, understand, and address key phenomena and challenges in the particular contexts of the region. Knowledge that is disseminated through academic programs that provide a steady stream of graduates who directly contribute to the development of sufficient human and material resources, effective organization, adequate legal and policy frameworks, and over time even favorable norms and values in society is needed.

Functioning universities exist even in the weakest states of Southern Africa and it is time for the international community to focus more long term and acknowledge their central role in breaking the vicious cycle of disasters and failed development. One good research-based academic program per country may be all it takes to start effecting change and transformation in the region. About 20–30 graduates from a program fully dedicated to DRR would literally double the pool of that kind of expertise in several of the countries after the first year. This would mean more qualified individuals to recruit to the many positions on national, provincial, and local level that are currently required in legislation and policy, but for which there are often not enough people with the necessary knowledge and skills. Such continuous stream of graduates would also provide enough people to maintain this pool of human resources even with the often high staff turnover of civil servants in the region. Having such dedicated programs would facilitate more general DRR courses to be integrated into other academic programs at the universities, which would over time facilitate the necessary cross-sectoral buy-in and momentum for accelerated and sustainable change.

Knowledge and experience of DRR exists in the region already, mainly in a few established dedicated research centers within universities and among them are often not more than a handful of overstretched professionals per country. Harnessing these capacities and sharing them through a network of mutual collaboration and support between universities would facilitate the development of the necessary research base and curricula in DRR. This will particularly be effective if such a network is based on an organized system of twinning new members with the growing number of universities with established research and educational programs. Examples of this already exist, such as Periperi U and SASDiR (see boxes above), and such are the initiatives that the international community should support with its DRR funds if its stakeholders are sincere in their wish to substantially reduce disaster losses and to facilitate sustainable development in Southern Africa.

REFERENCES

Andrade Perez, A., Herrera Fernandez, B., Cazzolla Gatti, R., Management, I.C.O.E., 2010. Building Resilience to Climate Change: Ecosystem-Based Adaptation and Lessons from the Field. IUCN, Gland, Switzerland.

Becker, P., 2009. Grasping the hydra: the need for a holistic and systematic approach to disaster risk reduction. Jàmbá: J. Disast. Risk Stud. 2 (1), 12–24.

Becker, P., Abrahamsson, M., Tehler, H., 2014. Resilience engineering perspectives. In: Nemeth, C., Hollnagel, E. (Eds.), An Emergent Means to Assurgent Ends: Societal Resilience for Safety and Sustainability, vol. 3. Ashgate, Farnham and Burlington.

Becker, P., Abrahamsson, M., Hagelsteen, M., 2013. Parallel structures for disaster risk reduction and climate change adaptation in Southern Africa. Jàmbá: J. Disast. Risk Stud. 5 (2), 5 pp.

Birkmann, J., 2006. Measuring Vulnerability to Natural Hazards: Towards Disaster Resilient Societies. United Nations University, Tokyo; New York.

Blaikie, P.M., Cannon, T., Davis, I., Wisner, B., 1994. At Risk: Natural Hazards, People's Vulnerability, and Disasters. Routledge, London.

Botha, D., van Niekerk, D., 2013. Views from the Frontline: a critical assessment of local risk governance in South Africa. Jàmbá: J. Disast. Risk Stud. 5 (2), 10 pp.

Botha, D., van Niekerk, D., Wentink, G., Coetzee, C., Forbes, K., Maartens, Y., Annandale, E., Tshona, T., Raju, E., 2011. Disaster Risk Management Status Assessment at Municipalities in South Africa. South African Local Government Association (SALGA), Pretoria.

Brauch, H.G., 2011. Coping with Global Environmental Change, Disasters and Security. Springerverlag Berlin Heidelberg.

Bryan, E., Deressa, T.T., Gbetibouo, G.A., Ringler, C., 2009. Adaptation to climate change in Ethiopia and South Africa: options and constraints. Environ. Sci. Pol. Elsevier 12 (4), 413–426.

CADRI, 2011. Basics of Capacity Development for Disaster Risk Reduction. Capacity for Disaster Reduction Initiative, Geneva.

Cardona, O.D., 2004. The need for rethinking the concepts of vulnerability and risk from a holistic perspective: a necessary review and criticism for effective risk management. In: Bankoff, G., Frerks, G., Hilhorst, D. (Eds.), Mapping vulnerability, Disasters, development and people. Earthscan Publishers, London, pp. 37–51.
Carlzon, J., 2008. Riv pyramiderna!: en bok om den nya människan, chefen och ledaren. Natur & Kultur, Stockholm.
Chagutah, T., 2009. Towards improved public awareness for climate related disaster risk reduction in South Africa: a participatory development communication perspective. Jàmbá: J. Disast. Risk Stud. 2 (2), 113–126.
Coetzee, C., Fourie, K., June 2014. 2014. Skills and capacity shortages in South African disaster risk management entities: Observations from the USAID/ACDS Disaster Risk Management Training Project. 2nd IRDR Conference, 7–9 June 2014, Beijing.
Coetzee, C., van Niekerk, D., 2013. Towards a disaster risk assessment methodology for communities underlain by dolomite. Disast. Prev. Manage. 22 (4), 351–365.
Comfort, L., Wisner, B., Cuter, S., Pulwarty, R., Hewitt, K., Oliver-Smith, A., Wiener, J., Fordham, M., Peacock, W., Krimgold, F., 1999. Reframing disaster policy: the global evolution of vulnerable communities. Environ. Hazards 1, 39–44.
Cuny, F.C., 1983. Disasters and Development. Oxford University Press, New York.
De Graaf, H.J., Musters, C., Chapin, T., 1998. Defining socio-environmental systems for sustainable development. Ecol. Econ. 26, 243–258.
Eade, D., 1997. Capacity-Building: An Approach to People-Centred Development. Oxfam, Oxford.
Field, C.B., Barros, V., Stocker, T.F., Dahe, Q., Dokken, D.J., Ebi, K.L., Mastrandea, M.D., Mach, K.J., Plattner, G.K., Allen, S.K., Tignor, M., Midgley, P.M. (Eds.), 2012. Managing the Risks of Extreme Events and Disasters to Advance Climate Change Adaptation (SREX). Cambridge University Press, Cambridge.
GNCSODR, 2009. Clouds but Little Rain.... Global Network of Civil Society Organisations for Disaster Reduction, Teddington.
GNCSODR, 2011. If We Do Not Join Hands.... Global Network of Civil Society Organisations for Disaster Reduction, Teddington.
GNCSODR, 2013. Views from the Frontline: Beyond 2015. Global Network of Civil Society Organisations for Disaster Reduction, Teddington.
Hagelsteen, M., Becker, P., 2013. Challenging disparities in capacity development for disaster risk reduction. Int. J. Disast. Risk Reduct. 3, 4–13.
Handmer, J.W., Dovers, S., 2007. The Handbook of Disaster and Emergency Policies and Institutions. Earthscan, London.
Hewitt, K., 2013. Disasters in "development" contexts: contradictions and options for a preventive approach. Jàmbá: J. Disast. Risk Stud. 5.
Haulle, E., 2012. Evaluating earthquake disaster risk management in schools in Rungwe Volcanic Province in Tanzania. Jàmbá: J. Disast. Risk Stud. 4 (1), 7 pp.
Held, D., McGrew, A., 2001. The Oxford companion to politics of the world. In: Krieger, J., Crahan, M.E. (Eds.), Globalization. Oxford University Press, Oxford and New York, pp. 324–327.
Housner, G.W., 1989. An international decade of natural disaster reduction: 1990–2000. Nat. Hazards 2 (1), 45–75.
Huxley, T.H., 1882. Science and culture, and other essays. In: Technical Education. D Appleton & Company, New York, pp. 73–93.
Ireland, P., 2010. Climate change adaptation and disaster risk reduction: contested spaces and emerging opportunities in development theory and practice. Clim. Dev. 2 (4), 332–345.

ISDR, 2005. Hyogo Framework for Action 2005–2015: Building the Resilience of Nations and Communities to Disasters. International Strategy for Disaster Reduction, Geneva.

Jansen Van Vuuren, I.D., 2013. A framework for rapid impact assessment with special reference to wildfires. Jàmbá: J. Disast. Risk Stud. 5 (2), 8 pp.

Kellett, J., Sweeney, H., 2011. Analysis of Financing Mechanisms and Funding Streams to Enhance Emergency Preparedness. Development Initiatives, Somerset.

Kotter, J.P., Cohen, D., 2002. The Heart of Change: Real-Life Stories of How People Change Their Organizations. Harvard Business Press, Boston.

Lechat, M.F., 1990. The international decade for natural disaster reduction: background and objectives. Disasters 14 (1), 1–6.

Lozano, R., Naghavi, M., Foreman, K., Lim, S., Shibuya, K., Aboyans, V., et al., 2013. Global and regional mortality from 235 causes of death for 20 age groups in 1990 and 2010: a systematic analysis for the global burden of disease study 2010. Lancet 380 (9859), 2095–2128.

McEntire, D.A., 2007. Disaster Response and Recovery: Strategies and Tactics for Resilience. Wiley, Hoboken.

Mills-Knapp, S., Traore, K., Ericson, B., Keith, J., Hanrahan, D., Caravanos, J., 2012. The World's Worst Pollution Problems: Assessing Health Risks at Hazardous Waste Sites. Blacksmith Institute, New York.

Moyo, D., 2009. Dead Aid: Why Aid is Not Working and How There is a Better Way for Africa. Farrar, Straus and Giroux, New York.

Nordhaus, W.D., 2006. The Economics of Hurricanes in the United States. NBER Working Paper Series. W12813.

OECD, 2009. The Paris Declaration on Aid Effectiveness (2005) and the Accra Agenda for Action (2008). OECD, Paris.

Ogallo, L., 2010. The mainstreaming of climate change and variability information into planning and policy development for africa. Procedia Environ. Sci. 1, 405–410.

Parnell, S., Walawege, R., 2011. Sub-Saharan African urbanisation and global environmental change. Global Environ. Change, Elsevier Ltd 21, S12–S20.

Pelling, M., Holloway, A.J., 2006. Legislation for Mainstreaming Disaster Risk Reduction. Tearfund, Teddington.

Pelling, M., Wisner, B., 2012. Disaster Risk Reduction: Cases from Urban Africa. Earthscan, London.

Prabhakar, S.V.R.K., Srinivasan, A., Shaw, R., 2009. Climate change and local level disaster risk reduction planning: need, opportunities and challenges. Mitig. Adapt. Strat Gl. 14 (1), 7–33.

Schipper, L., Pelling, M., 2006. Disaster risk, climate change and international development: scope for, and challenges to, integration. Disasters 30 (1), 19–38.

Schulz, K., Gustafsson, I., Illes, E., 2005. Manual for Capacity Development. Sida, Stockholm.

Smith, K., 2009. Environmental Hazards: Assessing Risk and Reducing Disaster. Routledge.

Spahn, H., Hoppe, M., Vidiarina, H.D., Usdianto, B., 2010. Experience from three years of local capacity development for tsunami early warning in Indonesia: challenges, lessons and the way ahead. Nat. Hazards Earth Syst. Sci. 10 (7), 1411–1429.

Thompson, M., Gaviria, I., 2004. Weathering the Storm: Lessons in Risk Reduction from Cuba. Oxfam America, Boston.

UCDP, 2012. UCDP Conflict Encyclopedia, Uppsala Conflict Data Program [URL document] February 11, 2013. from http://www.ucdp.uu.se/gpdatabase/search.php.

UNCSD, 2012. The Future We Want. United Nations, New York.

UNDP, 2009. In: Wignaraja, K. (Ed.), Capacity Development: A UNDP Primer. UNDP, New York.

UNDP, 2013. Human Development Report 2013: The Rise of the South – Human Progress in a Diverse World. UNDP, New York.

United Nations Development Programme, 2004. Reducing Disaster Risk. United Nations, New York.

United Nations Inter-Agency Secretariat of the International Strategy for Disaster Reduction (UNISDR), 2002. ISDR Background Paper for WSSD. UN, Geneva.

United Nations International Strategy for Disaster Reduction, 2004. Living with Risk. United Nations Publications.

United Nations International Strategy for Disaster Reduction, 2013. Global Assessment Report on Disaster Risk Reduction. United Nations Office for Disaster Risk Reduction, Geneva.

United Nations, 1994. Yokohama Strategy and Plan of Action for a Safer World. United Nations, Yokohama.

United Nations, 2000. "We the People": The Role of the United Nations in the 21st Century (No. A/54/2000). UN, Geneva.

United Nations, 2005. Hyogo Framework for Action 2005–2015: Building the Resilience of Nations and Communities to Disasters (Geneva).

United Nations, 1987. General Assembly Resolution 42/169. The International Decade for Natural Disaster Reduction. 3 February. UN, Geneva.

Van Niekerk, D., 2012. Transdisciplinarity: The Binding Paradigm for Disaster Risk Reduction. dspace.nwu.ac.za, Potchefstroom.

Van Niekerk, D., Coetzee, C., 2012. Chapter 17 African experiences in community-based disaster risk reduction. In: Shaw, et al. (Eds.), Community, Environment and Disaster Risk Management. Emerald Group Publishing, Bingley, pp. 333–349.

Van Niekerk, D., Wisner, B., 2014. Integrating disaster risk management and development planning: experiences from Africa. In: Lopez-Carresi, A., Fordham, M., Wisner, B., Kelman, I., Gaillard, J.C. (Eds.), Disaster Management: International Lessons in Risk Reduction, Response and Recovery. Routledge, New York.

Van der Waldt, G., 2013. Disaster risk management: disciplinary status and prospects for a unifying theory. Jàmbá: J. Disast. Risk Stud. 5 (2), 11 pp.

Van Wilgen, B.W., 2009. The evolution of fire management practices in savanna protected areas in South Africa. S. Afr. J. Sci. Academy of Science of South Africa 105 (9), 343–349.

Villholth, K.G., Tøttrup, C., Stendel, M., Maherry, A., 2013. Cartografia integrada do risco de seca das águas subterrâneas na região da Comunidade para o Desenvolvimento da África Austral (SADC). Hydrogeol. J. 21 (4), 863–885.

Webster, P.J., Holland, G.J., Curry, J.A., Chang, H.R., 2005. Changes in tropical cyclone number, duration, and intensity in a warming environment. Science 309 (5742), 1844–1846.

WHO, 2011. The Top Ten Causes of Death 2008. WHO, Geneva.

Wisner, B., Blaikie, P.M., Cannon, T., Davis, I., 2004. At Risk. Routledge, London.

World Bank, 2013. GDP Growth (Annual%), Indicators for Economic Policy & External Debt [URL document] September 15, 2013. from http://http://data.worldbank.org/indicator/NY.GDP.MKTP.KD.ZG.

World Meteorological Organization (WMO), 1997. Comprehensive Risk Assessment for Natural Hazards. WMO/TD No. 955. Geneva.

Young, T., Tucker, T., Galloway, M., Manyike, P., Chapman, A., Myers, J., 2010. Climate Change and Health in SADC Region. Strategic Evaluation, Advisory and Development Consulting, Cape Town.

Chapter 6

Understanding Rights-Based Approach in Disasters: A Case for Affirming Human Dignity

Supriya Akerkar[1] **and John Devavaram**[2]
[1] *Centre for Development and Emergency Practice (CENDEP), Oxford Brookes University, Oxford, UK,* [2] *Resource Centre for Participatory Development Studies, Madurai, India*

ABSTRACT
This chapter argues that a rights-based approach in disasters is committed to the reduction of social vulnerability and promotion of equity in society in its postdisaster responses. This in turn means going beyond the dominant framing of rights by the state or dominant groups in the society that may not respond fully to the experiences of rightslessness felt by the marginalized groups after a disaster. What is needed is the critical assessment of rights by such groups, also termed as subaltern groups, suggesting that, although marginalized, they also have a voice and perspective on disaster response that needs to be included. It is this critical assessment of rights by subaltern groups that could inform a nuanced rights-based approach in disasters. The chapter uses the case study of one such subaltern group, namely, the women widowed in the tsunami of December 2003 in Tamil Nadu, India, and their long journey to dignified recovery to exemplify this learning. It also shows that the attainment of social rights and affirmation of dignity and equity were important disaster recovery goals for these women. More generally, a rights-based approach demands working closely with the subaltern groups, listening closely to their critical assessment of rights, and working with them to exercise their agency to affirm their rights and dignity. This calls for a commitment by organizations to work on a sustained basis for longer periods of time after disasters.

6.1 INTRODUCTION

Issues in disaster response and reduction have been addressed mainly using a risk and vulnerability perspective to date. The two competing risk and vulnerability perspectives used are a natural hazards paradigm and a social vulnerability paradigm. The natural hazards paradigm has focused on interventions that reduce physical vulnerability through building of better structures that can withstand natural hazards (White, 1974). The social vulnerability paradigm (Hewitt, 1983; Susman et al., 1983; Wisner et al., 2004)

Hazards, Risks, and Disasters in Society. http://dx.doi.org/10.1016/B978-0-12-396451-9.00006-8

has argued for interventions that increase the coping capacities of the social groups to recover from the disasters, noting that disasters have differential impacts that arise from the differential capacities of the social groups to recover and absorb the impacts of disasters. This use of risk and vulnerability perspective has shed important insights on disaster dynamics and response mechanisms to induce better recovery and reduce future risks. However, the links between rights, vulnerability, and disasters have not been explored in these perspectives, which, if done, can provide new perspective on understanding the impacts of disasters and a new standpoint for actions in disaster response. This chapter addresses this knowledge gap by evaluating the connection between disasters, vulnerability, and rights.

The main assertion of this chapter is strategic disaster response needs a concept of a "subaltern agency," which is empowered to undertake critical assessments of rights in disasters. It is suggested that framing of rights is a discursive terrain, and dominant framing of rights, such as that by states or other dominant groups in the society may not respond to the experience of rightslessness felt by marginalized communities, also termed as "subaltern groups." The notion of "critical assessments" suggests a critical appraisal of the dominant framing of rights in any given context. These critical assessments by subaltern groups could then inform a more contextually nuanced rights-based approach in disaster recovery. The paper develops this insight by reflecting on the experiences of one such subaltern group in India, namely, women widowed in a tsunami to reflect on: what might a rights-based approach look like in disasters.

The chapter is structured as follows: (1) section two builds conceptual linkages between disaster vulnerability, recovery, and rights and introduces the idea of subaltern agency; (2) section three further critically evaluates the substance of "rights" and the need for their critical assessment by the subaltern agency; (3) section four presents the researched case study of the women widowed in the tsunami of December 2003 in India. In certain parts of India, widows face discriminatory social practices (Chen, 1998). It is through the window of the experiences of rightslessness faced by the women widowed in tsunami, and *their* critical assessment of rights, that the contours of the critical rights-based practice is configured as an example; (4) section five draws from the case study and the conceptual discussion from earlier sections to develop wider learnings on the critical rights-based approach to disasters, affirming the principle of human dignity.

6.2 DISASTERS, VULNERABILITY, AND RIGHTS: FORGING CONNECTIONS BETWEEN SUBALTERN AGENCY AND DIGNIFIED RECOVERY

Disaster response is by and large led by an emergency/disaster management cycle response, which includes immediate rescue, relief, recovery, and disaster preparedness or mitigation phases (Cutter, 2003). This is also evidenced in the ways in which most international development organizations have responded to major disasters of the decade: the tsunami of 2003 in the Indian Ocean that affected several countries, the Pakistan floods of 2010, the Haiti earthquake of 2010, East African drought crisis of 2011, and the Philippines typhoon in 2013. The adoption of the United Nations (UN)-led Hyogo Framework for Action 2005–2015 has led to a new emphasis on disaster risk reduction, which has outlined as one of its strategic goals: "The systematic incorporation of risk-reduction approaches into the implementation of emergency preparedness, response and recovery programs." Although this incorporation of risk reduction in the disaster management cycle puts a certain premium on all actors, governments and nongovernments, intervening in the postdisaster scenarios to be reflexive in their response strategies, it begs the interpretation of what is risk/vulnerability reduction, and how is it to be achieved. It is here that the different paradigms of understanding disasters that play a role in determining, *which* and *whose* risk and vulnerabilities ought to be reduced and *how*, by the actors involved in disaster response.

The divide in understanding disasters has been mainly led by two schools: the natural hazards and social vulnerability paradigms. Disasters as a result of natural hazards have emphasized the importance of reducing physical vulnerability through technological solutions whether by way of flood control methods, through river management, or water management strategies, or broadly speaking the strategies of natural resources management. In the context of other hazards such as earthquakes, the building of structures that could withstand seismic activity is the dominant focus of the disaster response work inspired by the natural hazards perspective (Tierney, 2007).

This natural hazards approach, which had emphasized reduction of physical vulnerability, was questioned by another paradigm, which has become influential since the 1980s, namely, the social vulnerability perspective, inspired through works by O'Keefe et al. (1976) and Hewitt (1983). In this analysis, people's susceptibility to risk is mediated through social, economic, and political processes of marginalization, which in turn have "nothing to do with nature as such, but are attributes of society" (Wisner et al., 2004, p. 6). Wisner et al. further suggest that "people's exposure to risk differs according to their *class* (which affects their income, how they live and where), whether they are *male* or *female*, what their *ethnicity* is, what *age group* they belong to, whether they are *disabled* or not, their *immigration status*, and so forth" (Wisner et al., 2004, p. 6). Class, gender, age, caste, occupations, patterns of conflict, and cooperation show the prevalent social, economic, and cultural differentiation within societal structures and shape their relations with the larger structures such as the state and their ability to influence and access resources for recovery (Oliver-Smith and Hoffman, 2002). In other words, it is the social vulnerability and people's differential abilities to prepare, cope, and recover from the disasters that make a hazard event a disaster for some, and not for others.

Now, different studies and evaluations following major disasters have pointed out that gender, class, caste, disability, age, and ethnicity variables affect the disaster outcomes as different social groups have differential access to resources for recovery (Enarson and Morrow, 1997; Peacock et al., 1997; Enarson, 1998; Fothergill, 1998; Enarson and Fordham, 2001; Fordham, 2003, 2004). Social constraints such as the restricted mobility of female-headed households—in some contexts in public spaces—also led to their being left out of humanitarian assistance (Bryne and Baden, 1995; Akerkar, 2007). Following these evaluations, for several organizations working on the ground, engagement with social vulnerability means addressing the needs of the vulnerable groups, which in reality means treating "people as a long list of vulnerable groups" (Sphere, 2011, p. 11). Such an approach is also termed as the checklist approach (Wisner, 2004). In principle, such an approach meets with the humanitarian principle of giving relief and recovery support based on impartial assessments of the needs of different social groups such as children, women, people with disability, and elderly persons.

As this chapter demonstrates, a relief and recovery response based on assessment of needs of different social groups is a necessary, albeit being only a starting point involved in the long drawn-out process of social vulnerability reduction. The shift in the international discourse through the Hyogo Framework for Action, which seeks to connect risk/vulnerability reduction and disaster response strategies, invariably means engaging with a long-term developmental agenda, developing a framework for such a response. For example, development as a political discourse necessitates a positioning by international nongovernmental organizations (NGOs) and government actors on the outcomes of the interventions beyond rescue/relief, such as whether the postdisaster interventions enhance social justice, that is, distributive outcomes, which reduces vulnerability of social groups and improves their capacity to resist or overcome risks. It is argued here that the rights-based approach in disaster response consciously works toward interventions that further equity-based distributive outcome and in turn furthers the goal of social vulnerability reduction. In other words, through disaster response strategies, a rights-based approach consciously aligns itself with emancipatory interests of reducing inequities after disaster. This means attending to the lived experiences of the subaltern groups, their experiences of rightslessness, showing solidarity with their voice, and agency in their struggle to achieve a dignified recovery after disasters. The next section expands on these issues as it discusses and develops the concept of the rights-based approach in general, and in the specific context of disasters.

6.3 RIGHTS-BASED APPROACH IN DISASTERS: THE NEED TO INCORPORATE THE IDEA OF "SUBALTERN" IN RIGHTS-BASED PRACTICE

The rights-based approach is often mobilized through the vocabulary of state–citizen social contract or relation, wherein the state as a duty bearer has the responsibility of upholding the rights of its citizens, the rights holders. This is reflected through a country's constitution, which enshrines the rights of its citizens and the principles of human rights, to which the country is signatory to. The universal human rights discourse gets its mandate through the UN-led declarations, covenants, and resolutions on civil, political, social, and economic rights, as well as those dealing with specific constituencies such as rights of children and women. These UN resolutions when ratified by the countries become obligations that states need to deliver on. These UN-led framing of rights and the constitutional rights inform the laws and the responsibilities that the state has toward its citizenship in a country. Although diverse ways exist in which the rights-based approaches are currently applied by the different NGOs, such as through legal processes and grassroots mobilizations, broadly speaking, it is to make the state accountable to its citizens using the state–citizen social contract as the basis for these rights-based obligations. It is within this framework that the rights-based approaches in disasters adopted by some agencies focus on enabling the survivors to access the right to protection and assistance from the state through legal redress and compensations (Sphere, 2011). Some agencies such as Action Aid International have actively followed such policies and practices after disasters.[1] Further, a translation of the checklist approach is also seen, from a rights perspective, by organizations with specific sectoral mandates, as working for the right to protection and assistance of specific groups such as children, women, people with disability, or elderly persons.

Yet, the rights-based discourse that is configured through the state–citizen relation also reveals a real paradox, as we know following Foucault's work, that the state and its laws are themselves regulatory sites that regulate social relations and produce particular identities and subjectivities through the discourse of rights. This has led Brown (2002) to suggest:

The regulatory dimension of identity-based rights emerges to the extent that rights are never deployed "freely," but always within a discursive, hence normative context in which "woman" (and any other identity category) is iterated and reiterated... The paradox, then, is that rights that entail some specification of our suffering, injury, or inequality lock us into the identity defined by our subordination, and rights that eschew this specificity not only sustain the invisibility of our subordination but potentially even enhance it.

(Brown (2002)), pp. 422–423

1. Personal experience of working with Action Aid International on emergency response in particular Gujarat earthquake and the tsunami response in India, Sri Lanka, Thailand, and Somalia.

The problem with rights arising from particular identities or even eschewed identities in general, Brown (2002) further argued, taking the case of gender justice:

We appear not only in the law but in courts and public policy either as (undifferentiated) women, or as economically deprived... but never as the complex, compound, and internally diverse subjects that we are. This feature of rights discourse impedes the politically nuanced, socially inclusive project to which feminism has aspired in the past decade.

(pp. 428–429)

This critique of rights, in particular highlights the double-edged nature of identity-based and checklist-based mobilization of right to assistance to vulnerable groups from the states in disasters. Although a focus on gender (or other identity) may ensure that both women and men are enabled in a right to assistance, it may not address a range of other experiences of subordination, violence, and suffering faced by these groups, locking them in a particular state-prescribed identity-linked rights experience. Further, disaster relief and recovery assistance is often given around particular issues of loss such as death, injury, shelter, or livelihoods, namely, to the household head, who in most cases is a male, except where the male-household head has died in a disaster. As such, it can be argued that the rights-based approach in disasters as currently followed by the state and organizations itself regulates and promotes a particular kind of postdisaster, gendered-identity politics. In particular, it reconstructs hegemonic ideologies of women's role in the household and subordination, which contributes to women's differential vulnerability to disaster.

The other issue with the state-citizenship-centric, rights-based approach is that in several countries, not only the state but also wider actors and institutions beyond the state, affect the extent to which rights are recognized and respected (Kabeer, 2002). As Kabeer argued,

The state, in its various manifestations, is clearly central in determining which needs and priorities are given the status of rights in their operationalization. However, beyond the state, a wider range of institutions, including those of the market and civil society, also contribute to the process through their recognition and respect for these rights.

Kabeer (2002), p. 21

In other words, an overtly legalistic or state-centric, rights-based approach has its limitations; it is not able to uncover and hence engage with the social field of everyday practice that constrains or authorizes people to exercise their agency to claim their rights. Recognizing that agency is embedded in the field of social relations means that the rights-based approach needs to develop strategies that expand the freedom and the capability of the agency to act in ways that protects and promotes its emancipatory interests. Challenging structures of power embedded in entrenched everyday social relations need a rights discourse that moves beyond "state" as the only site of redress and claim affirmation. This in turn also means giving the subjective experiences of the agency a due place in rights-based analysis and action (Jackson, 2003). Given that the subjectivities of the agency are produced through the everyday practice, a rights-based agenda needs to be reflexive, responding to the emerging issues and processes rather than working with a model of abstract rights vested in an individual.

Disaster response is often led by the "tyranny of the urgent" and the organizational work is commonly driven by the need to respond to the overt subject and sectoral needs, namely, through supply-based strategies. For example, Sphere standards (2011) outlines minimum standards of response, as agreed by major international development organizations. And while it includes the discourse of right to assistance, it has at its core, the technical standards to be achieved in different sectors such as food, water, and housing, and as such remains supply driven in its strategic response. The moot question that is rarely understood and answered in disaster response is: How can emerging subjectivities of disaster-affected persons be understood and tapped upon as a necessary resource to develop an ethical disaster response? From a rights-based perspective, this calls for processes that understand and engage with the emerging subjecthood and subjectivities of the disaster survivors, in order that solidarity and support can be extended to the subaltern groups in their struggle for dignified recovery.

To summarize from the above, the current practice of rights-based approaches in disasters is limited by its approach to understanding and engaging with rights, such that they are predominantly exercised within the domain of a state–citizen relation. Although such an approach might be able to assist survivors to receive compensations or assistance from the state, it does so by building upon identities regulated by the state's discourse on *who* can access compensations/assistance, which in turn has its own problems. Secondly, this approach is driven by working with the rights of an individual abstracted from the social relations, thus ignoring the role that subjecthood and subjectivities play in shaping agencies, the societal constraints and possibilities inhered through them that can potentially lead to an emerging and ever-dynamic rights agenda.

A critical rights-based practice therefore demands an *engagement in a critique of rights as a part of its practice*. This in turn requires a notion of subjects who consciously engage with the social world around them, that is, an *agency* that is able to develop such a critique of rights. This entails a concept of subaltern agency that can undertake a critique of rights by interrogating critically the rights discourse, and the sociopolitical structures that simultaneously enable as well as disenfranchise them in exercising their agency and affirming their dignity. The concept of subaltern represents an "agency of change" (Spivak, 1988, p. 3) and is "a name for the general attribute of subordination in South Asian society, whether this is expressed in terms of class, caste, age, gender and office or in any other way" (Guha, 1988, p. 35). This means that, although subordinated, the subaltern groups actively construct their worlds, are politicized, and construct critiques of the world around them. This concept of subaltern groups is a contextually sensitive and not an ascribed concept that denotes the active voice of the *excluded* groups, depending upon their positionality in the social relations at stake. It denotes the active voice of the socially excluded groups who aspire social inclusion as socially embedded beings than as autonomous or isolated unitary individuals (Santos and Rodriguez-Garavito, 2005). A complex play of identities may lead to a complex analysis of rights issues that defy a one-dimensional way of framing a rights-based response. Women widowed in disasters may find themselves in situations where intimate relations within the family and the extended community that provide them with protection and belongingness also simultaneously became spaces that exclude them from participation in family and community affairs. It is the complexity of experiences that the subaltern agencies have to navigate through, as they aspire a dignified recovery, sensitivity to which requires an emerging and critical right-based practice.

6.4 SUBALTERN AGENCY AND WOMEN WIDOWED IN THE TSUNAMI OF DECEMBER 2003 IN INDIA

The tsunami of December 2003 in the Indian Ocean led to a widespread loss of lives and property in various countries (See Figures 6.2 and 6.3). Countries affected include India and Sri Lanka in South Asia, Indonesia and Thailand in Southeast Asia, Maldives in Asia Pacific, and Somalia in Africa. In India, Tamil Nadu was one of the worst affected regions with several men and women losing their spouses (See Figures 6.1 and 6.3). Given its emancipatory interests, this case study focuses on the experiences of the women widowed in

FIGURE 6.1 Muthu, widowed after tsunami with her children, in Tamilnadu, India

FIGURE 6.2 Impact of tsunami with a boat hitting inlands disrupting railways

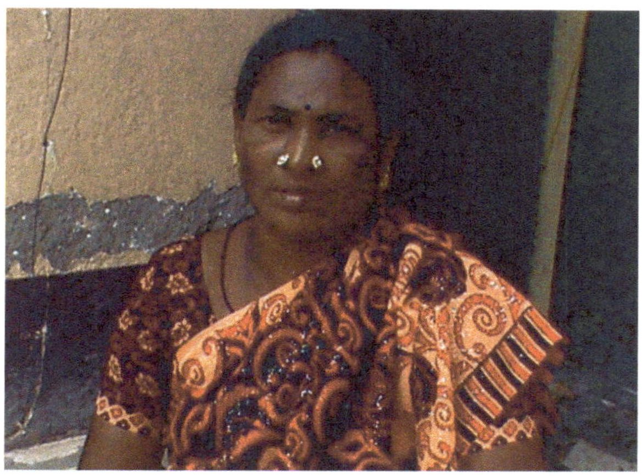

FIGURE 6.3 Balamma, widowed after tsunami in Tamilnadu, India

the tsunami in Tamil Nadu, given that the societal structures and beliefs marginalizes them from the mainstream society (Chen, 1998). Further, a study undertaken in 2004–2005, immediately after the tsunami, had revealed that women, and in particular widows, were facing specific challenges following the tsunami in accessing relief and rehabilitation materials (Akerkar, 2007). The longitudinal nature of our study, namely, revisiting tsunami survivors after several years in 2012–2013, means that we were able to develop insights into the challenges faced by this group of women over a prolonged period, as they continued their struggle for a dignified recovery.

6.4.1 Research Methodology Used

The study draws on a range of data: in-depth interviews, namely, the collection of narratives from 10 women widowed in the tsunami; focus group discussions (FGDs) and semistructured conversations with widowed women; and observations in the two selected villages for field work from Nagapattinum district, along with a review of secondary literature. Field interviews were done in two phases, one in January of 2012 and the other in July of 2013. Further, one of the researchers has been working with the tsunami survivors in the two selected field work villages since 2004 and has a personal knowledge of the field area,[2] whereas the other researcher[3] has worked between 2004 and 2006 with women affected by tsunami in Tamil Nadu. Given the active time spent by the researchers in these villages after tsunami, participant observation has been used to supplement data gathered through interviews. The two villages chosen for the field study were Nambikainagar and a neighboring village Pattinacheri. Nambikainagar is a newly built village space by one of the organizations, Society for People's Education and Economic Change (SPEECH),[4] involved in the relief and recovery, wherein 17 widows from affected villages were given a new house on this site, whereas Pattinacheri is one of the worst affected villages with a high number of deaths following the tsunami. Women participating in the study were between the age group of 28 and 45 years in 2012–2013. This in turn meant that at the time of tsunami, most of our respondents were young widows.

Given that the study wanted to *open up* the enquiry into the discourse of rights in critical ways, rather than working within the governmental framing of right to assistance after tsunami, the use of narratives, FGDs, and semistructured conversations with widowed women, and participant observations were deemed to be a suitable methodological approach. In the interviews and FGDs, the women were invited to share their experiences after the tsunami, and it is through their stories that the research has had an insight into their subjectivities and shared experiences of disaster response and recovery processes.

The main research questions that were sought to be understood through the collected data were the following:

1. How was "rightslessness" experienced by these widowed women in the posttsunami disaster scenario?
2. How did they *now* evaluate the disaster recovery outcomes, that is, several years since the tsunami? How much of agency (making a choice or a decision) were they able to exercise in relation to these outcomes? What factors enabled or disabled these women from exercising their agency?

---

2. Dr. John Devavaram is the Director of Resource Centre for Participatory Development Studies (RCPDS), one of the organizations that have worked with the tsunami survivors since 2004 in Tamil Nadu.
3. Dr. Supriya Akerkar was working between 2004 and 2006 in Tamil Nadu as women's rights advisor to Action Aid's tsunami reconstruction program.
4. RCPDS has worked closely with SPEECH in developing this initiative of giving housing to women widowed in the tsunami.

It was through an understanding into the above two questions that an inductive insight into a critical rights-based approach and practice after disaster was sought.

6.4.2 Findings of the Study

In this section, we summarize the main themes arising from our data analysis in relation to our research questions.

6.4.2.1 Forced Effacing of the Self-worth: On Being a Widow

A recurring theme was the difficult social transition that women had to make immediately after the death of their husband. Women's social rights were closely aligned with their relation to the males in the family. In the case of married women, their social rights were linked with that of their husbands. With the death of their husbands came the *forced* incognito living on them by their in-laws and in some cases their natal families. Emotions were still raw, after so many years, and some women broke down even as they narrated their experiences.

As U, aged 28 years, recalled, "I was very hurt by my in-law's family. I was now (after husband's death in tsunami) regularly side-lined in all family events, whereas earlier I was an important member of the family. But afterwards… my own mother also said: Don't go out alone. Don't take any responsibility now… Decisions were now taken by others. Earlier I was the leader and decision-maker for the family. Now I was completely set aside. At one point of time, I thought of committing suicide, but looking at the faces of the children, changed my mind, hardened my heart and decided to go on."

Others such as B, aged 35 years, from Nambikainagar now shared: "My parents had come to take me back after my husband's death… On the 15th day was the ritual for my dead husband. On that day, my in-laws told me: We have lost a son. Why don't you stay with us? So I stayed with them. But after 10 days, my in-laws started asking me: why are you taking a bath/shower everyday? Your husband is dead… are you looking at another man? Why do you want to dress up? Why do you want to be clean? Because of this, I came back to stay with my parents and my brother. But my sister-in-law had a negative attitude towards me. So my father put up a small shack for me and my children outside their house. I was mentally upset then, was just 28 years old then, and wanted to commit suicide. But my father and mother made me promise that I will not make such attempts."

These stories are symbolic of the painful social effacement and exclusion experienced by these widows in their own families, whether from having a say and everyday participation in the family affairs to an attempt to efface their sense of self-worth by denying them access to means that led to their well-being. B maintained that she appreciated the support of her parents but she also acutely felt the pain and affront on her pride by the circumstances that she found herself in both in her in-laws as well as her parent's house.

Although such an experience was common, some such as M, now aged 38 years, from Nambikainagar, with three children, who had lost her entire in-law's family and her husband (injured in the tsunami) after a month, had support of her father, who came to stay with her and her children, and helped her slowly rebuild her life. However, she is also quick to point out:

"Most women do not have capacity to be on their own. Society is not prepared to accept women without men. Other people can manage, but women who have lost husbands cannot face the society immediately as there is a lot of pressure on them. I managed to do so with support from my family and father."

Other social prejudices such as widows bringing bad luck or attributing the husband's death to the ill-luck of women (FGD, 2012) also contributes to the widows not being invited to the family or community events. K, aged 28 years, from Pattinacheri recalls:

"On the fifth day of my husband's death, people started talking about me. I was very worried then about my three children. Everybody in the family started saying: Because of her, the husband is dead, hearing which I felt very bad about."

Although most widows faced social exclusion, one of the widows, M, aged 45 years with three children, from Pattinacheri village, who belonged to a minority caste group had an opposite experience: "I have never experienced the issue of community not calling me for any event… I am always invited. In our community, we have very strong networks."

She attributed her inclusion to the homogenous caste nature of her village community, and further added: "We are a minority caste in this state… If women like me are neglected, then women can get married to men from other castes, which is a risk. So what they try to do is, create strong caste integrity… They say whatever happens within the community, we will look after her, including allowing a contact with some other man within the community… Also there are property issues, and if we seclude women from the rest of the community, then there is a likelihood of third person coming into our community, which we don't want."

Thus, as can be seen from M's story, her social rights and freedom of movement are preserved, but only so that these spaces can be carefully regulated and controlled through her caste formation. In that sense, although other women experienced a forced effacement of self-worth, with the family and the community singling out their identity of a woman without a man a "widow," in the case of M, her sense of self and widowhood was mediated mainly through her minority caste identity.

6.4.2.1.1 Facing Unwanted Advances Is a Part of Being a Widow

Five widows faced unwanted sexual advances from either their own families or village community, or at work place: "There are many instances of sexual advance of that sort, but I have managed to tackle them. This is a part of widowhood and there are many instances of people trying to sexually abuse me, even within my family… People tried to abuse me in the name of helping me, but I kept to myself and came out of it" (M, aged 38 years).

Others such as L, aged 40 years, faced pressure from the brother-in-law to marry him, whereas B, aged 35 years, was persisted on by her sister's husband to marry him; however, both stood their ground and refused.

6.4.2.1.2 "I Lost Everything, My Life in This Village… I Belong to This Village": Struggle to Access Resources for Recovery

Widows faced varied levels of challenges in accessing recovery assistance. Widows faced adverse pressures from their in-laws families and the communities in relation to access to recovery assistance. This hostility was felt most acutely by S, aged 38 years, who lived with her two children in Pattinacheri after her husband's death. She later remarried (the only one from our sample to remarry) about four-and-a-half years after the tsunami. S states: "My in-laws did not agree to my remarriage. Even my own village community did not accept my remarriage. I was told: You don't belong to this village, so get out of it. But I demanded and had to put up with a very strong struggle. When I was to be allocated a house, a permanent shelter, I was told that I was not entitled to the house, and they reallocated the house. I was told, I will not get the house. But I told them: I lost everything, my life in this village… So I belong to this village. If you don't want to give, then give it to me in writing. Me and children will go out and commit suicide… Finally I won and got a house in my village…"

Although S was successful after a prolonged struggle, A, aged 33 years, living with her two children, struggled to get either death compensation or housing support, which has been denied to her to date: "I had to struggle here all alone. Even the relief material and other assistance, I could not get, as I could not get the death certificate of my husband, as we were not registered as married. I was not officially declared a widow. Benefits therefore reached my in-laws' family and I did not receive anything. Whenever I asked for anything, they (government) told me to get a death certificate… I was unable to mobilise such certificate… After my husband's death, there were special occasions in my in-laws house, such as marriage… I tried to go for them, but was told to not to come in. I am not even informed of good events. Whenever I ask about the death certificate, they (in-laws family) say that it is not possible as my marriage was not recorded… I tried to get help of local NGOs, but nothing happened… I was finally given a certificate from the self-help group federation, which I have submitted to the government, but nothing has happened." A asks: "In the absence of any relief assistance, in absence of rehabilitation, and without my husband, how I will survive with my two children?"

A, today is surviving on labor work, and says: "but for the support of the three other widows from my village, I would not have survived. They supported me and helped me find this work, where they were also employed."

M has also not managed to get the death compensation of about 2000 US dollars, as her husband who was injured in the tsunami died after a month of hospitalization. As a result, M was given only about $180 US dollars, as injury assistance.

All these three women, M, A, and S, felt that their claims, although rightful, had been met with resistance due to the regulated governmental and community interpretations of their rights and claims. In all the three cases, the fluid transition of the identity of the women was not recognized or appreciated, as governmental rules and interpretations tried to freeze them into a singular frame, which highlighted their "non-widowedness." In the case of M, her identity was frozen as being a nonwidow immediately after the tsunami, her transition to widowhood not considered, although her husband died of tsunami injuries after 1 month. In the case of S, her identity was being frozen as a "remarried" woman and hence a nonwidow, whereas in the case of A, her identity as a woman who was de facto widowed, had borne children, was not recognized with governmental rules freezing her into a nonwife and hence a nonwidow claimant. Governmental law and its interpretations of recovery assistance and compensations produced and regulated the identity-based rights associated with "widowhood" in ways that froze these women's lives, and excluded them from accessing recovery assistance. Further, community and social regulation of the ideal of widowhood were also mobilized in ways that supported the governmental regulation. It is only prolonged fight for a dignified recovery by S that finally led to her gaining the housing assistance.

Apart from these, some widows also stated that their experience of being effaced as persons was also attended with their in-laws family wanting to control the relief and recovery assistance received from state and other players.

6.4.2.1.3 Finding Socially Acceptable Work

Lastly, one struggle that stood out in all stories is the financial difficulties faced by all the widows. Except for one widow, none of the other widows were working or had experience of working outside the house when their husband died, and with small children and sometimes, responsibility of extended family members, all widows faced financial crisis, and had to find socially acceptable work for themselves.

Finding socially acceptable work was a challenge. For example, A, who came from a fishing family says: "I was planning to do vending of the fish. However other widows told me that other people will look down upon you. They told me to take up my current employment. Vending is not considered good for widows and is looked down upon. Because in vending, we also come across many men. Also it means buying raw material—fish from men—bringing more contact with men. This lifestyle is looked down upon."

In spite of the social challenges faced, all widows were employed at the time of the interview, working in a local factory, local organizations, or doing agricultural labor work. Although the above were some of the main experiences of rightslessness faced by these widows, their life journeys till date have also revealed how they had exercised their agencies in this social environment. Indeed, in spite of all the hardships they had suffered, the overall evaluation of the widows of their life today was a realist appreciation of their own strength, the varied support they had received on the way, as each of them had embarked upon the long drawn journey to recovery, after the fateful day in December 2003.

6.4.2.1.4 "Become a Strong Person…" (B): Overcoming Adversities and Regaining Social Status

The idea of "becoming" denotes a person in making, and the stories of these women were about how they had become "strong" through their struggle for dignity and recognition of self after the disaster. In spite of all the adversities and experiences of rightslessness faced, they evaluated the current life they lived in positive terms, the main reason being that they felt socially accepted today. In B's words: "My life has changed a lot today. My son is studying in a polytechnic, my daughter is married already. People came seeking my daughter's hand in marriage from a distant place. And in her marriage, relatives and friends came all the way to participate. This showed that although I am a widow, I was on par with others, and am recognized with respect. Even my sister and her husband (who had pressured her earlier for marriage) came for the ceremony without earlier feelings… No more harassment. Their attitude has changed… I am so happy that people have accepted me…"

B also said that the single most support she valued, which enabled her to overcome adversity and make a dignified recovery today, was the social support, in her case the support of her parents: "It requires a will-power. Anybody who loses husband suddenly requires support from parents or somebody else… who can help her to become a strong person to overcome pressures of different kinds. There should be a shoulder to lean back upon, who can share her issues… More important is the social support system."

The women's stories highlighted important transitions in their lives: if widowhood was a painful transitional period, filled with trauma, forced social isolation, or economic and social struggle for basic life resources, each of them had a story about how they had made a *positive transition* after this period of trauma. The transition periods included events that brought meaningful changes in their lives. For example, for Sa, this transition was made when her in-laws, with whom she continued to stay after her husband's death, changed their outlook toward her work. Sa had taken up volunteering work with a local organization, which included going out to meet other women affected by the disaster, to enquire into their well-being, and do counseling work. Initially, her in-laws were unhappy and made innuendos about her moral character, and Sa shares how the transition came about:

"My father-in-law and mother-in-law were the key people in my life... But whenever I had to go out for counseling work, they would make a comment: 'Today she is going out. Is she going for counseling or something else?' Due to this I was in the house most of the time, but the counselors from the organization came to my house and spoke to my in-laws and told them about the nature of my work... After that my in-laws tell me: why don't you go and talk to other women? The best support I have received is that of my in-laws and other counselors." Being a counselor has also given Sa a sense of meaning and social status.

A recurring theme in the stories of transitions shared was how and what helped them regain social status and respect. Their stories suggest that women had received such support from various sources: whether friendship of other widows in the village, self-help groups, their parents, in-laws, or organizations with which they worked. The theme of social status appeared more important than even economic recovery as an indicator of positive recovery today. Indeed, the house that was mobilized from the government or an NGO, or the work and livelihood they earned, were important to them, but more so, because these resources gave them social respect and acceptance in the society. In other words, achieving social status was an important central theme in their stories of how they understood and experienced disaster loss and later recoveries.

S, the only widow who had remarried, highlights this social achievement, when she stated: "I am happy in my new remarried life. It is the social acceptability that I enjoy the best. Now the village has accepted me. My children feel confident and go out with other community children. When there are functions and celebrations in my house, (the) whole community participates. The whole family has social acceptance. I can now put kumkum on my forehead (red vermillion on the forehead, an indicator of a married woman in Hindu society)..., which I could not as a widow. I had felt then the pain of widowhood. But now things are changed. People accept me. Their social acceptance has brought me mental satisfaction."

S's story reasserts how the transition to widowhood, which led to a discriminatory treatment of these women, and forced effacement of self were affronts on their social rights and sense of dignity. To summarize, the stories of these widows revealed their understanding of what constituted a real disaster for them after the tsunami, what were the most painful and destructive losses, and what really constituted to them a real recovery. The loss of social status and acceptance was the most traumatic loss, and its recovery their most important achievement as they looked back at their experiences after the tsunami. In that sense, each story revealed in general two crucial transitions: one, a transition to the state of widowhood, in general a traumatic period, where they faced issues of social exclusion, and erosion of social rights. This was a period of loneliness and intense struggle, in many cases prolonged to several years, followed by other positive transition events that indicated their regained social status in their societies. The stories also showed how they had individually exercised their agency, and how family members and village community had played both a disabling and positive enabling role in both the loss and regaining of social status in these transitions, respectively. Other widows, counselors, and self-help groups were found playing a positive role in their transitions.

The stories of the widows indicate the importance of *achieved social status* as an indicator of their *recovery*, sense of well-being, self-worth, and dignity. When they were asked what advice they would give other organizations seeking to work in disasters, they unequivocally suggested that vulnerable groups and widows and their issues of *social acceptance* should be given priority. In other words, working for the "social rights" and "regaining of social status," aspects closely linked with the affirmation of their dignity, were for them the most important aspects of disaster recovery work in their societal context.

6.5 UNDERSTANDING THE RIGHTS-BASED APPROACH IN DISASTERS: SOME LEARNINGS

This chapter has asserted that a critical rights-based approach in disasters needs to go beyond the legal and governmental framing of right to assistance to the people affected by disaster. In order to do so, a critical rights-based practice needs a notion of *subaltern agency* that is able to reflect on the societal processes in the postdisaster scenarios, so as to critique the governmental or dominant societal framing of rights. The case study discussed above supports this assertion, as it goes on to show how a clear chasm existed between governmental and dominant societal framing of rights to assistance and disaster recoveries, and that of the subaltern groups. In this case, the subaltern groups were widows, whose own reflections and self-analysis have questioned such a framing of their rights. Their own life journeys and stories were signifiers of the struggles they individually undertook to reclaim their social rights and dignity. The stories show that intimate relations such as immediate families and village communities were both a source of support and a risk to their well-being, both playing a supportive as well as oppositional role in their stories of recoveries. But more generally, it showed that wherever possible the widows made a strategic use of the social resources available to them to make crucial positive transitions that helped them regain social respect, and status, albeit achieved after a prolonged period of struggle. In other words, the inductive evidence from the case study suggests that regaining social rights was a result of a long journey traveled by these women. In that sense, the rights-based approach needs a commitment of the organizations to be with the subaltern groups in their struggle for dignified recovery for long periods of time, whereas the current dominant practice of organizations is to leave the affected areas within short spans of relief and rehabilitation work. Commitment to long-term recovery work by organizations is crucial as

FIGURE 6.4 On the road after tsunami - a survivor

affirmation of rights of excluded groups require societal change, which is a slow and prolonged process that generally comes only with years of sustained engagement.

A rights-based approach thus demands that organizations working in disasters could well work closely with subaltern groups, understanding their stories and their critique of the framing of their rights by both government and dominant sections in the society; and enabling them with solidarity to exercise their agency to regain their rights. For subaltern groups, affirmation of their rights is closely aligned with the affirmation of their dignity (as this case study has also affirmed), and as such a rights-based approach or a critical rights-based practice in disasters must have at its core the principle of affirmation of human dignity (Figures 6.1–6.4).

REFERENCES

Akerkar, S., 2007. Disaster mitigation and furthering women's rights: learning from tsunami. Gender Tech. Dev. 11 (3).

Bryne, B., Baden, S., 1995. Gender, Emergencies and Humanitarian Assistance. Bridge, IDS, Sussex.

Brown, W., 2002. Suffering the paradoxes of rights. In: Brown, W., Halley, J. (Eds.), Left Legalism/Left Critique. Duke University Press, Durham and London.

Cutter, S., 2003. GI science, disasters and emergency management. Trans. GIS 7 (4).

Chen, M., 1998. Widows in India: Social Neglect and Public Action. Sage Publications, New Delhi.

Enarson, E., Morrow, B.H., 1997. A gendered perspective: the voices of women. In: Peacock, W.G., Morrow, B.H., Gladwin, H. (Eds.), Hurricane Andrew: Ethnicity, Gender and Sociology of Disasters. Routledge, London and New York.

Enarson, E., 1998. Battered Women in Disaster: A Case Study of Gendered Vulnerability. Disaster Preparedness Resource Centre, University of Columbia.

Enarson, E., Fordham, M., 2001. Lines that divide, ties that blind: race, class and gender in women's flood recovery in the US and UK. Aust. J. Emerg. Manage. 15 (4).

FGD, 2012. Focus group discussions held in January with women's group in Nambikai nagar.

Fordham, M., 2003. Gender, disaster and development: the necessity for integration. In: Pelling, M. (Ed.), Natural Disasters and Development in a Globalizing World. Routledge, London and New York.

Fordham, M., 2004. Gendering vulnerability analysis: towards a more nuanced approach. In: Bankoff, G., Frerks, G., Hilhorst, D. (Eds.), Mapping Vulnerability: Disasters, Development and People. Earthscan, London and Sterling.

Fothergill, A., 1998. The neglect of gender in disaster work: an overview of literature. In: Enarson, E., Morrow, B.H.M. (Eds.), The Gendered Terrain of Disasters. Praeger, Westport, Conn.

Guha, R., 1988. Preface. In: Guha, R., Spivak, G.C. (Eds.), Selected Subaltern Studies. Oxford University Press, Oxford.

Hewitt, K., 1983. The idea of calamity in a technocratic age. In: Interpretations of Calamity: From the Viewpoints of Human Ecology. Allen and Unwin, Boston, London and Sydney, pp. 3–32.

Jackson, C., 2003. Gender analysis of land: beyond land rights for women? J. Agrar. Change 3 (4).

Kabeer, N., 2002. Citizenship, affiliation and exclusion: perspectives from the south. IDS Bull. 33 (2).

O'Keefe, P., Westgate, Wisner, B., 1976. Taking the naturalness of natural disasters. Nature 260.

Oliver-Smith, A., Hoffman, S., 2002. Introduction: why anthropologists should study disasters. In: Hoffman, S., Oliver-Smith, A. (Eds.), Catastrophe & Culture: The Anthropology of Disaster. School of American Research Press and James Currey, Santa Fe and Oxford.

Peacock, W., Morrow, B., Gladwin, H. (Eds.), 1997. Hurricane Andrew. Routledge, London and New York.

Santos, S.B.D., Rodriguez-Garavito, 2005. Law, politics and the subaltern in counter-hegemonic globalisation. In: Santos, S.B.D., Rodriguez-Garavito (Eds.), Law and Globalisation from Below: Towards a Cosmopolitan Legality. Cambridge University Press, Cambridge.

Sphere Project, 2011. The Sphere Project: Humanitarian Charter and Minimum Standards in Disaster Response [Online]. Available at: www.sphereproject.org.

Spivak, G.C., 1988. Subaltern studies: deconstructing historiography. In: Guha, R., Spivak, G.C. (Eds.), Selected Subaltern Studies. Oxford University Press, Oxford.

Susman, P., O'Keefe, P., Wisner, B., 1983. Global disasters, a radical interpretation. In: Hewitt, K. (Ed.), Interpretations of Calamity. Allen and Unwin, Boston, London, Sydney.

Tierney, K., 2007. From the margins to the mainstream? disaster research at the crossroads. Ann. Rev. Soc. 33.

Wisner, B., Blaikie, P., Cannon, T., Davis, I., 2004. At Risk: Natural Hazard, People's Vulnerability and Disasters. Routledge, London and New York.

Wisner, B., 2004. Assessment of capability and vulnerability. In: Bankoff, G., Frerks, G., Hilhorst, D. (Eds.), Mapping Vulnerability. Earthscan, London.

White, G.F., 1974. Natural Hazards: Local, National, Global. Oxford University Press, New York.

Chapter 7

Reactive to Proactive to Reflective Disaster Responses: Introducing Critical Reflective Practices in Disaster Risk Reduction

Nibedita S. Ray-Bennett[4], Anthony Masys[2], Hideyuki Shiroshita[3] and Peter Jackson[1]

[1] *University of Leicester, Leicester, UK,* [2] *Centre for Security Science (CSS), Defence R&D Canada, Ottawa, ON, Canada,* [3] *Kansai University, Japan,* [4] *Civil Safety and Security Unit, University of Leicester, Leicester, UK*

ABSTRACT

Beck (1992) described the modern or postindustrial landscapes as a "risk society." The fundamental characteristic of this risk society is its interconnectedness and interdependence that makes systems and networks highly complex so much so that they are often vulnerable to abrupt failures. The landscape of risk society is highly prone to social, natural, and technical hazards. The "dangers" created from the combination of these hazards in this interconnected world are hyper/hybrid-risks illustrated through the increase in "natural"/environmental and naturally triggered tec3hnical disasters in general. In such a context, this chapter calls for a "reflective response," one that is based on critical reflective practices and systems thinking to counter hyper-risks and develop organizational resilience. Some methodologies and methods to promote reflective response are discussed. Usefulness of a reflective response and disaster risk reduction are also explored. A charter of reflective response is suggested.

7.1 INTRODUCTION TO "NATURAL" AND NATURALLY TRIGGERED TECHNICAL DISASTERS AND THEIR IMPACT WORLDWIDE

In 2010, 263 million people were affected by "natural" or environmental disasters, which are about 110 million more than in 2004, the year of the Asian tsunami (DFID, 2013). The number of people affected by disasters is

Hazards, Risks, and Disasters in Society. http://dx.doi.org/10.1016/B978-0-12-396451-9.00007-X

increasing every year with a marked debilitating effect on the developing countries as compared with developed countries (Coppola, 2011; DFID, 2013). Between 1980 and 2000, 53 percent of the deaths due to disasters occurred in poor countries, although these countries accounted for only 11 percent of the world's "at-risk" population (UNDP, 2004; quoted in Coppola, 2011). It is likely that more people, particularly in developing countries, will be affected by environmental disasters in the coming decades (Coppola, 2011). It is known that the achievement of the Millennium Goals by 2015 is much harder for those countries repeatedly affected by disasters (DFID, 2005; UN, 2007).

Naturally triggered technical hazards (NATECH) are also on the rise (Cruz, 2012; Ozunu et al., 2011). Cruz (2005, p. 2) highlighted some key NATECH disasters, revealing the risks, hazards, and consequences associated with such events:

- Asian tsunami on December 26, 2004 caused extensive damage to several industrial plants, triggering hazardous materials releases.
- Banda Aceh, the Pertamina oil depot in Kreung Raya Bay, reportedly leaked about 8,000 kL of oil.
- The Kocaeli earthquake (magnitude of 7.4) in Turkey in 1999 triggered unprecedented hazardous materials releases resulting in off-site consequences to the population. More than 6 million kilograms of toxic acrylonitrile was released into air, water, and soil from an acrylic fiber plant (Cruz, 2005, p. 2).

The most recent case includes the devastating earthquake named Great East Japan aka Tohoku on March 11, 2011, which triggered a giant tsunami that reached the height of 40.5 m in the city of Miyako and killed more than 15,844 people. The tsunami also sparked nuclear accidents, including the meltdown of three nuclear reactors in the Fukushima Daiichi Nuclear Power Plant. The destruction of the Fukushima Daiichi Nuclear Power Plant resulted in massive radioactive contamination of the Japanese mainland. It resulted in radioactive cesium entering the ecosystem and becoming ubiquitous, contaminating water, soil, plants, and animals. It has been detected in a large range of Japanese foodstuffs, including spinach, tea leaves, milk, beef, and freshwater fish up to 200 miles from Fukushima (Starr, 2013). The contamination from the radioactive cesium affected 11,580 square miles (30,000 sq km) of the land surface of Japan (Starr, 2013). Estimates of the total economic loss range from US$250 to US$500 billion, and over 150,000 people were displaced.

To counteract some of the effects of environmental as well as NATECH disasters, the governmental and multilateral organizations have adopted several United Nations international conventions, including the Hyogo Framework for Action (HFA or the Framework) (2005–2015) for disaster risk reduction (DRR), Millennium Development Goals (2000–2015) to eradicate poverty, and the Intergovernmental Panel for Climate Change (2007) for climate change and adaptation, in order to promote disaster resilient communities.

According to the HFA, resilience can be built by adopting the five priorities for actions (PfA) (for details, see UN, 2005). These PfA are seminal because they urge the national and international organizations to shift from reactive to proactive disaster response through preparedness and mitigation. It also recognizes the mutual importance of integrating disaster into development paradigm (UN, 2005; Collins, 2009). It urges the involvement of different stakeholders, the adoption of a multisectoral approach, and the devolution of disaster management activities (UN, 2005). As a result, the HFA has led to an increasing level of "risk appetite" worldwide that previously was unforeseen (Ray-Bennett, 2009, 2012).

However, the Framework is tacit in explaining how to build the capacities and resilience of those organizations that apply the principles of DRR into practice in their own context. The Framework places much emphasis on the capacity building of the "at-risk" community rather than that of the organizations and professionals who are involved in the day-to-day "messy" (Schön, 1983) businesses of averting environmental and technological risks. As a result, building organizational capacity either of the national or local authorities, nongovernmental organizations, or environmental agencies has received far less attention by the DRR community (for instance, see DFID, 2006, 2012). The Framework also fails to capture the complexities of the postmodern world and those of the organizations (Argyris, 1960) in order to understand NATECH disasters in particular. This chapter aims to address this gap. In doing so, "reflective response" is introduced to the dominant crisis and disaster management practices, one that is based on "critical reflective practices" and "systems thinking" in order to promote organizational resilience (Ray-Bennett et al., forthcoming).

A "resilient organization" is understood as a "learning organization" that does not unravel in "predictable" and "unpredictable" disasters. A resilient organization is realized only when the "communities of practice" are supported and nurtured as part of the overall goal of resilience. Such an organization will also be in a better position to respond and mitigate disaster risks both internally as well as externally for the "at-risk" community. "Communities of practice" means "where participants share understandings about what they do and what the doing means for them and for their communities" (Lave and Wenger, 1991; quoted in Welsh and Dehler, 2004, p. 21). According to Turner (1999), communities of practice are influential because they exchange experience and can be brought together to build webs of relationships in order to develop organizational capacities and learning (Turner, 1999) relative to environmental and NATECH disasters.

7.1.1 Scope for Reflective Response

The fundamental characteristic of today's world is its interconnectedness and interconnectivity that lends itself to a "risk society" (Beck, 1992) and a "world risk society" (Beck, 2009). Beck (1992) described the modern or postindustrial landscapes as a "risk society." The industrial society is the distribution of "goods," whereas the risk society is the distribution of "bads" or dangers. Risk society is the by-product (or residue) of the industrial society that produces "risk society's risk" and "dangers" in conjunction with science and technology. Although the interconnectivity in risk society supports and enables global prosperity, it also increases vulnerability to transnational risk (Beck, 2009). More and more we see the consequences of disasters extend beyond borders having regional and global reach, as described earlier through the case of NATECH disasters such as the case of Great East Japan aka Tohoku.

The landscape of the risk society is then highly vulnerable to social, natural, and technical hazards. The dangers and "bads" created from these hazards have the potential to spread both locally and globally due to the nature of interconnectedness. Also these "bads" have the potential to jeopardize livelihoods, natural resources, and ecosystem services of the current and future generations, as observed in the case of Japan. No longer can we view impacts as discrete events and as the sum of their local consequences. The mental models we use to conceptualize risk must address the complex and dynamic interconnectivity and interdependence that exists in systems. Consequently, what has emerged from the interconnected world are "risks" or hyper/hybrid-risks that are unpredictable and unknown (Helbing, 2013; Taleb, 2007; Beck, 2009).

These risks are hybrid because "a number of basic features that may have often been regarded as mutually exclusive" (Beck, 2009) by the dominant risk, crisis, and disaster theories are inclusive and fused. Hyper- or hybrid-risks are connected to several systems such as the society, environment, organization, and the like (Helbing, 2013). In contrast, the dominant risk framework compartmentalizes risks and hazards into three types: natural, social, and technological (Jones and Hood, 1996; Beck, 1992). These are useful analytical distinctions. But in the context of hyper-risks, they overlap and intersect to produce hybrid hazards known as quasi-natural hazards or NATECH hazards (Jones and Hood, 1996; David et al., 2007). As a result, the conventional distinction between these three types of hazards has come under serious challenge from the disaster sociologists and geographers in light of recent disasters such as the hurricane Katrina (David et al., 2007) and the Tohoku earthquake.

The emerging and systemic risks and inherent uncertainty associated with surprising events require understanding of a variety of possibilities in order to facilitate management of these hyper-risks. Many of the systemic risks that characterize the disaster risk often arise from unanticipated consequences of interactions within and between different types of systems. In this light, a new paradigm is required to support DRR embedded in hyper-risks, one that will not only develop anticipatory measures for risk management but also prepare for the unpredictable and the "unknown" by building organizational resilience (Wildvasky, 1988; quoted in Hood and Jones, 1996; Taleb, 2007) for hyper-risks. This merits the question, how can we build organizational resilience to comprehend hyper-risks such as conjoined environmental and NATECH disasters. Several ways exist to promote organizational resilience. This chapter proposes a "reflective response," one that is based on "critical reflective practices" and "systems thinking."

7.2 THE PERSPECTIVE OF "REFLECTIVE RESPONSE" IN AN INTERCONNECTED WORLD

In 2005, the HFA urged the DRR community to shift from a reactive to proactive disaster response. In order to promote organizational resilience, against the odds of environmental and NATECH disasters, "reflective response" is proposed. At the outset, "reflective response" is defined as a combination of individual, organized, and critical reflections and reflective strategies embedded in an organization's context. Reflective response has its roots in management, education, psychology, and social care studies and practices. Reflective response is not a stand-alone response; it is rather complementary to reactive and proactive disaster responses. The three tenets—individual, collective, and critical reflections—are mutually inclusive to build organizational resilience for hyper-risks, and they are discussed below.

7.2.1 Individual Reflection

Reflection is quintessential to all human beings. It is the very essence of human existence. Reflection is also the most important building block and driver of human learning (Higgins, 2011). The ability to think and reflect separates human beings from animals (Freire, 1972). Reflection is also the connection between thoughts and actions (Thorson and DeVore, 2013). Then, if we are to use reflection as a mode to improve the profession and practices of DRR, we need to reflect within reflection as to how this can be achieved. Why reflection is so important? How does reflection get its credibility and power? What is the nature of this reflection that takes one above the day-to-day reflection we do at home, whereas driving, shopping, and the like, to a superior level at which we improve our practices and, in Freire's (1972) words, have the "potential to transform the world"? To explore these questions, the theories on "reflection" and the ways in which it has evolved conceptually are revisited.

John Dewey, the American philosopher and an educationalist, is arguably considered to be the father of modern reflection theory not only for learning and education but also for management studies (Reynolds and Vince, 2004). For Dewey, thought and action were inextricably interlinked. Inspired by Dewey's work, Schön (1983) analyzed the gap that exists between theory and practice and of the practitioners dealing with "messy problems" of the day-to-day workplaces through the model of "reflective practice" and "reflective practitioner" (Karban and Smith, 2010). Schön brings to the fore the idea that reflection is not only retrospective but also an element of the experience (Reynolds and Vince, 2004). More concretely, he made the distinction between "reflection-in-action" and "reflection-on-action" as a way of thinking about a situation while engaged within it, in order to reframe and solve some breakdown in the smooth running of experience (Plager, 1994; Schön, 1983). Reflection-in-action is based on a rapid interpretation of the situation, whereas "reflection-on-action" occurs after the event to improve future action (Ghaye and Ghaye, 1998; Schön, 1983).

The concept of "sense making" in organizations (Weick, 1995) is another important aspect of reflection. Sense making means "making of sense" (Weick, 1995, p. 4) "about an activity or a process" that requires "interpretation" (Weick, 1995, p. 13). According to Weick (1995, p. 15), sense making has a "strong reflexive quality" to the process of interpretation because "people make sense of things by seeing a world on which they already imposed what they believe."

According to Johns (2009), reflection is a balance between the right side of the brain (which is concerned with creativity, imagination, perception, intuition, synthesis, wonder, and spirit) and the more dominant left side of the brain (which is concerned with qualities of the mind associated with analysis, reason, rationality, and logic). Reflection is then a balanced approach that requires a shift in thinking and new ways of responding. To him, reflection is a way of being within everyday practice (Johns, 2009, p. 10).

7.2.2 Organized Reflection

The theories of "organized reflection" or collective reflection emerged as a critique to the theories on reflection (Reynolds and Vince, 2004). Organizational theorists argued that "reflection" did not fully capture the critical perspective of reflection in the context of organization and management studies (Antonacopoulou, 2004; Kayes, 2004; Nicolini et al., 2004; Reynolds and Vince, 2004; Welsh and Dehler, 2004). According to them, reflection has to go beyond the individual to draw lessons from the power relations (Kemmis, 1985) existing within and between communities of practice in order to advance professions and management practices (Welsh and Dehler, 2004). Reynolds, in particular, spearheaded this initiative by taking reflection to a "higher level" and introducing a "critical element into the process" (Welsh and Dehler, 2004, p. 18). They argued that reflection has to be "less about the individual reflective practitioner and more about organizing reflection." (Reynolds and Vince, 2004, p. 1). Based on this tenet, Reynolds and Vince (2004) conceived the concept of "organized reflection," one that takes account of "social and political processes at work in the organization of reflection." The emphasis is "placed [...] on creating collective and organizationally focused processes for reflection" (Reynolds and Vince, 2004, p. 11). Organizing reflection is also linked closely to developing strategies for organizational development (see Antonacopoulou, 2004; Kayes, 2004; Nicolini et al., 2004; Reynolds and Vince, 2004; Welsh and Dehler, 2004).

7.2.3 Critical Reflection

"Critical thinking" is analogous to critical reflection. "Critical thinking is a complex process of deliberation which involves wide range of skills and attitudes" (Cottrell, 2005, p. 2) "to think in critically analytical and evaluative ways [means] using mental process such as attention, categorisation, selection and judgment" (Cottrell, 2005, p. 1). Skepticism and reasoning are two important aspects of critical thinking (Ennis, 1987; quoted in Cottrell, 2005).

According to Vaughn (2010, pp. 4–5), critical thinking is

the systematic evaluation or formulation of beliefs, or statements, by rational standards [...]. Critical thinking applies not just to some of your individual beliefs, but to all of them together. It applies to your worldview, the vast web of fundamental ideas that help you make sense of the world, what some people call a philosophy of life. We all have a worldview, and most of us want the beliefs that constitute it to be true and coherent [...].

Since critical thinking can be acquired, Vaughn (2010) posits that it can lead to enhancement of knowledge, understanding, and, in some cases, empowerment. This is because critical thinking enables "problem solving, active learning, and intelligent self-improvement" (Vaughn, 2010, p. 5).

In the context of management and organization theories, the frameworks of "organizational learning" and "learning organizations" are central to critical reflection (Gould, 2004). Organizational learning is learning that takes place at an organizational level and as a result gains new knowledge (Argyris and Schön, 1996; quoted in Fook, 2004). Organizational learning understands organization itself as an entity,

which has a life over and above its individual elements [...] [to] bring about organisational learning, specific aspects of the organisation need to be work with

Argyris and Schön, 1996; quoted in Fook, 2004, p. 67.

The framework of a "learning organization," on the other hand, engages with systemic thinking, teamwork, and work-based learning of a practitioner(s) within the wider organizational context (Senge, 1990; Gould, 2004).

"Learning organization" is different from "organizational learning." Organizational learning talks about the processes through which learning takes place, whereas learning organization is about characteristics of an organization that learns (Gould, 2004). A learning organization focuses on the different parts of an organization by making connections between individual learning for organizational change (Senge, 1990). In a learning organization, "people are continually discovering how they create their reality. And how they can change it" (Senge, 1990, p. 13) being a part of the world rather than a separate entity.

Learning organization goes beyond the confines of an organization to connect with the global world to promote learning at individual, collective, as well as organizational levels. Learning organization also values human agencies that are capable of making decisions for personal and professional development as well as for the organization. According to Senge (1990), systems thinking is then played out through the paradigm of interdependency, complexity, and wholeness.

Culture is another key aspect to learning organization (Baldwin, 2004). By reflecting critically on the broad concept of culture, it enables one to grasp the worldviews that people share based on their race, class, gender, caste, disability, and ageism, for instance. These shared and collective perspectives have historical, social, and political dimensions. If these are understood correctly, learning organizations can promote "a culture of learning" in project teams as well as in developing effective services (Baldwin, 2004, p. 43) for the interconnected world.

However, Gould (2004) argues that although organizational learning and learning organizations exhibit different characteristics, in reality they overlap. In the context of DRR, the frameworks of organizational learning and learning organizations are equally important to understand the disparate and diverse practices that exist worldwide.

7.3 METHODOLOGY AND METHODS TO PROMOTE REFLECTIVE RESPONSE

According to Hickson (2011, p. 830), "there are a myriad of methods, tools, techniques and frameworks for reflective practice and these methods can be used individually, in a group, online or by following a self-help book." Reflective response is an underresearched topic in DRR; therefore, more empirical research is required to identify the appropriate methodology and methods that can develop organizational resilience akin to the interest of the communities of practice.

7.3.1 Interpretive Approach

However, some of the methodologies and methods that can promote critical reflective practices have been deduced. These include Habermas's (1973) "historical hermeneutic" sciences. This is a methodology that refers to the "science of interpretation and explanation" (Mezirow, 1981, p. 5) rather than relying solely on empirical analytic approach (Mezirow, 1981). This methodology includes descriptive social science, history, ethnographic, legal, and other qualitative methods (Mezirow, 1981; Habermas, 1973) that can interpret the complexity of disaster risks.

Other methodologies can include Kemmis's (1985) "emancipatory action research" and Baldwin's (2004:50) "cooperative inquiry." Both of these methodologies are action research and have the potential to promote the double dialectic of thought and action, and the individual and society in a collaborative and participatory way. Cooperative inquiry also has the potential to capture the critical and reflective cyclical processes of learning (Baldwin, 2004).

Outcome mapping (OM) is another methodology

that can be used to create planning, monitoring, and evaluation mechanisms enabling organisations to document, learn from, and report on their achievements. It was developed by the International Development Research Centre (IDRC) in Canada in support of development efforts around the world. It was designed to assist in understanding an organization's results, while recognizing that contributions by other actors are essential to achieving the kinds of sustainable, large-scale improvements in human and ecological well-being toward which the organization is working

Earl et al. (2001, p.viii)

The application of this methodology to the disaster management domain is particularly relevant to the notion of building resilience. Two of the key features or principles of OM that resonate with systems thinking are its recognition of the importance of embracing different worldviews and perspectives as well as the acceptance of nonlinear (complex) causality. More details on this methodology are available at IDRC (2001).

7.3.2 Tools of Systems Thinking

There is a myriad of system-thinking tools that can build the capacities of the communities of practice, such as casual loop diagrams (CLD) (Sterman, 2000; Senge, 1990; Senge et al., 1994), which provide a language for articulating our understanding of dynamic, interconnected situations. Systems dynamics (Sterman, 2000) is another product of CLD to support modeling of the dynamic behavior of systems. Assumption-based planning (Dewar et al., 1993) is an essential tool for explicitly capturing the assumptions underlying plans that if invalid, could derail the plans and operations stemming from it. It recognizes how "load-bearing vulnerable assumptions" require explicit analysis. "Scenario planning" is a useful tool to tackle black swans or the "unknown" or "unpredictable risks" (Masys, 2012). Scenario planning represents a set of tools for foresight, facilitating discourse with the purpose not to predict, but to facilitate a change in the mindset of the people who use them: in short, to think the unthinkable (Farber and Lakhtakia, 2009, p. S13). Numerous approaches exist to scenario planning in the literature such as Schwartz's Eight-Step Scenario Building Model (Schwartz, 1996), Schoemaker's 10-Step Scenario Building Model (Schoemaker, 1993), Avin's 12-Step Scenario Building Model (Avin, 2007), and the 6-step processes for the development of scenarios by JISC (2007).

7.3.3 Tools of Critical Reflective Practices

The tools of critical reflective practices can further complement reflective response at individual and collective levels, which are as follows: Gibbs's (1988) "experiential learning cycle," Pearson and Smith's (1985) "briefing and debriefing," and Bolton's (2005) reflexivity as an important strategy. As part of reflexivity, "mindfulness" is an invaluable approach (Bolton, 2005; Johns, 2009). Mindfulness is a conscious "exclusion of other elements of life" apart from the one that is being attended to (Johns, 2004; quoted in Bolton, 2005). This is achieved when senses and awareness are turned into action: the opposite of multitasking (Epstien, 1999; quoted in Bolton, 2005). Being mindful develops communication according to Bolton (2005). Practical wisdom and observation are other skills quintessential of reflective practice and of a practitioner (Bolton, 2005).

Powell's (1985) autobiographical learning such as keeping diaries, journals, and portfolios on a regular basis are other methods in order to reflect and develop strategies; Knights (1985) highlighted the importance of "listening and of a listener" in order to support others through the process of learning and reflection. Related to this is the importance of effective supervisor or manager to develop a culture of listening, learning, and listener. Antonacopoulou's (2004) "feedback" and (Argyris, 1960) "feed-upon" are other important methods of learning and changing and embracing complexities in order to minimize disaster risks.

7.4 THE USEFULNESS OF REFLECTIVE RESPONSE IN DRR

In light of the above discussions, it is worth questioning the practical usefulness of reflective response. What can an organization learn from reflective response? What value can it have to the communities of practice and management? This chapter posits the following:

1. Improve management practices and philosophy

The increasing complexity of today's interconnected and interdependent systems as it pertains to DRR calls for greater understanding and development mechanisms for coping with uncertainty and shocks. In this light, reflective response can lead to improved management practices because it promotes not only anticipatory measures but also resilience. Through the capacity building of a "learning organization," it prepares for the "unknown" hyper-risks. This preparation has the potential to develop active learning, one that "is knowing about something and then taking remedial action to rectify the deficiencies that have been uncovered" (Toft and Reynolds, 2005, p. 660) but also forethought. Forethought means being prepared for the unknown hyper-risks through scenario building and the like, in order to build a safe and resilient culture.

The hyper-risks that characterize the complexity associated with NATECH disasters make it a challenge to untangle the interactions and relations between environmental disasters and technological accidents (Cruz, 2012). If the conjoint effects of NATECH disasters are not identified, assessed, and planned for by both industrial facilities and government agencies in charge of industrial risk and emergency management, they will be ill prepared to respond (Cruz and Okada, 2008, p. 200). Such events as the Great East Japan earthquake and the hurricane Katrina highlighted the need for greater awareness and action regarding NATECH. From the interdependencies that characterize the NATECH disaster regime emerges the possibility of cascading disasters (domino effects) resulting from failing mitigation measures and strategies that did not consider the hyper-risks.

In the complex space associated with risk, crisis, and disaster management, decision-making calls for judgments under uncertainty, ambiguity, and time pressure. As described by Dekker (2011, p. 13):

Reasoning in complex systems is governed by people's local understanding, by their focus of attention, goals, knowledge. In complex systems, of course, that which can work really nicely locally, can make things fail globally. Optimizing decision and conditions locally can lead to global brittleness.

2. Developing human and organizational resilience to hyper-risks

Strategic reflection creates an opportunity to share, compare, and explore mental models, thus opening up consideration of a variety of possibilities. Enabling resilience therefore requires a constant sense of unease that prevents complacency. As argued by Hollnagel and Woods (2006),

[...] it requires a realistic sense of abilities of where we are. It requires knowledge of what has happened, what happens, and what will happen, as well as what to do. A resilient system must be proactive; flexible; adaptive; and prepared. It must be aware of the impact of actions, as well as the failure to take action.

Reflective practice also has the potential to

- identify the sensitive and "forgotten" areas of practices;
- provide for the analysis of hesitation, skills, and knowledge gaps;
- provide relief from stress;
- identify learning needs; and
- disseminate experience and expertise to and from colleagues and result in an increased confidence in professional practice (Bolton, 2005).

Furthermore, reflective practices can lead to the elimination of "second-order errors[1]" of management (Gould, 2004), latent errors, and organizational breakdown (Reason, 1997; Toft and Reynolds, 2005) and evolve, adapt, and generate learning and leadership (Baldwin, 2004; Senge, 1990) to address the fast-changing landscapes of hyper-risks and disasters in which organizations and communities of practice operate.

1. Second-order errors are those errors caused by management decisions.

3. Bridging the gap between theory and practice

Of equally significant importance, reflective practices can lead to a reduction in the gap between theory and practice by informing theories based on messy practices (Fook, 2004). This is explained in Fook's words:

reflective practices can contribute to the possible improvement of practice by closing the gap between espoused theory and enacted practice; the learning of knowledge generating capacities; the potential for on-going evaluation of practice; and the integration of theory, practice and research.

Fook, 1996; quoted in Fook, 2004

Reflective practices are an on-going combination of reflection and critical reflection processes undertaken at workplaces to meet the everyday organizational challenges. In the context of this chapter, this was extended to comprehend hyper-risks and NATECH incidents. Reflective practices and reflective responses are a "dynamic developmental process" (Bolton, 2005) that occurs at the interface of individual, collective, and critical levels and can be applied to support complexity management. Similarly to reflective practice, mitigating hyper-risks is an on-going process. As a result, both mitigation and reflective practices are mutually inclusive. It is therefore posited that a combination of mitigation and reflective practices will enable disaster mitigation to attain a critical edge in order to progress the multidisciplinary DRR concept and practice.

7.5 CONCLUSION: A CHARTER FROM REFLECTIVE RESPONSES

To conclude, this chapter has argued for "reflective response" that is based on "critical reflective practices" and "systems thinking" as a means with which to counter environmental and NATECH risks. The fundamental characteristic of today's world is the interconnectedness and interconnectivity that generates wealth as well as "dangers" and "bads" of the postmodern societies in conjunction with science. The landscape of risk society is highly vulnerable to social, natural, and technical hazards. The dangers and "bads" created from the combination of these hazards and hyper-risks are illustrated through the case of NATECH disasters in general.

The caveat of the dominant crisis and disaster theories and practices is that they fail to capture the complexities of the postmodern world and those of the organizations involved in DRR practices. Most importantly, there is a lack of understanding on NATECH disasters and hyper-risks. Hyper-risks are embedded in complex processes. They are unpredictable with a likelihood of transborder cascading effect. The case of NATECH incidents in Japan or Turkey illustrates some aspects of hyper-risks.

To tackle hyper-risks, this chapter has argued for developing organizational resilience by adopting "reflective response." Reflective response is understood as a combination of individual, collective, and critical reflections and reflective strategies to build the capacity of learning organizations and that of the communities of practice. Reflective response is complementary to the dominant reactive and proactive disaster responses. Building organizational resilience is seminal in this interdependent world.

Currently, the dominant risk, crisis, and disaster theories and practices are inadequate to offer a space to comprehend the true nature of hyper-risks and NATECH disasters. Ozunu et al. (2011, p. 1319) argue that "the consequences of these conjoint events are much more substantial for the health of people, environment and property than those posed by each hazard alone." In this light, a new paradigm is much needed to support the DRR practices worldwide.

To realize reflective response, a charter is suggested:

1. Include complexity/systems thinking in the HFA so that effective responses can be designed to address the complexity associated with conjoint NATECH disaster

The charter begins by leveraging the HFA's strategic goals and the five PfA:

PfA 1: Make DRR a priority
PfA 2: Know the (hyper)risks and take action
PfA 3: Build understanding and awareness
PfA 4: Reduce (hyper)risk
PfA 5: Be prepared and ready to act

Organizational resilience thereby becomes the central node emerging from the reflective practice properties of organizational learning, sense making, and system mindset (Figure 7.1).

Mental models incorporate our biases, values, learning, experiences, and beliefs about how the world works. Systems thinking (Senge, 1990), characterized by seeing wholes and interconnections, is critical to understanding complex disaster etiology associated with NATECH. As described in

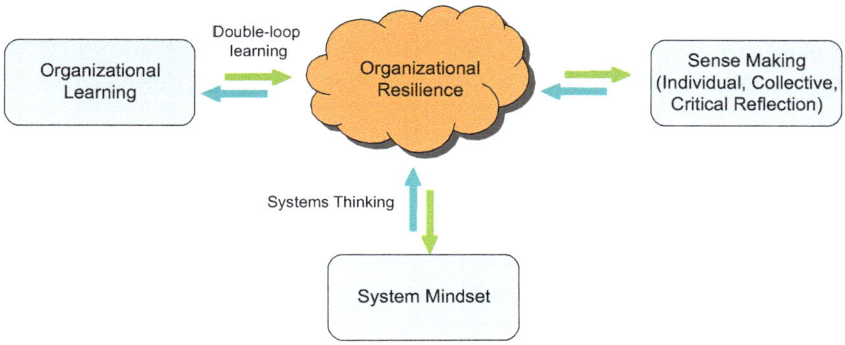

FIGURE 7.1 Organizational resilience enabled through reflective practices.

Masys (2010), the systems lens can enable decision-makers to see beyond events and detect underlying patterns as well as the forces and causal relationships that hold these patterns in place. Van der Merwe (2008, p. 220) argues, "...a systems worldview, together with tools and techniques to make structure visible, is important for building quality scenarios." Scenario planning as a reflective practice has been shown to support the challenge of mental models (Chermack and Nimon, 2013).

Foresight and imagination that recognize the inherent interdependencies in such a complex hyper-risk landscape support understanding of the vulnerabilities, triggers, and consequences, thereby influencing decision-making. Reflective practice of scenario thinking can support the challenges of mental models. It is evident that effective scenario planning did not take place in the context of Fukushima, which the authors have discussed elsewhere (see Ray-Bennett et al., forthcoming). In the context of scenario planning, the major purposes of changing mental models can be seen as opening up perceptions, transitioning toward a learning organization, and as a result, decisions can encompass a wider range of thought (Chermack and Nimon, 2013).

2. Build organizational resilience by involving the communities of practices and practitioners

Reflective practice at the organizational level should allow members to "...critically evaluate their own thinking, but also, to investigate the shared, collective assumptions and expectations, as well as the institutionalized rules and routines" (Hilden and Tikkamaki, 2013, p. 82). Within the context of DRR and NATECH, this provides a methodology to address many of the gaps identified regarding the hyper-risks. Reflective practices that support mindfulness afford organizations involved in DRR within the context of NATECH a "heightened awareness in critical and complex situations which require novel responses" (Zundel, 2012). As described in Weick and Sutcliffe (2007), mindfulness (as an offshoot of reflective practice) supports a preoccupation with failure and a reluctance to simplify interpretations, coupled with sensitivity to weak signals and the ability to respond locally and in real time. In this sense, reflection is about engaging in analysis, considering alternatives, and seeing things from various perspectives to better understand the NATECH hyper-risks and potentially cascading effects. As noted from the Fukushima case study and the studies on NATECH accidents conducted by Cruz (2012) and Cruz and Krausmann (2008), the requirement to test and challenge assumptions, theories, and ideas emerges as a key enabler to support DRR.

3. Leverage the principles of high reliability organizations (Weick and Sutcliffe, 2007) to support mindfulness as part of the reflective practice

The increasing complexity of today's interconnected and interdependent systems as it pertains to DRR has resulted in calls for greater understanding and development mechanisms for coping with uncertainty and shocks to the system. Strategic reflection creates an opportunity to share, compare, and explore mental models, thus opening up consideration of the variety of possibilities. Enabling resilience therefore requires a constant sense of unease that prevents complacency, as argued by Hollnagel and Woods (2006). In supporting resilience, the ability to learn and change requires the application of double-loop learning. This learning is rooted in an organization's ability to frame and then reframe, or to change the way in which they think about or view a concept or idea (Senge, 1990). Double-loop learning requires the organization to be cognizant of such organizational paradigms, mindsets, and mental models that shape decision-making. Chermack and Nimon (2013, p. 816) argue that "most critical is the ability to change one's mental model in order to view the future of an organisation through a new lens. Such reflective practice requires dialogue and conversation about strategic issues decision makers are facing," a reflective capacity that was not apparent in the Fukushima case study (Ray-Bennett et al., forthcoming).

4. Limitation of reflective response

In light of the above, certainly a need exists for increased awareness regarding the interdependencies of NATECH accident and their subsequent cascading effects. Cruz (2012, p. 79) shows that although such awareness has improved, "…this is not reflected sufficiently in the laws and regulations of individual countries, and a general framework for the governance of NATECH risk is lacking."

This brings one of the major caveats of reflective responses to the fore. Reflective responses are likely to encounter challenges at governmental and nongovernmental organizations with regard to time, bureaucracy, and centralized decision-making processes—all of which underpin power, politics, and authority. Practitioners and communities of practice do not work in a social vacuum (Argyris, 1960; Kemmis, 1985), and reflective practices can bring some of the nuances of messy businesses and power dynamics of bureaucracy and management to the fore through lack of awareness and inactions (Gardner et al., 2006; Fook, 2004).

In this light, more empirical research is required to document the "messy" practices of DRR practitioners in order to inform theory from practice, but also to identify the motivating factors that can promote reflective responses.

ACKNOWLEDGMENT

This research was partly funded by the Professional Development Innovation Fund (PDIF) at the University of Leicester. Special thanks are due to the Fund Managers of PDIF, Ms Elain Crewe, and Dr Natalie Cozier. An earlier version of this chapter has appeared as an Input Paper Prepared for the Global Assessment Report on Disaster Risk Reduction 2015 for the UNISDR. Civil Safety and Security Unit (CSSU) at the University of Leicester runs a short course, **Practising Reflection: Reflective Practices in DRR**. More information on this course can be found at www.le.ac.uk/cssu.

REFERENCES

Antonacopoulou, E.P., 2004. The dynamics of reflexive practice: the relationship between learning and changing. In: Reynolds, M., Vince, R. (Eds.), Organizing Reflection. Ashgate Publishing Ltd, Hampshire, UK.

Argyris, C., 1960. Understanding Organizational Behaviour. Travistock Publications, London, UK.

Avin, U., 2007. Using scenarios to make urban plans. In: Hopkins, L.D., Zapata, M. (Eds.), Engaging Our Futures: Tools for Effective Planning Practices, Lincoln Institute of Land Policy, Cambridge, MA.

Baldwin, M., 2004. Critical reflection: opportunities and threats to professional learning and service development in social work organizations. In: Gould, N., Baldwin, M. (Eds.), Social Work, Critical Reflection and the Learning Organization. Ashgate Publishing Ltd, Hampshire, UK.

Beck, U., 1992. Risk Society: Towards a New Modernity. Sage Publications, London, UK.

Beck, U., 2009. What Is Globalization? Polity Press, Cambridge, UK.

Bolton, G., 2005. Reflective Practice: Writing and Professional Development. Sage Publications, London, UK.

Chermack, T.J., Nimon, K., 2013. Drivers and outcomes of scenario planning: a canonical correlation analysis. Eur. J. Training Dev. 37 (9), 811−834.

Collins, A.E., 2009. Disaster and Development. Routledge, London.

Coppola, D.P., 2011. Introduction to International Disaster Management. Butterworth-Heinemann, Boston, USA.

Cottrell, S., 2005. Critical Thinking Skills: Developing Effective Analysis and Argument. Palgrave Macmillan, New York.

Cruz, A., 2005. NaTech disasters: a review of practices, lessons learned and future research needs. In: Paper Presented at the 5th Annual IIASA-DPRI Forum, Integrated Disaster Risk Management: Innovations in Science and Policy, Beijing, China, 14−18 September 2005, p. 10.

Cruz, A.M., 2012. Challenges in Natech risk reduction. Revis. Ingen. 37, 79−87.

Cruz, A.M., Krausmann, E., 2008. Results of the Workshop: Assessing and Managing Natechs (Natural-Hazard Triggered Technological Accidents). EUR Report 23288 EN. http://www.preventionweb.net/files/9623_LBNA23288ENC002.pdf (accessed 06.01.14.).

Cruz, A.M., Okada, N., 2008. Methodology for preliminary assessment of Natech risk in urban areas. Nat. Hazards 46, 199−220.

David, L.S., Overfelt, D., Picou, J.S., 2007. The Sociology of Katrina: Perspectives on a Modern Catastrophe. Rowman and Littlefield Publishers, INC, New York, USA.

Dekker, S., 2011. Drift into Failure: From Hunting Broken Components to Understanding Complex Systems. Ashgate Publishing, Aldershot, Hampshire.

Dewar, J.A., Builder, C.H., Hix, W.M., Levin, M.H., 1993. Assumption-based Planning: A Planning Tool for Very Uncertain Times. RAND Corporation, California, USA.

DFID, 2005. Disaster Risk Reduction: A Development Concern. Policy Briefing Paper. DFID, London, UK.

DFID, 2006. How Resilient Is Your Business to Disaster. DFID, London, UK.

DFID, 2012. Minimum Standards for Embedding Disaster Resilience in DFID Country Offices. DFID, London, UK.

DFID, 2013. Helping Developing Countries Deal with Humanitarian Emergencies. DFID Publication, London. Available online: www.gov.uk/government/publications.

Earl, S., Carden, F., Smutylo, T., 2001. Outcome Mapping: building learning and reflection into development programs. International Development Research Centre, Ottawa, Ontario Canada.

Farber, D., Lakhtakia, A., 2009. Scenario planning and nanotechnological futures. Eur. J. Physics 30 (4), S3–S15.

Fook, J., 2004. Critical reflection and organizational learning and change: a case study. In: Gould, N., Baldwin, M. (Eds.), Social Work, Critical Reflection and the Learning Organization. Ashgate Publishing Ltd, Hampshire, UK.

Freire, P., 1972. Pedagogy of the Oppressed (M. Bergman Ramos, Trans.). Penguin Books, London, UK.

Gardner, F., Fook, J., White, S., 2006. Critical reflection: possibilities for developing effectiveness in conditions of uncertainty. In: White, S., Fook, J., Gardner, F. (Eds.), Critical Reflection in Health and Social Care. Open University Press, Maidenhead, UK.

Ghaye, A., Ghaye, K., 1998. Teaching and Learning through Critical Reflective Practice. David Fulton Publishers, London, UK.

Gibbs, G., 1988. Learning by Doing: A Guide to Teaching and Learning Methods. Further Education Unit. Oxford Polytechnic, UK.

Gould, N., 2004. Introduction: the learning organization and reflective practice – the emergence of a concept. In: Gould, N., Baldwin, M. (Eds.), Social Work, Critical Reflection and the Learning Organization. Ashgate Publishing Ltd, Hampshire, UK.

Habermas, J., 1973. Theory and Practice (J. Viertel, Trans.). Polity Press, Cambridge, UK.

Helbing, D., 2013. Globally networked risks and how to respond. Nature 497, 51–59.

Hickson, H., 2011. Critical reflection: reflecting on learning to be reflective. Reflective Practice 12 (6), 829–839.

Higgins, D., 2011. Editorial. Reflective Practice 12 (5), 583–584.

Hilden, S., Tikkamaki, K., 2013. Reflective practice as a fuel for organizational learning. Admin. Sci. 3, 76–95.

Intergovernmental Panel on Climate Change (IPCC), 2007. Climate change 2007: synthesis report. In: Contribution of Working Groups I, II and III to the Fourth Assessment Report of the Intergovernmental Panel on Climate Change. http://www.ipcc.ch (accessed 06.01.14.).

JISC, 2007. Scenario planning. Available at: www.jiscinfonet.ac.uk/tools/scenario-planning.pdf

Johns, C., 2009. Becoming a Reflective Practitioner, third ed. Wiley-Blackwell, Oxford, UK.

Jones, D., 1996. Anticipating the risks posed by natural perils. In: Hood, C., Jones, C. (Eds.), Accident and Design: Contemporary Debates in Risk Management. UCL Press, Oxon, UK.

Jones, D., Hood, C., 1996. Introduction. In: Hood, C., Jones, C. (Eds.), Accident and Design: Contemporary Debates in Risk Management. UCL Press, Oxon, UK.

Karban, K., Smith, S., 2010. Developing critical reflection within an inter-professional learning programme. In: Bradbury, H., Frost, N., Kilminster, S., Zukas, M. (Eds.), Beyond Reflective Practice. New Approaches to Professional Lifelong Learning. Routledge, London, UK.

Kayes, C.D., 2004. The limits and consequences of experience absent reflection: implications for learning and organising. In: Reynolds, M., Vince, R. (Eds.), Organizing Reflection. Ashgate Publishing Ltd, Hampshire, UK.

Kemmis, S., 1985. Action research and the politics of reflection. In: Boud, D., Keogh, R., Walker, D. (Eds.), Reflection: Turning Experience into Learning. RoutledgeFalmer, London, UK.

Knights, S., 1985. Reflection and learning: the importance of a listener. In: Boud, D., Keogh, R., Walker, D. (Eds.), Reflection: Turning Experience into Learning. RoutledgeFalmer, London, UK.

Masys, A.J., 2010. Opening the Black Box of Human Error: Revealing the Complex Aetiology of Fratricide. VDM Publishing, Saarbrücken, Surrey. pp. 59–84.

Masys, A.J., 2012. Black swans to grey swans- revealing the uncertainty. Int. J. Disaster Prevention Manage. 21 (3), 320–335.

Mezirow, J., 1981. A critical theory of adult learning and education. Adult Education 32 (1), 3–24.

Nicolini, D., Sher, M., Childerstone, S., Gorli, M., 2004. In search of the 'structure that reflects': promoting organisational reflection practices in a UK health authority. In: Reynolds, M., Vince, R. (Eds.), Organizing Reflection. Ashgate Publishing Ltd, Hampshire, UK.

Ozunu, A., Senzaconi, F., Botezan, C., Stefanescu, L., Nour, E., Balcu, C., 2011. Investigations on natural hazards which trigger technological disasters in Romania. Nat. Hazards Earth Syst. Sci. 11, 1319–1325.

Pearson, M., Smith, D., 1985. Debriefing in experience-based learning. In: Boud, D., Keogh, R., Walker, D. (Eds.), Reflection: Turning Experience into Learning. RoutledgeFalmer, London, UK.

Plager, K., 1994. Hermeneutic phenomenology: a methodology for family health and health promotion study in nursing. In: Benner, P. (Ed.), Interpretive Phenomenology. Sage Publications, London, UK.

Powell, J.P., 1985. Autobiographical learning. In: Boud, D., Keogh, R., Walker, D. (Eds.), Reflection: Turning Experience into Learning. RoutledgeFalmer, London, UK.

Ray-Bennett, N.S., 2009. Caste, Class and Gender in Multiple Disasters: The Experiences of Women-headed Households in an Oriya Village. VDM Verlag, Sarbrucken, Germany.

Ray-Bennett, N.S., 2012. Risky talk and talking risks: the way forward or backward?. In: International Disaster and Risk Conference 2012: Integrative Risk Management in a Changing World- Pathways to a Resilient Society, 26–30 August 2012, Davos, Switzerland.

Ray-Bennett, N.S., Masys, T., Shiroshita, H and Jackson, P. 2014. Hyper-risks in a hyper connected world: A call for critical reflective response to develop organisational resilience. Input Paper Prepared for the Global Assessment Report on Disaster Risk Reduction 2015. Global Assessment Report on Disaster Risk Reduction. UNISDR.

Reason, J., 1997. Managing the Risks of Organizational Accidents. Ashgate Publishing Ltd, Aldershot, UK.

Reynolds, M., Vince, R., 2004. Organizing reflection: an introduction. In: Reynolds, M., Vince, R. (Eds.), Organizing Reflection. Ashgate Publishing Ltd, Hampshire, UK.

Schoemaker, P., 1993. Multiple scenario development: its conceptual and behavioral foundation. Strategic Manage. J. 14 (3), 193–213.

Schön, D.A., 1983. The Reflective Practitioner: How Professionals Think in Action. Ashgate Publishing Ltd, London, UK.

Schwartz, P., 1996. The Art of the Long View: Planning for the Future in an Uncertain World. Doubleday, New York, NY.

Senge, P.M., 1990. The Fifth Discipline: The Art and Practice of The Learning Organization. Century Business, London, UK.

Senge, P.M., Kleiner, A., Roberts, C., Ross, R.B., Smith, B.J., 1994. The Fifth Discipline Fieldbook. Doubleday, New York, USA.

Starr, S., 2013. Costs and Consequences of the Fukushima Daiichi Disaster. http://www.psr.org/environment-and-health/environmental-health-policy-institute/responses/costs-and-consequences-of-fukushima.html (accessed 06.01.13.).

Sterman, J.D., 2000. Business Dynamics: Systems Thinking and Modeling for a Complex World. McGraw-Hill Publishing, Boston, USA.

Taleb, N.N., 2007. The Black Swan: The Impact of the Highly Improbable. Penguin Books Ltd, London, UK.

Thorson, C.A., DeVore, S., 2013. Analyzing reflection on/for action: A new approach. Reflective Practice: International and Multidisciplinary Perspectives 14 (1), 88–103.

Toft, B., Reynolds, S., 2005. Learning from Disasters: A Management Approach. Palgrave Macmillan, Eastbourne, UK.

Turner, C., 1999. What are "communities of practice"? In: Senge, P., Kleiner, A., Roberts, C., Ross, R., Roth, G., Smith, B. (Eds.), A Fifth Discipline Resource: The Dance of Change: The Challenges of Sustaining Momentum in Learning Organizations. Nicholas Brealey Publishing, London, UK.

UN, 2005. Building the resilience of nation and communities to disasters: Hyogo framework for action 2005–2015. In: World Conference on Disaster Reduction, 18–22 January, 2005, Hyogo, Japan. http://www.unisdr.org (accessed 23.12.13.).

UN, 2007. Human Development Report 2007/8: Human Solidarity in a Divided World. Palgrave Macmillan, Basingstoke, UK.

Van der Merwe, L., 2008. "Scenario-based strategy in practice: a framework". Advances in Developing Human Resources Vol. 10 (No. 10), 216–239.

Vaughn, L., 2010. The Power of Critical Thinking: Effective Reasoning about Ordinary and Extraordinary Claims, third ed. Oxford University Press, Oxford.

Weick, K.E., 1995. Sensemaking in Organiszations. Sage Publications, London, UK.

Weick, K.E., Sutcliffe, K.M., 2007. Managing the Unexpected: Resilient Performance in an Age of Uncertainty, second ed. John Wiley and Sons Inc, California, USA.

Welsh, M.A., Dehler, G.E., 2004. P(l)aying attention: communities of practice and organized reflection. In: Reynolds, M., Vince, R. (Eds.), Organizing Reflection. Ashgate Publishing Ltd, Hampshire, UK.

Woods, D.D., 2006. Essential characteristics of resilience. In: Hollnagel, E., Woods, D.D., Leveson, N. (Eds.), Resilience Engineering: Concepts and Precepts. Ashgate Publishing Ltd, Hampshire, UK.

Zundel, M., 2012. Walking to learn: Rethinking reflection for management learning. Management Learning 44 (2), 109–126.

Section 2

Hazards in Social, Technological and Political-Economic Change

Chapter 8

Vulnerability, Coping and Loss and Damage from Climate Events

Kees van der Geest and Koko Warner

Institute for Environment and Human Security, United Nations University, Bonn, Germany[1]

ABSTRACT

"Loss and damage" is an emerging concept in the fields of climate-change adaptation and disaster risk reduction. It results from inadequate efforts to reduce greenhouse gas emissions and insufficient capacity to adapt to climatic changes and cope with impacts. This chapter uses original data from study sites in four African and Asian countries that were surveyed in the context of the first ever multisite study of loss and damage in vulnerable communities. National researchers conducted fieldwork in rural areas, looking at impacts of drought (Gambia), floods (Kenya, Nepal), and cyclones (Bangladesh). Methods included a household survey ($n = 1,431$) about climate-related stressors, vulnerability, impacts, coping strategies, and residual loss and damage. The survey data are used to investigate how rural households attempt to cope with impacts of climatic stressors and how successful they are in avoiding loss and damage. A multidimensional vulnerability index (MDVI) is created to analyze differences between more and less vulnerable households in the uptake and effectiveness of seven types of coping strategies. The study reveals that vulnerable households used a more diverse portfolio of coping measures. They were more likely to reduce expenses, modify food consumption and rely on social networks to cope with impacts of climatic stressors. No significant differences were found between more and less vulnerable households in the uptake of migration, asset sales and reliance on relief. The impacts of climate-related stressors in vulnerable communities are beyond the majority (69 percent) of households' capacity to cope. This is especially true for more vulnerable

1. The data presented in this chapter were gathered in the context of the Loss and Damage in Vulnerable Countries Initiative (www.lossanddamage.net). The Climate and Development Knowledge Network (CDKN) funded this project. The authors wish to thank Dr Sidat Yaffa in the Gambia, Denis Opiyo Opondo in Kenya, Dr Golam Rabbani in Bangladesh, and Dr Prakash Korala and Dr Kenneth Bauer in Nepal for their great efforts in conducting the household survey on which this chapter is based.

Hazards, Risks, and Disasters in Society. http://dx.doi.org/10.1016/B978-0-12-396451-9.00008-1

households, but over half the households with a low MDVI also experience losses and damages.

8.1 INTRODUCTION

8.1.1 Loss and Damage

"Loss and damage" is a new concept in the fields of climate-change adaptation and disaster risk reduction. Loss and damage results from inadequate efforts to reduce greenhouse gas (GHG) emissions and insufficient capacity to adapt to climatic changes and to cope with impacts of climate change-related events. The topic emerged in climate negotiations after the establishment of a work program on loss and damage at the 16th UNFCCC Conference of the Parties in Cancun, Mexico in December 2010. The concept has gained further interest from 2012 onward, as a mandate was given to establish institutional arrangements to address loss and damage. At COP19 (November 2013), the Warsaw International Mechanism was established to promote "implementation of approaches to address loss and damage associated with the adverse effects of climate change…in a comprehensive, integrated and coherent manner" (UNFCCC, 2014). The mechanism creates a policy space to discuss and address the negative consequences of climate change if society's efforts to mitigate and adapt are not sufficient. An important question that needs to be answered over the next few years is what can be done to support vulnerable people, communities, and societies who are already feeling the negative consequences of climate change (Kates et al., 2012; Warner, 2013; Roberts and Huq, in press).

Loss and damage is already a significant—and in some places growing—consequence of inadequate ability to adapt to changes in climate patterns across the world (Warner et al., 2012, 2013; Huq et al., 2013). It undermines sustainable development and can impede progress in improving human well-being. Yet, there is currently a lack of empirical evidence of the circumstances under which households manage climatic stressors, the resulting societal impacts, and the loss and damage that results from not being able to adjust sufficiently. Policymakers and practitioners need better information about both the challenges and the potential solutions.

Although a tendency exists among legal scholars, in the media, and among some parties in the climate negotiations to reduce the topic of "loss and damage" to compensation and liability, the authors of this chapter link loss and damage more explicitly to adaptation limits and constraints. The authors acknowledge that some losses and damages are unavoidable, and that these need to be addressed with a separate set of policy measures (social protection, safety nets, resettlement, etc.). However, we emphasize that much loss and damage could be avoided if efforts to reduce GHG emissions are boosted, if more adaptation funding becomes available, if more effective

TABLE 8.1 Avoidable and Unavoidable Loss and Damage

Nature of Loss and Damage	Ways to Address Loss and Damage
Avoidable	• Reduction of greenhouse gas emissions • More effective adaptation • Improved disaster risk reduction • Increased resilience and coping capacity
Unavoidable	• Social protection and safety nets • Resettlement without undermining resilience
Unavoided	• Removing constraints to adaptation

adaptation policy is designed, if sustained progress in comprehensive climate risk management is achieved, and if people's resilience and coping capacity increases (Table 8.1).[2] A third category is unavoided loss and damage, which concerns impacts of climate-related stressors that can be avoided through mitigation and adaptation, but which have not been avoided because the appropriate measures were not adopted due to financial, technical, or other constraints (Verheyen and Roderick, 2008). This category moves the concept from an insecure future to present-day realities of vulnerable people. The way to address such losses and damages would be to understand and remove the constraints to actions required to avoid loss and damage.

Definitions of the term "loss and damage" vary. For this research project, a working definition of loss and damage was used that includes households' inability to respond adequately to climate stressors and the costs and adverse effects associated with the adaptation and coping measures themselves:

Adverse effects of climate-related stressors that households have not been able to avoid through adaptation.

In our working definition, we used the term households because these were our units of analysis. However, the term "households" can be replaced by "actors" (e.g., companies, governments, communities, etc.) to widen the applicability of the definition. The working definition for our local case studies does not explicitly mention reduction of GHG emissions because this was less relevant for the empirical work in vulnerable communities. However, we do recognize that emission reductions play a crucial role in avoiding dangerous climate change and its impacts.

2. For discussions of avoidable and unavoidable loss and damage, see Huq et al. (2013) and Pinninti (2014).

The working definition we used links the concept of loss and damage firmly to the emerging research field of adaptation limits and constraints (Burton, 2009; Adger et al., 2009; Dow et al., 2013; Preston et al., 2013), which for the first time has a chapter in the Fifth Assessment Report of the Intergovernmental Panel on Climate Change (Klein et al., 2014). Loss and damage from sudden-onset events as well as slow-onset processes have been examined, but the focus in this chapter is on concrete events and less on gradual changes. Household measures to avoid loss and damage include risk reduction, coping strategies, and adaptation (Figure 8.1). The focus in this chapter is on the coping measures that households adopt in the aftermath of climate-related events. The climate-related events include droughts (Gambia), floods (Kenya and Nepal), and cyclones (Bangladesh). A multi-dimensional vulnerability index (MDVI) is used to study differences in the uptake of coping strategies between more and less vulnerable households. The next section introduces a framework connecting the key concepts used in the study.

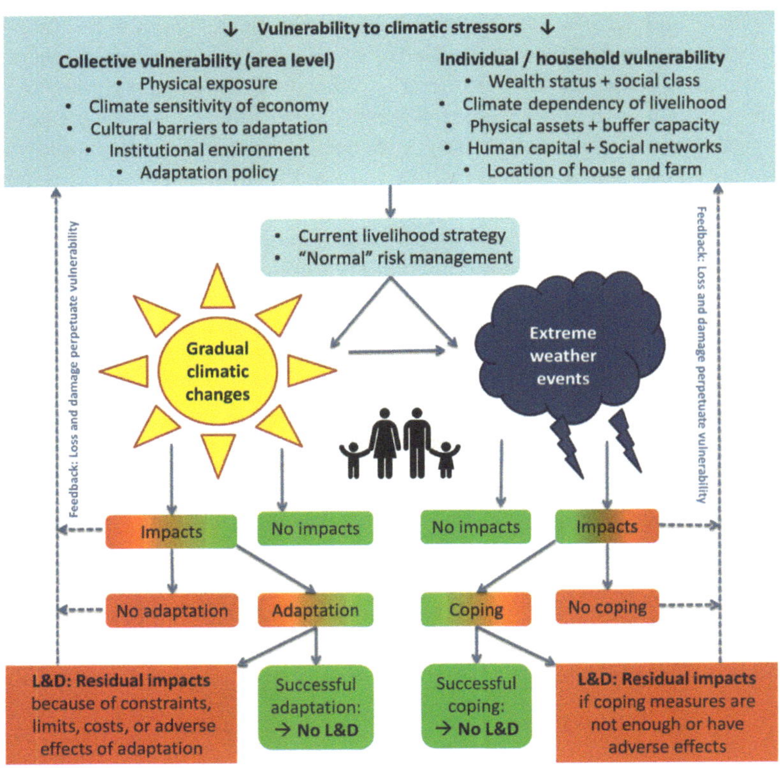

FIGURE 8.1 Conceptual framework: linking loss and damage (L&D) to vulnerability, risk management, and adaptation.

8.1.2 Conceptual Framework

The framework discussed in this section connects loss and damage from climate-related stressors to vulnerability, risk management, impacts, coping strategies, adaptation, and limits and constraints of adaptation (Figure 8.1). The framework results from progressive insights of working on loss and damage in vulnerable communities in the past 2 years (Warner et al., 2012, 2013; Warner and van der Geest, 2013) and previous work on impacts of and adaptation to climate change in dryland West Africa (van der Geest, 2004, 2011; van der Geest and Dietz, 2004). This work, in turn, builds on a longer tradition of studying livelihoods in risk-prone environments that emerged in the 1990s (Chambers, 1989; Davies, 1996; Blaikie et al., 1994; Scoones, 1998; Ellis, 1998).

The blue box in the upper part of Figure 8.1 shows the vulnerability context of households and communities that shapes households' livelihood strategies and the measures they put in place to reduce the risk of being adversely affected by climatic and other stressors. The framework distinguishes collective vulnerability—resulting from area-level variables that are the same for all households in a given community—from individual or household vulnerability (Adger, 1999). When a region experiences changes in the climate or when extreme weather events hit, some households will experience impacts (such as a crop failure or damage to properties), although others may not. This depends on their vulnerability profile—particularly their exposure—and household risk management. When the household experiences no impact, no loss and damage occurs (hence the green color of the "no-impact box"). When the household is affected by the climatic stressor, it may incur or avoid residual loss and damage depending on whether effective measures are adopted to adjust (hence the red-green color of the impact and coping/adaptation boxes). If there is nothing the household can do to adjust, it will incur loss and damage (hence the red color of the no-adaptation and no-coping boxes). If coping or adaptation measures are adopted, these may or may not be effective in avoiding residual loss and damage, depending on the household's adaptive capacity and the magnitude of the climatic stressor (or in other words adaptation constraints and limits[3]). If measures are insufficient, costly, or "erosive" in the longer term, households incur loss and damage (Warner and van der Geest, 2013). Lastly, a feedback loop connects loss and damage back to the household's vulnerability profile. This is because the losses and damages incurred render the household more vulnerable in the face of ongoing climatic changes and future extreme events.

3. IPCC's Fifth Assessment Report defines adaptation constraints as "factors that make it harder to plan and implement adaptation actions" and adaptation limits as "the point at which an actor's objectives… cannot be secure from intolerable risks through adaptive actions" (See also Dow et al. (2013).).

The framework distinguishes two types of household responses to climate-related stressors: "coping" and "adaptation." Many studies use these terms synonymously (Birkmann, 2011). This is problematic because they involve different types of responses to different types of stresses (van der Geest and Dietz, 2004). Coping strategies are short-term responses to the impacts of sudden or unusual events. By contrast, adaptation refers to longer-term adjustments to more permanent changes in the climate.[4] Besides coping and adaptation, a third type of response involves the preventive measures (risk reduction) that households adopt in response to normal characteristics (including variability) of the climate and environment and *in anticipation* of unusual events.[5]

The three groups of climatic stressors and household responses are shown in Table 8.2 with some nonexclusive examples. There are multiple linkages between different types of household responses to climatic stressors. Firstly, the success of ex-ante preventive measures determines the need for and success of ex-post coping strategies. Secondly, short-term coping measures can evolve into more permanent livelihood adaptations when they become recurrent. Thirdly, when households change their preventive measures in response to changes in perceived risk, they are adapting.[6]

8.2 METHODS

The case studies of the Loss and Damage in Vulnerable Countries Initiative (www.lossanddamage.net) focused on the losses and damages that climate-related stressors are already causing in vulnerable communities. Researchers in Bangladesh, Bhutan, the Gambia, Kenya, Micronesia, and Nepal conducted a total of 2,068 household interviews and over 100 focus groups and key informant interviews. They looked at a wide range of climate-related stressors such as droughts, floods, changing rainfall patterns, sea level rise, cyclones, salinity intrusion, and coastal erosion. Each case study looked at extreme weather events as well as slow-onset changes. The overall *research question* was how does the impact of climate-related stressors lead to loss and damage among households in vulnerable communities.

For this chapter, household data from four of the case studies are used (Bangladesh, the Gambia, Kenya, and Nepal). Findings from the Bhutan case

4. For the more elaborate definition of adaptation we used in the case studies, see Moser and Ekstrom (2010). Their definition recognizes that adaptation measures are often adopted in response to a mix of climatic and nonclimatic changes and aim to meet more than climate goals alone.
5. The relationship between preventive strategies, coping, and adaptation is described in detail in van der Geest (2004: 20–29).
6. For an overview of linkages between prevention, coping, and adapting, see van der Geest and Dietz (2004). The framework is inspired by the early work of Susana Davies (1996) on "adaptable livelihoods" in Mali.

TABLE 8.2 **Different Climatic Stressors Require Different Household Responses**

Climatic Stressor	Household Response
Climate variability • "Normal" uncertainties • "Normal" risk of extreme weather events	Preventive measures • Physical protection • Risk spreading • Creating buffers • Building safety nets
Climate-related events • Floods • Droughts • Cyclones/storms	Coping strategies • Rely on social networks • Food aid and other relief • Alternative income • Selling assets
Climatic changes • Changes in "average" conditions • Changes in risk of extreme weather events	Adaptation • Agricultural change • Livelihood diversification • Migration • Changes in "normal" risk management

study are omitted because the study yielded data of limited use with regard to impacts of climate-related *events*. The study looked at glacier lake outburst floods (GLOFs), which had affected only very few households in the area. Besides looking at GLOFs, the study also investigated impacts of longer-term changes in monsoon patterns on rice production, which did yield insightful findings (Kusters and Wangdi, 2013). However, these results are beyond the scope of this chapter. The Micronesia case study looked at loss and damage from coastal erosion and storm surges. Findings are omitted here because the study area was very different from the other four cases in terms of socioeconomic development. Household incomes were about 10 times higher, and most households did not depend on natural resources for their livelihood activities (Monnereau and Abraham, 2013).

Table 8.3 shows the study areas included in this chapter. The sites had diverse ecologies (coastal, floodplain, dryland savannah, hill), but crop cultivation was the main economic activity in all the four areas. In Bangladesh and Nepal, rice was the main crop, whereas farmers in the Gambia and Kenya mostly cultivated millet, maize, and sorghum. The sample size for the household questionnaire varied between 300 and 400 households per study area.

8.2.1 Fieldwork

The research presented here generated original data from the perspective of vulnerable people who experience climate-related stressors, using a systematic

TABLE 8.3 The Case Study Areas

Country	District/ Region	Ecological Zone	Main Crops	Stressor	Sample Size
Bangladesh	Satkhira	Coastal	Rice, vegetables	Cyclone	360
The Gambia	North Bank	Savannah	Millet, groundnut, maize	Drought	371
Kenya	Budalangi	Floodplain	Maize, sorghum, beans	Flood	400
Nepal	Udayapur	Hill, valley	Rice, maize, mustard	Flood	300

assessment approach that employed a variety of methods, including a household survey, focus group discussions, in-depth interviews with a selected number of questionnaire respondents, and expert interviews. In addition, local meteorological and other relevant data were gathered and compared with local perceptions of changes in climatic stressors. Household data were gathered in the following domains: experience with climate-related stressors, impacts in households, the current adaptation and coping measures, and residual loss and damage.

The principal data source for the current chapter is the questionnaire survey that was conducted in the case study sites. A template questionnaire, designed by the project's science coordinator at United Nations University Institute for Environment and Human Security, was used, but national research teams adapted the template for each case study to suit its thematic focus and the characteristics of local livelihood systems and environments. The questionnaire had 10 pages, and interviews usually took 45 min to 1 h. The questionnaires had four sections. The first section focused on socioeconomic and demographic characteristics of the household and their sources of food and income. The information gathered in this section was used to distil multidimensional vulnerability indicators. Second and third sections of the questionnaire focused on impacts of extreme weather events and slow-onset processes, household responses (coping and adaptation), and residual losses and damages. Open questions were combined with closed questions to optimize the balance between listening to the voices of vulnerable people and being able to quantify how widespread different impacts and responses are. Fourth section contained open questions about differences in vulnerability between households in the communities, between men and women, and between children and adults. In this section, respondents were also asked to share their ideas about ways to address loss and damage.

Climate-related losses and damages usually have nonclimatic causes as well. Floods, for example, can be caused by extreme rainfall, whereas at the same time, deforestation and unsustainable land use practices can contribute to the severity of floods. Droughts can become more frequent because of regional climatic changes but are also part of the natural system of climatic variability. It is usually not possible to attribute losses and damages entirely to climate change (Huggel et al., 2013; Wrathall et al., in press).[7] That was also not the objective of the research project this chapter reports on. Rather, the objective was to explore situations in which households face adaptation constraints and lack coping capacity, and to assess the consequences for people's livelihoods and sustainable development pathways. The study and its methods should be treated as points of departure for further research on loss and damage in vulnerable communities.

8.2.2 Analysis

The results of this research project presented so far[8] have been mostly descriptive. By contrast, the analysis presented in this chapter goes beyond the descriptive to study how household vulnerability influences the uptake of different coping strategies in the aftermath of climate events and the ability of these measures to avoid loss and damage.

An important step was to determine for each household in the sample whether it incurred loss and damage or not. No attempt was made to quantify loss and damage in monetary terms. In line with the conceptual framework in Figure 8.1, households incurred loss and damage when

- climate-related stressors affected them and they did not/could not adopt any measures to mitigate the effects and
- the measures that they adopted to mitigate the effects of climate-related stressors were not enough to avoid loss and damage or had adverse effects.

The next step was to assess the effect of two groups of independent variables on household loss and damage: (1) a group of multidimensional vulnerability indicators and (2) the coping strategies that people adopted in the aftermath of climatic events.

The conceptual framework for this study distinguished collective vulnerability (area-level) and individual vulnerability (household-level). To measure household vulnerability, we used an index that builds on the Alkire Foster method for measuring the multiple dimensions of poverty (Alkire and Foster, 2011). The method was originally developed by researchers at the Oxford

7. See Allen (2003), Lott et al. (2013), and Otto et al. (2013) for discussion of and attempts to estimate the "fraction of attributable risk" of climatic events to external drivers of climatic change such as GHG emissions (Poster presentation by Rachel James at CCDA-3, Addis Ababa).

8. Warner and van der Geest (2013), Warner et al. (2012, 2013), Bauer (2013), Kuster and Wangdi (2013), Monnereau and Abraham (2013), Opondo (2013), Rabbani et al. (2013), Yaffa (2013).

Poverty and Human Development Initiative to measure deprivations in health, education, and living standard based on household surveys. To adjust the multidimensional poverty index to an MDVI, household variables were selected that represent household vulnerability to climate-related stressors (see Section 4).[9] Another departure from the Alkire Foster method is that instead of using just two vulnerability options (vulnerable and nonvulnerable) for each indicator, we used a five-point scale on each of the 10 indicators we identified.

8.3 DESCRIPTIVE CASE STUDY FINDINGS

Across the research sites, it was found that households struggled to manage impacts of climatic stressors on their livelihoods, assets, and culture. Despite their efforts to cope with the impacts of extreme weather events and to adapt to slow-onset climatic changes, many could not avoid incurring residual impacts. Some of the most notable impacts were on household food production and livelihoods, raising questions about the ability of adaptation measures, both formal and informal, to stem the interacting negative impacts of climate change and vulnerable societies.

Results from the case studies revealed four different "loss and damage pathways" (Warner and van der Geest, 2013). Loss and damage from climate-related stressors occurs when

- measures to cope or adapt are not enough to avoid loss and damage,
- measures have costs (including noneconomic) that are not regained,
- despite short-term merits, measures are erosive and undermine longer-term livelihood sustainability, and
- no measures are adopted or possible at all.

Each of the case studies has a different story to tell, but all point to the same key finding: Loss and damage is happening now, despite adaptation efforts.[10]

Satkhira, a coastal district in Bangladesh, faces the threat of sea level rise and cyclones. Both result in saltwater intrusion, which severely impacts rice cultivation, the mainstay of the local economy. To adapt, farmers have planted new saline-tolerant rice varieties. This worked well until 2009, when cyclone Aila hit and caused a sudden and drastic increase of salt content in the soil. Almost all farmers lost their complete harvest that year.

The North Bank Region of the Gambia experienced a severe drought in 2011, affecting almost all farmers in the area, many of whom lost their entire harvests. They tried to cope by looking for additional income (e.g., sale of property) to buy food. However, the majority had to reduce their food intake,

9. See, e.g., Hahn et al. (2009) work on the livelihood vulnerability index.
10. These excerpts of case study findings have earlier been published in a blog of the Thomson Reuters Foundation. See http://www.trust.org/item/20131120115704-706cv/.

for example, by changing from three to two meals a day because their coping measures were insufficient.

In December 2011, River Nzoia in Western Kenya broke its dikes and wreaked havoc in Budalangi Division. Crops were washed away, livestock drowned, houses were severely damaged, and there was an outbreak of waterborne diseases. For survival, many households were forced to adopt erosive coping strategies such as the sale of productive assets and taking children out of school to earn a meager income in the informal sector. Erosive coping measures have severe implications for future livelihood security.

Nepal is particularly susceptible to climate-related disasters such as floods, landslides, and debris flows. In the Udayapur district, floods have become more severe over the past two decades, destroying crops and damaging houses. In addition, the study found that food prices increased sharply in the aftermath of floods. Although households expend much effort on preventive and coping measures, these have not been enough to counteract adverse effects.

8.4 VULNERABILITY

To measure household vulnerability, we used an index that builds on the Alkire Foster method for measuring the multiple dimensions of poverty (Alkire and Foster, 2011). Based on the household survey data, 10 multidimensional vulnerability indicators were defined, including, for example, education level, land and livestock ownership, livelihood diversity, income, house quality, and food security. For each variable, four thresholds were chosen, dividing the sample population in five vulnerability groups on each indicator. For example, for education, the groups are "no formal education," "attended literacy classes," "primary school," "secondary school," and "tertiary education." For quantitative indicators (e.g., land size, livestock ownership, and total income), the sample population for each case study is divided, based on quintiles, and household scores are assigned accordingly.[11] The vulnerability indicators selected for the analyses in this chapter are listed and described in Table 8.4.

Each household in the sample has a score of 1−5 on each indicator. The overall vulnerability is calculated as the average score and ranged from 1.2 to 4.7. A higher score on the index indicates high vulnerability to climatic stressors. For the whole sample, the average score is 3.02 with a standard deviation of 0.64. The distribution closely resembles the normal curve (see Figure 8.2). The MDVI is used in the following sections to relate the uptake of coping measures to households' vulnerability levels.

Figure 8.3 shows vulnerability profiles for the four case study areas. The multidimensional vulnerability indicators illustrate that different study areas

11. The upper and lower bounds of quintiles for each country are listed in the appendix.

TABLE 8.4 Vulnerability Indicators

Indicator	Description and Thresholds
1. Education	Level of education of household head: none = 5; literacy training = 4; primary = 3; secondary = 2; tertiary = 1.
2. Dependency ratio	(Dependent household members (aged <18 and >65)/adult household members (aged 18–65)) × 100 (score based on quintiles; see appendix for threshold values).
3. Land ownership	Size of land owned by the household (score based on quintiles).
4. Livestock ownership	Expressed in tropical livestock units (score based on quintiles). See appendix for conversion factors.
5. Nonfarm income	Most vulnerable (5) if household has no nonfarm income activities at all; less vulnerable (4–1) according to amount of nonfarm income (score based on quartiles).
6. Livelihood diversity	Number of the following livelihood sources: crop cultivation, livestock keeping, fishing, trees, farm labor, nonfarm income, remittances, other (mostly pension, rent).
7. Total cash income	Total amount of cash income from all sources.
8. House quality	Based on floor material (vulnerable if earth, mud, cow dung) and own perception of house quality (better, average, worse). See appendix.
9. Sanitation	Based on household access to clean drinking water and private toilet facilities. See appendix.
10. Food security	Based on months of food shortage in the past year and years of food shortage in the past decade.
Overall vulnerability	Average score on the 10 indicators, allowing for one missing value.

Note: More details and threshold values for quantitative indicators in appendix.

are vulnerable for different reasons. For example, households in the Gambian study site were well endowed in terms of land and livestock but had low levels of education and poor access to nonfarm income. Selling livestock to buy food is an important way to deal with drought-induced crop failures. Hence, having a good stock of domestic animals made households less vulnerable. By contrast, relatively few households in the Gambian case study area had viable nonfarm activities to fall back on when their harvest failed. In the densely populated Asian study sites (Bangladesh and Nepal), land scarcity was a key problem, limiting households' possibility to attain self-sufficiency in food production. However, food insecurity was less common here because more households had reliable nonfarm income sources.

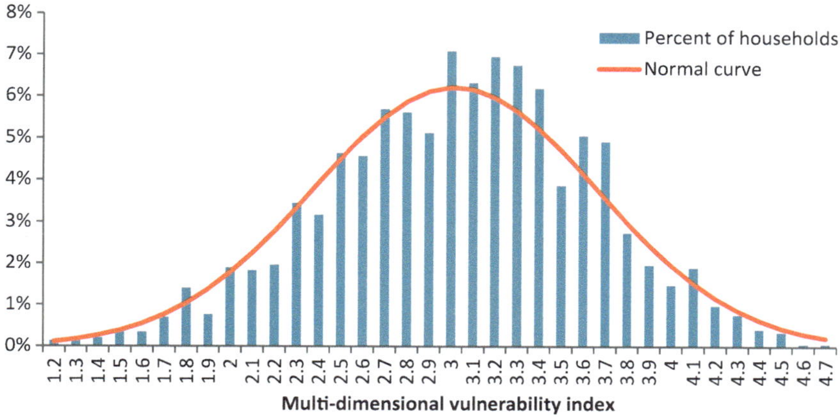

FIGURE 8.2 Distribution of household scores on the multi-dimensional vulnerability index. Note: Normal curve based on an average of 3.02 and standard deviation of 0.64.

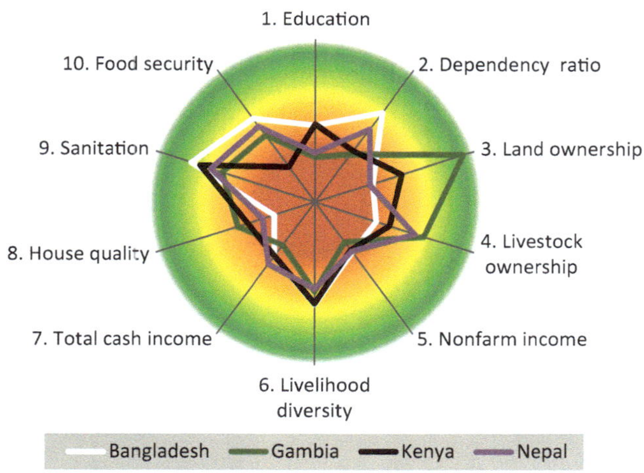

FIGURE 8.3 Household vulnerability profile by indicator and country. Note: The closer to the center of the radar (red), the more vulnerable.

8.5 IMPACT OF CLIMATE EVENTS

Each of the four study sites had been hit by climate-related events in the past years. In the North Bank Region of the Gambia, a severe drought hit in 2011; the Kenyan study site, at the shore of Lake Victoria, experienced extreme flooding that same year when River Nzoia broke its dike; a cyclone (Aila) caused havoc in southwest Bangladesh in 2009; and households in Udayapur district in Nepal had to deal with recurrent floods over the past decade. We asked respondents whether their households had been affected by these events, and if so, whether they would qualify the impacts as “moderate” or "severe." In a follow-up question, respondents were asked about impacts on specific aspects of their household economies such as crop cultivation, livestock, fishing, trees, trade, food prices, housing, properties, and other.

The first graph of Figure 8.4 shows the proportion of households reporting "moderate," "severe," and "no impacts" of climate stressors (y-axis) and relates this to the household scores on the MDVI (x-axis). The graph shows that the majority of households reported severe impacts, and only very few indicated that they did not experience any impacts at all. Vulnerable households were significantly ($p < 0.01$) more likely to report "severe" impacts. It could be argued, however, that the more unexpected finding is that approximately two-thirds of the *least* vulnerable households also reported severe impacts. The second graph of Figure 8.4 shows the proportion of households reporting impacts on specific household activities and assets. Almost 90 percent of surveyed households experienced adverse effects of climate-related events on their crops. Over 80 percent reported that they were affected by high food prices, which made it more difficult to cope with crop losses and attain food security. Qualitative information about the "other" category revealed that it would have been worthwhile to add a separate category for health impacts, which included waterborne diseases in the flood-afflicted study sites, skin and reproductive health problems in Bangladesh, and problems related to undernutrition in the Gambia.

8.6 COPING STRATEGIES

This section looks at the coping measures households adopted to deal with impacts of climate-related events. It shows which measures were most

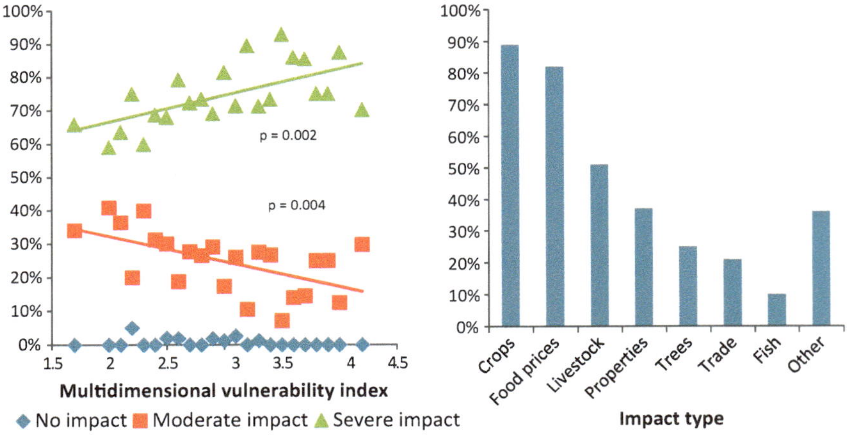

FIGURE 8.4 Impact of climatic events by household vulnerability (left) and impact type (right)—percent of households. Note: The questionnaire used in the Nepal case study did not distinguish moderate and severe impacts. Hence, the graph is based on data from the Gambia, Kenya, and Bangladesh only.

common across the four study sites and investigates whether there is a relation between vulnerability levels and the uptake of certain coping measures. As explained in the conceptual framework section, we differentiate coping strategies and adaptation. Coping refers to short-term responses to the impacts of events. By contrast, adaptation refers to longer-term adjustments to more permanent changes in the climate. The focus here is on coping measures. A cross-country analysis of adaptation measures made less sense because these differed much more widely. To adapt to increasing flood risks requires very different measures than to adapt to drought. This is different for coping measures. It does not differ so much whether households lose their harvest in a flood or a drought; to gain access to food and survive, similar measures are adopted. The survey instrument inquired about eight different types of coping measures (see Table 8.5).

Figure 8.5 relates the uptake of coping measures to household scores on the MDVI. More vulnerable households were more likely to cope by reducing expenses ("economize"), modifying food consumption, and relying on social networks. No significant differences were found between more vulnerable and less vulnerable households in the uptake of migration, asset sales, and reliance on relief. Reliance on nonfarm income was slightly more common among less vulnerable households, but the relation was not statistically significant ($p = 0.15$). "Other" coping measures were more often used by less vulnerable households. A qualitative analysis of this category revealed that most of these households borrowed money to deal with impacts of climate-related events.

Figure 8.6 shows the number of coping measures households adopted in the aftermath of climatic events. Most commonly, households used three to five different types of coping strategies. The figure further shows that less vulnerable households needed a less diverse set of coping measures to deal with impacts of droughts, floods, or cyclones.

8.7 LOSS AND DAMAGE

This section looks at households' ability to avoid loss and damage and relates this to vulnerability levels and the types of coping measures households adopted. As explained in the methods section, households incurred loss and damage when

- climate-related stressors affected them and they did not/could not adopt any measures to mitigate the effects and
- the measures that they adopted to mitigate the effects of climate-related stressors were not enough to avoid loss and damage or had adverse effects.

Figure 8.7 shows that across the four study sites, over two-thirds (69 percent) of households incurred loss and damage from climatic events. The proportion ranged from 62 percent in Nepal to 76 percent in Kenya. The

TABLE 8.5 Coping Measures and Their Uptake

Coping Measure	Uptake	Description
Reliance on social networks	48%	Asking food or other support from relatives, friends, or neighbors.
Reliance on support from organizations	54%	Food aid, camps, medical care, etc.
Engaging in other income-generating activities	50%	The climate-related stressors mostly affected crop production. A way to gain access to food and income was to (temporarily) intensify nonfarm activities or engage in new ones.
Migration	37%	This could involve temporary migration of individual household members, usually aimed at earning an income to support the family, or the movement of entire households (displacement), mostly in response to floods.
Asset sales	33%	Selling properties, mostly livestock, to buy food to deal with crop losses.
Reducing expenses	69%	When food is scarce because of harvest losses, people can avoid spending money on nonfood items such as school fees, health care, house maintenance, investments, etc.
Modifying food consumption	73%	Eating cheaper or less nutritious food, reducing the number of meals, reducing portion sizes. It is debatable whether this is a coping measure or a sign that other coping measures are failing.
Other	18%	Respondents were asked whether they had done anything else to deal with impacts of climatic stressors. The most common "other" measure was reliance on loans.

figure further shows that there is a clear and significant relation between households' score on the MDVI and their likelihood of incurring loss and damage. More vulnerable households are—as one would expect—more likely to experience impacts of climate-related events that are beyond their coping capacity. However, even households with a very low MDVI are more likely than not to incur loss and damage.

Figure 8.8 looks at the relation between the uptake of different coping measures and households' ability to *avoid* loss and damage. As most households adopted several coping measures, the graphs cannot be interpreted as assessments of the effectiveness of individual coping measures. A low bar in the left graph indicates that relatively few households that

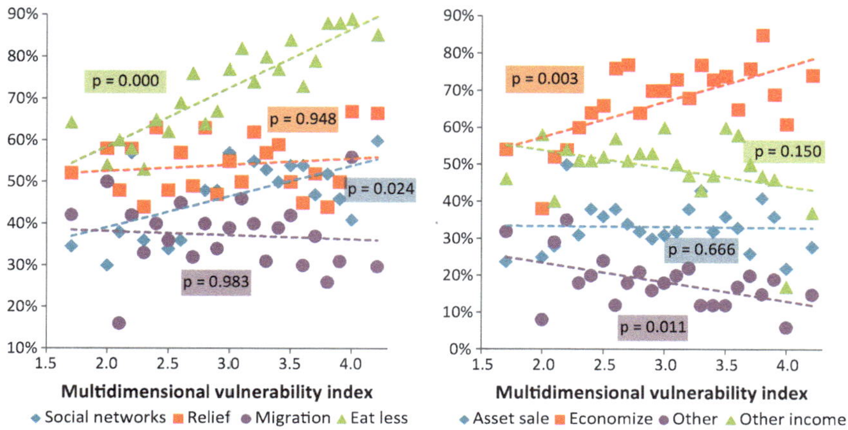

FIGURE 8.5 Uptake of coping measures by vulnerability—percent of households. Notes: (1) The coping measures were plotted in two separate graphs because of layout considerations. (2) The lower the p-value, the more significant the relation between uptake and households' multidimensional vulnerability index score.

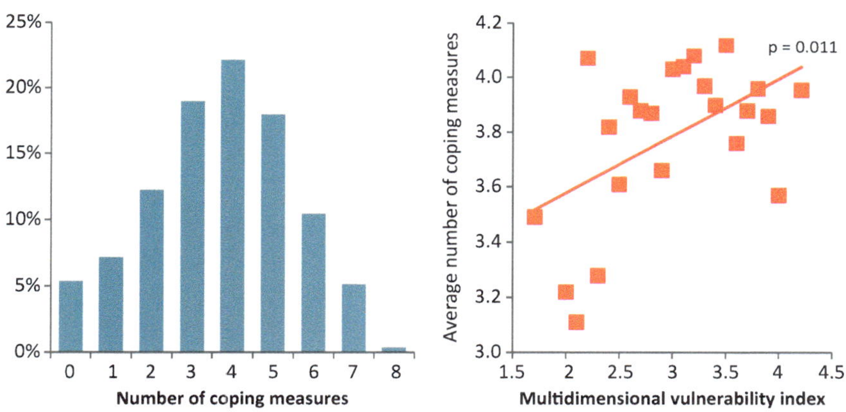

FIGURE 8.6 Number of household coping measures by vulnerability.

adopted that measure were successful in avoiding losses and damage. This was the case for reliance on relief, migration, and modification of food consumption ("eat less"). By contrast, households that relied on nonfarm income, reducing expenses, and "other" measures (mostly taking loans) to cope with climatic events were more likely to avoid losses and damages. However, the differences are quite small. The graph on the right looks at more vulnerable and less vulnerable households separately (threshold set at MDVI = 3.0). It shows that the association of coping measures with households' ability to avoid loss and damages differs between more

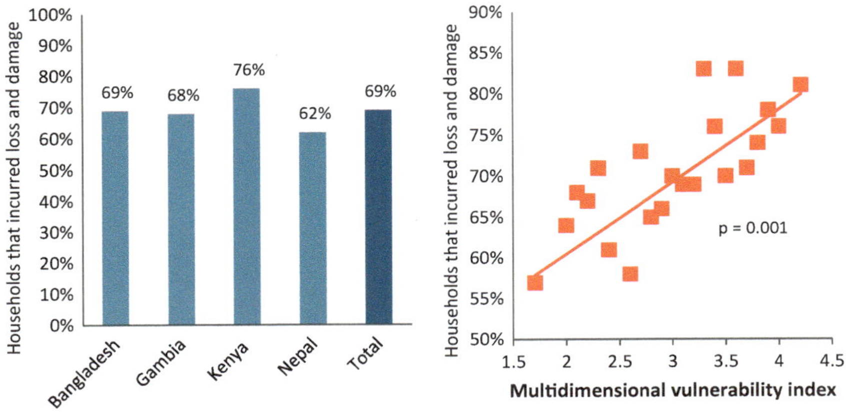

FIGURE 8.7 Loss and damage by vulnerability.

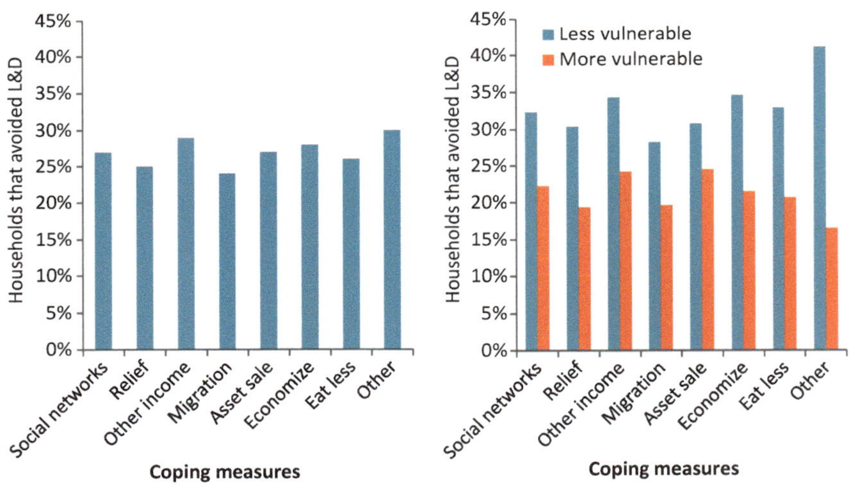

FIGURE 8.8 Coping measures, vulnerability, and the ability to avoid loss and damage (L&D). Notes: The cutoff between more vulnerable and less vulnerable households was set at multidimensional vulnerability index = 3.0.

vulnerable and less vulnerable households. This is clearest in the category "other," which mostly consisted of taking loans. Less vulnerable households that adopted this coping measure were most likely to avoid loss and damage, while the contrary was the case for more vulnerable households, many of whom indicated that the loans helped them to deal with short-term impacts, but had erosive effects in the longer term because they had great difficulties in paying back the loans.

8.8 CONCLUSIONS

'Loss and damage from climate change impacts' is an emerging theme in the international climate negotiations and associated fields of research. It results from inadequate efforts to reduce greenhouse gas emissions and insufficient capacity to adapt to climatic changes and cope with impacts of climatic events. Empirical research about the climate-related losses and damages that actors across the world are experiencing is still very limited. This chapter uses original data from study sites in four African and Asian countries that were surveyed in the context of the first ever multi-sited study of loss and damage in vulnerable communities.

The study uses a multi-dimensional vulnerability index (MDVI) to analyse differences between more and less vulnerable households in the uptake and effectiveness of measures to cope with impacts of climatic events. The results show that virtually all households surveyed experienced adverse effects of climate-related stressors, but more vulnerable households reported 'severe' impacts more often. Over 95% of surveyed households adopted at least one coping measure in response. Most commonly, households used 3 to 5 different types of coping strategies. Vulnerable households needed a more diverse set of coping measures to deal with impacts of droughts, floods or cyclones. Some coping measures, such as reducing expenses, modifying food consumption and relying on social networks were significantly more common among vulnerable households. By contrast, reliance on non-farm income and taking loans was more common among less vulnerable households. No significant relation was found between household vulnerability and selling assets, migration and reliance on relief to cope with climate impacts.

Over two thirds (69.4 percent) of surveyed households were not able to avoid residual loss and damage because their coping measures were not efficient enough, costly or erosive in the longer term. The percentage was highest among households that relied on migration, relief and modifying food consumption to cope with impacts of climatic stressors. A strong and significant relation was found between households' score on the multi-dimensional vulnerability index and their ability to avoid loss and damage. More vulnerable households are - as one would expect - more likely to experience impacts of climate-related events that are beyond their capacity to cope. However, even in the group of least vulnerable households over fifty percent incurred loss and damage.

APPENDIX: THRESHOLDS FOR VULNERABILITY INDICATORS

1. Education

Same for all countries: education level of the household head.

1 = Tertiary

2 = Secondary
3 = Primary
4 = Literacy course
5 = None

In "other" category: technical/vocational = 2; madrasa/monk = 4.

2. Dependency Ratio

Calculated as: (dependent household members (aged <18 and >65)/adult household members (18–65)) × 100. The higher the dependency ratio, the more vulnerable the household.

	1	2	3	4	5
Bangladesh	0–20	22–33	38–63	67–100	>117
Gambia	0–63	64–91	100–136	138–200	>206
Kenya	0–43	50–83	100–150	157–233	>250
Nepal	0–22	25–50	56–88	100–140	>150
All	0–32	33–65	67–100	108–178	>180

3. Land Size

Size of owned land (hectares).

	5	4	3	2	1
Bangladesh	0–0.07	0.07–0.20	0.20–0.34	0.35–0.74	>0.80
Gambia	0–3.00	3.5–5.00	5.50–8.75	9.00–13.5	>14.0
Kenya	0–0.28	0.3–0.40	0.45–0.73	0.81–1.21	>1.42
Nepal	0–0.05	0.07–0.17	0.20–0.34	0.36–0.68	>0.71
All	0–0.14	0.14–0.40	0.40–0.87	0.89–3.75	>4.00

4. Livestock

Livestock owned, expressed in tropical livestock units. Conversion factors: horse, 0.8; cow, 0.7; donkey, 0.5; pig, 0.2; sheep/goat, 0.1; poultry, 0.01.

	5	4	3	2	1
Bangladesh	0–0.07	0.08–0.26	0.27–0.90	0.92–1.76	>1.80
Gambia	0–0.75	0.80–1.77	1.80–3.05	3.06–5.00	>5.02
Kenya	0–0.01	0.02–0.43	0.44–1.76	1.80–3.05	>3.15
Nepal	0–0.47	0.50–1.61	1.64–2.50	2.52–3.63	>3.65
All	0–0.13	0.14–0.90	0.92–2.00	2.04–3.47	>3.50

5. Nonfarm Income

Total income derived from nonfarm-income-generating activities in the past 12 months, expressed in US$. The category of most vulnerable households (5) has no nonfarm activities (NFA) at all. Category 4 is populated by households that have very low-yielding or occasional NFA. This includes households with zero income from these activities in the past 12 months.

	5	4	3	2	1
Bangladesh	No NFA	12–245	270–441	444–736	>854
Gambia	No NFA	10–81	97–161	193–387	>406
Kenya	No NFA	0–88	94–228	234–527	>527
Nepal	No NFA	0–147	170–452	475–1,130	>1,220
All	No NFA	0–129	129–293	294–687	>702

6. Livelihood Diversity

Same for all countries: Number of livelihood sources out of the following list: crop cultivation, livestock keeping, fishing, trees, farm labor, nonfarm income, remittances, and other (mostly pension, rent).

1 = 6–8 sources
2 = 5 sources
3 = 4 sources
4 = 3 sources
5 = 0–2 sources

7. Total Income

Total cash income over the past 12 months, expressed in US$. Calculated as the sum of income from all sources mentioned under Livelihood Diversity.

	5	4	3	2	1
Bangladesh	0–208	209–449	453–753	770–1,242	>1,245
Gambia	0–147	148–316	319–568	574–1,081	>1,084
Kenya	0–211	215–410	415–716	722–1,334	>1,337
Nepal	0–136	139–525	565–1,028	1,051–1,865	>1,876
All	0–170	171–407	410–736	737–1,355	>1,356

8. House Quality

Same for all countries: Based on floor material (vulnerable if earth, mud, or cow dung) and own perception of house quality (better, average, worse).

	Is Floor of the House Made of Earth, Mud, or Dung?	
Own Perception of House Quality	**No**	**Yes**
Better	Vulnerability = 1	Vulnerability = 3
Average	Vulnerability = 2	Vulnerability = 4
Worse	Vulnerability = 3	Vulnerability = 5

9. Sanitation

Same for all countries: Based on household access to clean drinking water and private toilet facilities.

Source of Drinking Water	Does House Have a Private Toilet or Latrine?	
	Yes	No
Tube, pump, borehole	Vulnerability = 1	Vulnerability = 3
Surface water/well + other sources	Vulnerability = 2	Vulnerability = 4
Only surface water/well	Vulnerability = 4	Vulnerability = 5

10. Food Security

Same for all countries: Based on months of food shortage in the past year and years of food shortage in the past 10 years.

Months of Food Shortage in Past Year	Years of Food Shortage in the Past Decade		
	0	1–3	4–12
0	Vulnerability = 1	Vulnerability = 2	Vulnerability = 3
1–3	Vulnerability = 2	Vulnerability = 3	Vulnerability = 4
4–12	Vulnerability = 3	Vulnerability = 4	Vulnerability = 5

REFERENCES

Adger, N.W., 1999. Social vulnerability to climate change and extremes in coastal Vietnam. World Dev. 27 (2), 249–269.

Adger, W.N., Dessai, S., Goulden, M., Hulme, M., Lorenzoni, I., Nelson, D.R., Naess, L.O., Wolf, J., Wreford, A., 2009. Are there social limits to adaptation to climate change? Clim. Change 93, 335–354.

Allen, M., 2003. Liability for climate change. Nature 421, 891–892.

Alkire, S., Foster, J., 2011. Counting and multidimensional poverty measurement. J. Public Econ. 95 (7), 476–487.

Bauer, K., 2013. Are preventive and coping measures enough to avoid loss and damage from flooding in Udayapur district, Nepal? Int. J. Global Warming 5 (4), 433–451.

Birkmann, J., 2011. First- and second-order adaptation to natural hazards and extreme events in the context of climate change. Nat. Hazards 58, 811–840.

Blaikie, P., Cannon, T., Davis, I., Wisner, B., 1994. At Risk: Natural Hazards, People's Vulnerability, and Disasters. Routledge, New York.

Burton, I., 2009. Climate change and the adaptation deficit. In: Schipper, E.L.F., Burton, I. (Eds.), Earthscan Reader on Adaptation to Climate Change. Earthscan, London, pp. 89–95.

Chambers, R., 1989. Editorial introduction: vulnerability, coping, and policy. IDS Bull. 20 (2), 1–7.

Davies, S., 1996. Adaptable Livelihoods: Coping with Food Insecurity in the Malian Sahel. MacMillan Press, London.

Dow, K., Berkhout, F., Preston, B.L., Klein, R.J., Midgley, G., Shaw, M.R., 2013. Limits to adaptation. Nat. Clim. Change 3 (4), 305–307.

Ellis, F., 1998. Household strategies and rural livelihood diversification. J. Dev. Stud. 35 (1), 1–38.
van der Geest, K., 2004. "We're Managing!" Climate Change and Livelihood Vulnerability in Northwest Ghana. African Studies Centre, Leiden.
van der Geest, K., 2011. The Dagara Farmer at Home and Away: Migration, Environment and Development in Ghana. African Studies Centre, Leiden.
van der Geest, K., Dietz, T., 2004. A literature survey about risk and vulnerability in drylands, with a focus on the Sahel. In: Dietz, T., Rueben, R., Verhagen, J. (Eds.), The Impact of Climate Change on Drylands. Kluwer, Dordrecht, pp. 117–146.
Hahn, M.B., Riederer, A.M., Foster, S.O., 2009. The livelihood vulnerability index: a pragmatic approach to assessing risks from climate variability and change—a case study in Mozambique. Global Environ. Change 19 (1), 74–88.
Huggel, C., Stone, D., Auffhammer, M., Hansen, G., 2013. Loss and damage attribution. Nat. Clim. Change 3 (8), 694–696.
Huq, S., Roberts, E., Fenton, A., 2013. Loss and damage. Nat. Clim. Change 3 (11), 947–949.
Kates, R.W., Travis, W.R., Wilbanks, T.J., 2012. Transformational adaptation when incremental adaptations to climate change are insufficient. PNAS 109 (19), 7156–7161.
Klein, R., et al., 2014. Chapter 16: adaptation opportunities, constraints, and limits. In: Climate Change 2014: Impacts, Adaptation, and Vulnerability. Contribution of Working Group 2 to the IPCC Fifth Assessment Report.
Kusters, K., Wangdi, N., 2013. The costs of adaptation: changes in water availability and farmers' responses in Punakha district, Bhutan. Int. J. Global Warming 5 (4), 387–399.
Lott, F., Christidis, N., Stott, P., 2013. Can the 2011 East African drought be attributed to human-induced climate change? Geophys. Res. Lett. 40 (6), 1177–1181.
Monnereau, I., Abraham, S., 2013. Limits to autonomous adaptation in response to coastal erosion in Kosrae, Micronesia. Int. J. Global Warming 5 (4), 416–432.
Moser, S.C., Ekstrom, J.A., 2010. A framework to diagnose barriers to climate change adaptation. PNAS 107 (51), 22026–22031.
Opondo, D., 2013. Erosive coping after the 2011 floods in Kenya. Int. J. Global Warming 5 (4), 452–466.
Otto, F., Jones, R., Halladay, K., Allen, M., 2013. Attribution of changes in precipitation patterns in African rainforests. Phil. Trans. B 368, 20120299.
Preston, B.L., Dow, K., Berkhout, F., 2013. The climate adaptation frontier. Sustainability 5, 1011–1035.
Pinninti, K.R., 2014. Climate Change Loss and Damage: Economic and Legal Foundations. Springer, New York.
Rabbani, G., Rahman, A., Mainuddin, K., 2013. Salinity induced loss and damage to farming households in coastal Bangladesh. Int. J. Global Warming 5 (4), 400–415.
Roberts, E., Huq, S. Coming full circle: the history of loss and damage in the UNFCCC process. Int. J. Global Warming, in press.
Scoones, I., 1998. Sustainable Rural Livelihoods: A Framework for Analysis. IDS Working Paper 72. Institute of Development Studies, Brighton.
UNFCCC. 2014. Report of the Conference of the Parties, on Its Nineteenth Session, Held in Warsaw from 11 to 23 November 2013, Addendum, Part Two: Action Taken by the Conference of the Parties at Its Nineteenth Session. FCCC/CP/2013/10/Add.1.
Verheyen, R., Roderick, P., 2008. Beyond Adaptation: The Legal Duty to Pay Compensation for Climate Change Damage. WWF-UK.
Warner, K. 2013. The Warsaw International Mechanism: A legitimate policy space for loss and damage widens and deepens. http://www.ehs.unu.edu/file/get/11501.pdf

Warner, K., van der Geest, K., Kreft, S., Huq, S., Kusters K., de Sherbinin, A. 2012. Evidence from the Frontlines of Climate Change: Loss and Damage to Communities Despite Coping and Adaptation. Loss and Damage in Vulnerable Countries Initiative. Policy Report. Report No. 9. Bonn: United Nations University Institute for Environment and Human Security (UNU-EHS).

Warner, K., van der Geest, K., Kreft, S. 2013. Pushed to the Limits: Evidence of Climate Change-Related Loss and Damage when People Face Constraints and Limits to Adaptation. Report No.11. Bonn: United Nations University Institute for Environment and Human Security (UNU-EHS).

Warner, K., van der Geest, K., 2013. Loss and damage from climate change: local—level evidence from nine vulnerable countries. Int. J. Global Warming 5 (4), 367—386.

Wrathall, D., Oliver-Smith, A., Fekete, A., Gencer, E., Lepana, M., Sakdapolrakal, P. Problemizing loss and damage. Int. J. Global Warming, in press.

Yaffa, S., 2013. Coping measures not enough to avoid loss and damage from drought in the North Bank region of the Gambia. Int. J. Global Warming 5 (4), 467—482.

Chapter 9

Flood Shelters in Bangladesh: Some Issues From the User's Perspective

M. Aminur Rahman[1], Fuad H. Mallick[1], M. Shahjahan Mondal[2] and Mohammad Rezaur Rahman[2]

[1] *Postgraduate Programs in Disaster Management (PPDM), BRAC University, Dhaka, Bangladesh,* [2] *Institute of Water and Flood Management (IWFM), Bangladesh University of Engineering and Technology (BUET), Dhaka, Bangladesh*

ABSTRACT

Flood is an integral part of Bangladesh and so is the suffering of the people living in flood-prone areas. Unlike cyclone shelters, very few designated flood shelters exist in Bangladesh; this is because of the difference in the nature of these two phenomena. Flood shelters in three locations along three major rivers were visited, and communities were consulted to address some selected issues regarding the shelters. Three major types of flood shelters were identified for the comparison, which are community shelter, school-cum-shelter, and individual homestead. The range of issues included accessibility to shelters, land availability, protection of lives and livelihoods, basic facilities and services, safety and security of women and children, willingness and priority of potential users, and maintenance and cost-effectiveness. The primary objective of this chapter was to investigate the benefits and shortcomings of the types of flood shelters that are in use and check if any particular type of shelter could be recommended as solution. The findings show that each of the three aforementioned types of flood shelters has their pros and cons, and considering the need and aspiration of communities, one particular type of flood shelter might not be suitable across different geographical and social contexts.

9.1 INTRODUCTION

Emergency shelters or transitional shelters are a common issue in the aftermath of disasters. Worldwide, millions of houses are damaged or destroyed by disasters every year. People are forced to leave their homes and seek refuge, sometimes for short time and sometimes for a protracted period

Hazards, Risks, and Disasters in Society. http://dx.doi.org/10.1016/B978-0-12-396451-9.00009-3

based on the nature of the disaster and the extent of the damage done to the housing structures. With the varying levels of affordability of the affected population as well as the external assistance provided, the time for reconstruction of houses might take even months. Such delays sometimes lead to reluctant relocation of the affected communities. Yet, on the issue of livelihoods, some communities cannot avoid living in vulnerable locations, whereas economic constraints do not allow them to invest in stronger housing construction (IFRC, 2008).

Floods are an integral part of Bangladesh, and the suffering of the people living in low-lying areas is a recurrent phenomenon. When a flood is prolonged, households are forced to leave and search for a place to carry on with their lives and livelihoods. Thus, the issue of flood shelters comes into being.

The notion of "shelter" in Bangladesh prompts "cyclone shelter" to come to mind. At present, around 2,000 designated shelters are distributed all over the coastal districts, where people take refuge during cyclones. In addition, a few hundred more are under construction to keep up with the need of the rising population growth. A significant amount of research work on design, construction, and usefulness of cyclone shelters has been carried out, and a considerable number of initiatives have been taken for making cyclone shelters both more useful and more user-friendly. In spite of some limitations, cyclone shelters have been playing a great role in saving lives in Bangladesh.

Cyclone and flood, both affect housing structures severely, rendering them unusable without repair or reconstruction. However, in the case of floods, the homesteads remain submerged for a relatively long period of time before the repair or reconstruction work can be started.

Unlike cyclone shelters, not many designated places exist to house flood-affected people in Bangladesh. The reason, it can be argued, is the difference in nature of cyclones and floods. Cyclones require people to seek shelter for a relatively short time, whereas floods might sometimes make people homeless for weeks. Cyclone shelters are designed considering very little space for an individual (0.186 square meter/person) (converted from BUET-BIDS, 1993) and hence allow a high density of shelter seekers. But, as flood-affected people need shelters for days, sometimes for weeks, the occupant density should be conducive to their prolonged stay for which a substantial number of shelters are required. As floods are common across the country, it is indeed a challenging task to ensure designated flood shelters for all the affected people in a land-scarce and resource-strapped country such as Bangladesh.

Flood-hit people tend to remain in or near their own homesteads, though already submerged, for as long as possible (Hossain et al., 1987 cited in Paul (1997)). This is partly because of the difficulties of moving their belongings. This is further discouraged by the lack of designated places where they could seek refuge. For this, flood-affected people build "machan" or raised platform with bamboo and straw inside the house

(Paul, 1984 and Rasid, 1993 cited in Paul (1997)) at a level higher than that of the anticipated flood and move onto that "machan" with all their belongings. As the floodwater keeps on rising and crosses the anticipated level, very often flood-affected people are seen taking refuge on the roof of the house with the hope that the water level would go down soon. When even living on the roof becomes impossible, people move to nearby highlands and build temporary housing structures with the materials they have carried with them dismantling their regular houses.

Despite all the limitations and challenges and the inadequacy in number compared with the requirements, some initiatives have been taken to ease the sufferings of the flood-hit people. Some structures are built primarily to provide shelter to flood victims and some others with a different primary use that serve as shelters at the time of a flood. This chapter discusses the benefits and shortcomings of some of these primary and secondary flood shelters on the basis of some selected issues from the users' perspective. It is expected that the findings and analysis on these selected issues will help better understand the problems and challenges associated with flood shelters and will also prove useful for future flood shelter initiatives in Bangladesh.

9.2 FLOOD SHELTERS TYPOLOGY IN BANGLADESH

Purpose-built flood shelters in Bangladesh, though a comprehensive listing is not available as is the case for cyclone shelters, can be broadly categorized into two groups: community flood shelters and school-cum-shelters. Flood-proof individual homesteads also serve as flood shelters for the individual family.

The **community flood shelter** is a large earthen mound designated to function as a shelter in the first place. It can accommodate families within the range of 150–250 (found from the survey). In community shelters, people usually stay in makeshift houses. Some community flood shelters have structures erected on them where schools and other community functions are held during normal time. The other normal-time uses of community flood shelters include playground and local market place. Usually, such kind of shelters provides toilets and water supply for the shelter seekers, and they can bring along their livestock.

The **school-cum-shelter** is a building on stilts that functions as a school in normal time. During flood, the ground floor is used for sheltering cattle and people take refuge upstairs. After the 2004 flood, the government of Bangladesh, with the financial assistance from the World Bank, undertook a major rehabilitation program for secondary and higher secondary education institutions in the country. Under this initiative, 181 new three-storied school buildings-cum-flood shelters were constructed, keeping the ground floor partly open on one side and constructing the teachers' room in other side by elevating the plinth on the highest flood level. Another 179 school buildings were extended to allow flood-affected people to take shelter. The sites for construction of new schools were selected on the basis of their proneness to flood—either slow-onset or rapid-onset in nature. For extension work, schools that were badly damaged in the flood were selected (EPC, 2007).

The disruption of academic activities of schools because of turning them into shelters is perceived as a matter of concern. The reality is that during flood time, the schools are rendered unable to continue their regular activities, as the students and the teachers might be the victims of the flood. School authorities reported that in order to compensate for the flood-affected working days, the common practice is to cut off the designated vacation and arrange extra classes for students. This has been widely accepted in all the areas visited. Except for the case of extraordinary floods that stay over a month, as it was in the years 1988 and 1998, the number of days for which schools are turned into shelters is not too many to adjust with the designated vacation for the schools.

The flood-proof **individual homestead** is basically a house on a raised plinth. The plinth level is raised considering the perceived highest flood level and in a manner so that the household assets, such as cattle, can be accommodated and homestead gardening can continue. When a large parcel of land is raised to construct a number of homesteads, it is called a cluster village.

Homestead raising is an effective way of making houses flood proof; but the community is not always economically capable of raising the plinth level to a considerable height. The Chars Livelihoods Programme (CLP) financed the flood proofing of approximately 72,000 individual homesteads in the chars of the northern Jamuna (CLP Website). In 2006–2007, CARE Bangladesh, under the program titled Strengthening Household Ability to Respond to Development Opportunities (SHOUHARDO), raised some 3,671 homesteads in char areas of the Jamuna and Teesta rivers (Tod et al., 2008). The beneficiaries were the poor and the extreme poor who are highly vulnerable to flood.

9.3 APPROACH AND METHODOLOGY

This research was carried out based on field visits to three different parts of the country. The areas were selected from three major river basins—Faridpur from the Padma basin, Sirajganj from the Jamuna basin, and Lalmonirhat from the Teesta basin. The nature of floods in these basins is different. The Teesta basin experiences rapid-onset floods of 3–4 days duration, whereas the Padma and Jamuna basins experience slow-onset floods of relatively long duration in the range of 2–3 weeks. The *chars*[1] in the Teesta basin are more stable than those in the other basins, as it appeared from the discussion with the local people. Field visits were made during 7–8 July 2010 to Faridpur, on 9 July 2010 to Sirajganj, and during 22–25 July 2010 to Lalmonirhat. There was a flood at the time of visit to Lalmonirhat.

1. The *chars* are riverine islands formed as a result of river erosion and silt deposition and are surrounded by water throughout the year.

TABLE 9.1 Types and Numbers of Shelters in the Region

	Faridpur	Sirajganj	Lalmonirhat
School-cum-shelter	1	2	4
Community flood shelter	2	1	1
Cluster village	–	1	–

Table 9.1 shows the types and number of shelters visited in all the three areas excepting the individual homestead type.

Information for the research was collected basically through key informant interviews and focus group discussions (FGDs). In addition, transect walks and group consultations were also used as tools for information collection.

Nineteen people were interviewed as key informants in all the three locations. These included male and female shelter users (four), school or college teachers-cum-shelter managers (five), boatman (one), community leaders (three), members of the local government (three), member of the Upazila[2] Disaster Management Committee (one), and the local NGO staff (two). A total of four FGDs were conducted in Faridpur and Lalmonirhat (two in each location), where the flood-affected people participated. Transect walks were taken through Paikpara and Gunapara villages in Sirajganj Sadar to inquire about matters related to raised cluster and individual homesteads. In addition, group consultations were made with the local-project-implementing NGOs in all the three locations, held prior to visiting the shelters to discuss the planning and implementation issues. Furthermore, a number of short consultations were held with local people present at the time of visits at different spots, including tea stalls.

9.4 FINDINGS AND ANALYSIS ON THE SELECTED ISSUES

9.4.1 Accessibility

In terms of physical accessibility, all community shelters in *chars*, irrespective of their locations, have been found having almost the same level of difficulties. The shelters visited in Faridpur, Sirajganj, and Lalmonirhat (Figure 9.1) have been sufficiently raised, but except for the one in Lalmonirhat (Hatibandha), others did not have any designated access routes. It is well understood that since the houses in *chars* are constructed in a dispersed manner, it is not feasible to ensure access from all targeted households to the community

2. Upazila (sub-district) is the lowest level of official government administration. In Bangladesh, there are 483 Upazilas (BBS, 2009).

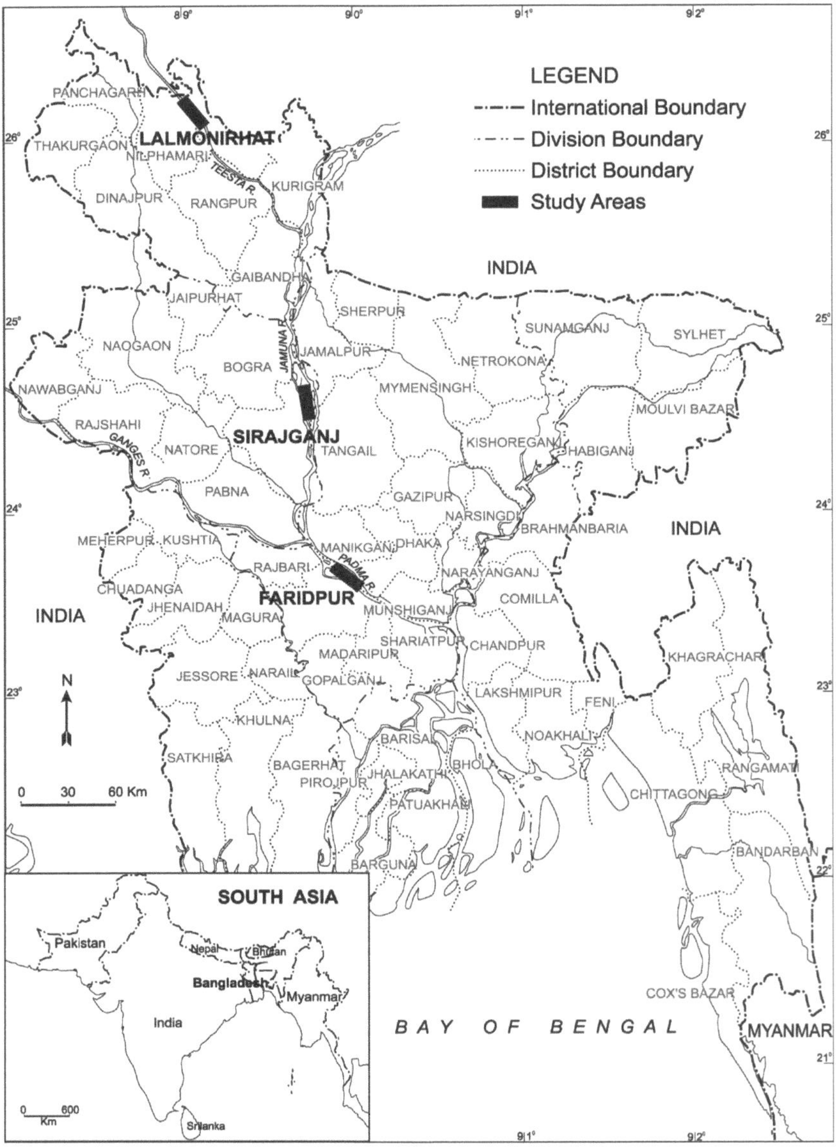

FIGURE 9.1 Map showing the study locations.

shelters. Even when the provision of a designated access from a certain point can be ensured, without regular maintenance they might be rendered unusable.

The difficulties of physical accessibility associated with community shelters in *chars* can be reduced to an acceptable level by the provision of boats as part of the shelter management activities. In Faridpur and Sirajganj, it has been observed that boats are quite available for transporting people and their belongings. But in Lalmonirhat, the situation was a bit different; very few boats were found available for transportation, and the fare seemed pretty high for poor households; this came out from the statements of some flood-affected families who failed to move their physical assets, mostly corrugated iron sheets and bamboo thatch, to a safe place because there were no boats available at the time of emergency. The unavailability of boats in the *chars* of the River Teesta can be attributed to the nature of flood (rapid-onset, short-duration flood). Unlike Faridpur and Sirajganj, where flood occurs for relatively longer period of time, boats are not a part and parcel of the daily life of the people living in *chars* of the Teesta.

From the accessibility perspective and assuming that reinforced cement concrete structures (referring to the "school-cum-shelter" type) are not feasible in *chars* from a cost–benefit perspective, the individual homestead raising can be regarded as the best available option.

Accessibility is not an issue of concern for shelters in *chars* only, rather shelters constructed on the mainland might fail to serve their purpose because of a lack of discretion regarding accessibility. The school-cum-shelter in Saadipur (Faridpur) though apparently a sound structure, did not have a well-designated access path, which needs to be constructed above the highest flood level (Figure 9.2). People from the catchment of the shelter complained of the suffering they undergo during flood in order to get to the shelter because there is no raised access path to it.

9.4.2 Land Availability

While consulting the community, several people reported that they had to shift their houses more than 10 times in their life so far. This indicates the unstable nature of *chars*, which poses a threat to individual homestead and community shelter equally. From this point, neither individual homestead raising nor

FIGURE 9.2 Access path to the community shelter-cum-school in Hatibandha. *Source: Authors.*

community shelter provision can be considered a permanent solution. But, since the whole char population cannot be shifted to shelters on mainland during a flood, either of the options, or a combination of both, has to be chosen keeping in mind the nature and type of assistance that the community would require during flood.

Finding land for individual homestead raising is not an issue, but in the case of a community shelter where a good number of people can seek shelter, this may be difficult. An individual landowner might be willing to hand over a piece of land for community welfare, but not as much land as is required. However, land for community shelters is made available through some formal arrangements among the landowner, the community, and the implementing agency.

From the land availability point of view, school-cum-shelter seems to be the most suitable option. School compounds have been reported as the first place where flood victims seek shelter. School management committees and teachers have accepted this practice well and consider it as part of their social responsibilities to provide shelter to the people. Moreover, school-cum-shelter is a good option in terms of ensuring the most effective normal-time use.

Although school-cum-shelter can be regarded as the best available option from land availability perspective, it has to be noted that there are not as many schools in *chars* as necessary to ensure refuge to all shelter seekers. Furthermore, the unstable nature of *chars* might require proper justification of investing on schools in *chars*. The average life expectancy of a char is considered to be less than 10 years, and therefore, the investment in infrastructure on chars might not be justified through normal cost–benefit analysis (CLP, 2005a).

9.4.3 Protection of Lives and Livelihoods

A sufficient number of flood shelters do not exist to accommodate the entire flood-affected population of the country. Perhaps, it is not practical or necessary as long as the concern is saving lives. During floods, the majority of the affected people find their way to nearby embankments or the national highway that is raised well above the highest flood level. The reported number of deaths that can be directly attributed to flooding is not significant compared with deaths occurring in the aftermath. Most of the deaths relating to flood actually take place after the flood, when the sources of potable water are all contaminated and when there is an acute crisis of water and food. This is necessary for discussion because just keeping provision of shelters during flood is not sufficient to save lives. Ensuring food and safe water when floodwater starts receding and the flood-affected people start leaving the makeshift shelters for their homes can contribute more to saving lives.

When structural viability is important for safety against floods, the conventional school-cum-shelter can be regarded as the best option. However, such safe structures might not be found within convenient distances for all flood victims. From this point of view, community shelters are more preferable. Another reason identified in favor of this type of shelter is that the affected people can take along their livestock and other physical assets to these places.

The main income-generating activities of char dwellers are agriculture and cattle rearing. A report of CLP shows that the average number of livestock ownership in CLP intervention areas is 2.9 per household (CLP, 2005b). For some families living on the chars visited, cattle are the main asset they have, since the land is yet not ready for agricultural pursuits and saving these cattle is as important to them as saving their own lives. Though school-cum-shelters and community shelters have enough space to accommodate cattle, sometimes households fail to get there before it is too late. When the water level has risen significantly, people might not consider it safe to transport their cattle to safe places, as there have been reported cases of livestock losses from drowning. The same report of CLP as stated earlier finds that more than 30 percent of households in chars reported loss of livestock due to flood and erosion. To avoid such loss, people rather prefer staying in their submerged house along with their livestock despite undergoing inhumane suffering.

For raised homesteads, when the plinth is extended beyond the house structure, it helps to keep the livestock safe and also allows space for homestead gardening, which meets some nutritional needs for the family during flood (Figure 9.3). Besides, an individual raised homestead extends its benefits to the neighbors whose homesteads are not sufficiently raised. The neighbors are allowed to keep their livestock and other physical assets safely at a nearby home that has a raised plinth above the flood level and can even seek shelter for the family temporarily until they find another suitable place to move to. This is possible because of their social network and the culture of fellow feelings that is commonly found among village people.

For areas like Hatibandha that are subject to rapid-onset flood, warning cannot be issued well ahead of the event, and individual homestead raising is the only preferred solution. People reported that sometimes floods occur in the

FIGURE 9.3 Raised homestead in Hatibandha with provision of space for homestead gardening and livestock. *Source: Authors.*

middle of the night when all are asleep, and they hardly get time to move to any nearby community shelters. It gets further difficult, since boats are not readily available for transportation.

However, raised homesteads have certain limitations from the viewpoint of protection of livestock. *Char* dwellers reported looting of their cattle by armed gangs even in broad daylight during floods. Usually, they come by engine boats and take away the cattle by force. Most of the time, the household members can do little to resist them, as they come in groups. Floods facilitate the quick movement of the thieves by engine boats, and no help is available from neighbors because of the inundation and threat.

9.4.4 Basic Facilities and Services

Tube wells are the most common source of potable water in rural Bangladesh, and as they go under water during floods, safe water becomes a crucial issue. The school-cum-shelter at Saadipur in Faridpur had a tube well on raised ground, but it was submerged during flooding, as it was not sufficiently raised considering the highest flood level. As a result, people had to carry drinking water from a distance. There are four toilets on each of the two floors that accommodated almost 100 people during 2007 flood, but the problem was with the flushing of the toilets, since the overhead water reservoir could not be filled up in absence of an electricity supply. This was managed by carrying water upstairs, which is a labor-intensive task. All the community shelters visited in the three sites had in-situ provisions of potable water and toilets. The only exception was the one in Paikpara, Sirajganj, which was newly constructed and the provisions are yet to be installed.

Toilets and tube wells in shelters ensure the basic sanitation and drinking water services, but a chance of spreading communicable diseases still exists because of the high density of shelter seekers. However, health services in those places could be provided more easily compared with the raised individual household type.

Most of the raised individual homesteads visited had both tube wells and toilets instilled on the raised plinth. But people living on raised homesteads expressed their concern that they are more likely to be left unattended by the relief delivery team. From this point of view, school-cum-shelter and community flood shelter are more convenient options. In a community shelter, a good number of people can be provided with relief assistance over a short period of time, as is the case with medical services.

9.4.5 Safety and Security of Women and Children

People on raised homesteads are left marooned and are less likely to receive social support at the time of emergency. While interviewing a female-headed household in Sirajganj, the household head expressed her interest to move to a community shelter, though her homestead was sufficiently raised. The reasons she underscored included safety of her children. Understandably, during flood, the risk of death of children from drowning becomes high. Although the adults of a household are more likely to remain busy securing the safety of their belongings and ensuring food for the family, it becomes difficult on their part to look after the children around the clock. But at the community shelter or school-cum-shelter, it is not difficult to find someone to look after the children. In community shelters, children from all the households that took shelter in it might engage themselves in playing games, provided that such provision exists. Thus, the safety and security of children can be better ensured in a community shelter or a school-cum-shelter than in individually raised homesteads.

The safety of women and adolescent girls in terms of harassment/assault is also considered as a matter of concern. Literature extensively suggests that in communal shelters, they are rendered vulnerable to abuse by men. Interestingly, at the time of consultation, they did not seem to be much concerned about it, although they regarded raised homesteads as a preferred option.

9.4.6 Willingness and Priority of Potential Users

After consultation with the flood-affected people in Faridpur, Sirajganj, and Lalmonirhat, it is evident that they prefer raised homestead to the other two options. This is because raised homesteads save both their lives and their belongings. Sometimes it is difficult to shift all their belongings, and they might not be allowed to move to a communal shelter with all their belongings; this was the case in Hatibandha. In a discussion with a Union Council member, who is also a member of the Union Disaster Management Committee, it was revealed that the shelter management committee did not encourage people to come to the *killa* (raised earthen mound) with their dismantled house, fearing that they might not want to leave when flood is over.

For floods of unprecedented severity and prolonged periods, the limitations with raised homesteads become evident. These include the chances of running out of household savings, the fear of failing to avail relief assistance, and living without communal support. Hence, communal shelters cannot be ruled out from the preferences of the community.

Flooding in *char* areas is usually accompanied by soil erosion. The strong wave action during floods aggravates erosion, and the affected households sometimes have to think beyond a mere temporary refuge. In Hatibandha of Lalmonirhat district, this was more evident. Some people were seen carrying the dismantled parts of their house and searching for a place to settle for the coming days, though a raised earthen plinth (*killa*) meant for community shelter stands unused nearby (Figure 9.4). This is also attributable to the fact that the region is prone to rapid-onset type of flood, which usually does not prolong over many days, as is the case in Faridpur and Sirajganj.

FIGURE 9.4 Flood victims shifting their house in Hatibandha. *Source: Authors.*

9.4.7 Maintenance

When floodwater starts receding, any obstruction to receding water delays the drainage process and exacerbates the flood situation. This is particularly true in situations where drainage channels are obstructed. In chars, usually designated drainage channels do not exist, as not many raised roads exist. Furthermore, homesteads are sparsely distributed. Therefore, the chance of drainage congestion due to obstruction of receding water is limited. No cases of significant drainage congestion in or around a shelter were reported in any of the field visit sites.

Rainfall runoff is responsible for damage to the earthen plinth of flood shelters. If rainwater stands on a flood shelter, the earthen plinth becomes slushy and eventually gets washed away. It was observed during the field visits that appropriate slopes (top and side) were not maintained during the construction except for the raised community shelter in Sirajganj. Gullies were prevalent in those shelters. Except for the school-cum-community shelter in Sirajganj, none of the community shelters had longitudinal, lateral, or side gutters for drainage of rainwater. Such slopes and gutters are necessary to let the water run down sideways during rains.

The school in Saadipur, Faridpur, was structurally sound when the construction was completed, but the damage that it incurred during the first flood it encountered has not been overcome yet. As the earthen plinth of the structure was not properly stabilized, floodwater washed it away, rendering the building "unsafe" in the perception of the people. The toilets of the building remain unused, as the sewerage lines got exposed and subsequently damaged. The building keeps the provision of an overhead water tank, but no means to pump water up there.

Raised individual homesteads, as observed in the districts of Faridpur, Sirajganj, and Lalmonirhat, act as individual family flood shelters. The common practice seen among households to increase soil stability and protect slopes from erosion is planting grasses and trees on the slopes. Banana trees are most commonly planted by residents, though these are not considered as an ideal solution, as they leave large holes in the soil when they die (Parkinson, 2006). Still, since the residents do the repair and maintenance by themselves, this is not to be considered much of a problem.

As stated earlier, from the multipurpose use perspective, school-cum-shelter is a preferred option. As schools have year-round activities, the maintenance can be intrinsically done. But one problem as identified during the field visits is the unavailability or insufficiency of funds for maintenance operations; from this perspective, raised individual homesteads offer a better choice.

9.4.8 Cost-effectiveness

It was found out during the survey that depending on the area of the homestead, the cost of raising the plinth level ranges between Taka[3] 10,000 to Taka 17,000, as observed in the three aforementioned districts. For community shelters, the cost varies between Taka 250,000 (in Hatibandha, Lalmonirhat) to Taka 2,100,000 (in Paikpara, Sirajganj), whereas a standard school-cum-shelter requires approximately Taka 3,000,000 (Singimari, Hatibandha). Taking the lowest figure, the cost of sheltering a household comes to around Taka 10,000 in case of raised homestead, Taka 2,500 per household in case of community shelter, and Taka 30,000 per household in case of school-cum-shelter (keeping apart the regular schooling function). However, from the cost-effectiveness point of view, it is very difficult to name one single type of shelter as most appropriate. Each of these types of shelter has some distinct advantages that cannot be replaced by others. Moreover, the perception of the notion "effectiveness" might vary from community to community when there are "direct"/"indirect," "tangible"/"intangible" issues. Still, community flood shelters can be considered as more cost-effective than the other two options, especially considering their potential for normal multipurpose use.

9.5 CONCLUSIONS

As Bangladesh is a floodplain country, it is very difficult to avoid impacts of floods only through flood control interventions. Moreover, the negative environmental consequences of flood control interventions are now becoming more and more evident. Thus the idea of "flood management" comes in the place of "flood control."

The provision of flood shelter can be considered as part of the flood management approach. Yet, it is to be noted that the needs and aspirations of people who require such shelters might vary across the floodplains. Chowdhury et al. (1997) underscored the need for giving due consideration to the socioeconomic condition and cultural heritage of the floodplain inhabitants

3. 1USD = Approximately 78 Taka (as of April 2014).

for the attainment of the objectives of an intervention. The nature and extent of the problems associated with floods sometimes vary across different river basins and across communities. It is evident from the discussion above that one particular type of flood shelter might not be suitable for all people for whom flood is a concern. Careful investigation into the nature of problems people face during floods can help better understand what type of shelters should be recommended in a particular area.

ACKNOWLEDGMENT

This chapter is an outcome of the Asia Regional Writeshop (workshop on academic writing) organized by the Stockholm Environment Institute (SEI) and the UN-ISDR, held in Bangkok from 20 to 24 September 2010. The first author expresses his sincere gratitude to Professor Mohammad Rezaur Rahman for nominating him to the writeshop and the organizers for the opportunity to participate. Special gratitude is due to Mr Terry Cannon, who has been his mentor for the chapter writing and extended his support painstakingly to improve the quality of this write-up over the months after the writeshop.

The field survey for this chapter was carried out as part of a research project commissioned by the DIPECHO Partners in Bangladesh (DPB) to the Institute of Water and Flood Management (IWFM) of Bangladesh University of Engineering and Technology (BUET) and the Postgraduate Programs in Disaster Management (PPDM) of BRAC University. The authors express their gratitude to Dr Humayun Kabir of the University of Dhaka for his kind help with the preparation of the map.

REFERENCES

BBS, 2009. Statistical Pocket Book of Bangladesh. Bureau of Statistics, Bangladesh.

BUET-BIDS, 1993. Multipurpose Cyclone Shelter Programme. Government of Bangladesh and UNDP/World Bank.

Chars Livelihoods Programme, 2005a. Chars Livelihoods Programme Inception Report. CLP.

Chars Livelihoods Programme, 2005b. A Profile of Households Living on Mainland and Island chars. CLP.

Chowdhury, J.U., Rahman, R., Salehin, M., 1997. Flood Control in a Floodplain Country: Experiences of Bangladesh. Islamic Educational Scientific and Cultural Organization (ISESCO), Morocco.

Chars Livelihoods Programme, (website): http://www.clp-bangladesh.org/index.php?option=com_content&task=view&id=131&Itemid=153 (accessed 28.06.11.), http://www.clp-bangladesh.org/index.php?option=com_content&task=view&id=82&Itemid=79 (accessed 11.10.11.).

EPC, 2007. Flood Rehabilitation Programme in Secondary and Higher Secondary Education Institutions Project, Project Completion Report. Education Engineering Department, Ministry of Education, Government of the People's Republic of Bangladesh (unpublished report).

Hossain, M., Islam, A.T.M., Saha, S.K., 1987. Floods in Bangladesh: Recurrent Disaster and People's Survival. University Research Centre, Dhaka.

IFRC, 2008. Global Shelter Programme 2009/2010. International Federation of Red Cross and Red Crescent Societies, Geneva.

Parkinson, J., 2006. Assessment of CLP Progress and Recommendations for Future Project Activities (Output 1). Report prepared for Maxwell Stamp Plc -Management Agency of the Chars Livelihoods Programme (CLP).

Paul, B.K., 1984. Perception of and agricultural adjustments to floods in Jamuna floodplain, Bangladesh. Human Ecol. 12, 3–19.

Paul, B.K., 1997. Flood research in Bangladesh in retrospect and Prospect: a review. Geoforum 2, 121–131.

Rasid, H., 1993. Preventing flooding or regulating flood levels?: case studies on perception of flood alleviation in Bangladesh. Nat. Hazards 8, 39–57.

Tod, I., Afroz, D., Ali, M.S., 2008. Assessment of the Effectiveness of Homestead Raising and Mound Protection Works Implemented by the SHOUHARDO Program, USAID, GoB and CARE Bangladesh, Dhaka (unpublished report).

Chapter 10

Cyber-Security Hazards in Society

Maitland Hyslop
Disaster and Development Network (DDN), Northumbria University, Newcastle, UK

ABSTRACT
Cyber-security hazards in society are viewed in the context of tensions, disasters, and development opportunities. The tension is essentially that between the physical and spiritual world as we know it and the challenge given by virtual cyberspace to this traditional understanding. The hazards can clearly cause disasters but can also create development opportunities. Beginning with a brief historical review of what is known, what is not known, and definitions of terms, the chapter moves on to look at the very different environments and difficulties created within cyberspace in the political, economic, social, technological, environmental, and legal arenas. Additionally, life, military, organizational, critical infrastructure, criminal, and moral hazards are explored. The analysis reveals some familiar, some surprising, and some unknown features of these hazards. The reluctant conclusion is that some form of regulation is required to ensure that society as a whole is protected from cyber-security hazards.

10.1 INTRODUCTION

The simplicity of the title of this chapter belies a variety of complex problems. Humans, generally, are struggling to come to terms with these, where they are aware of them at all. Cyber-security and hazards can, for the purposes of this chapter, be described in terms of life, politics, military, organizational, critical infrastructure, economic, social, technical including virtual/Internet, environmental, legal, and criminal. Society can be described in part in sociological or political terms. These descriptions, along with the disaster and development problems associated with each, are briefly examined in this short chapter. The lasting and lengthy nature of the challenge should not be underestimated. As Alexander (2010) tended to say, "We live in the new Baroque Age; an age of the tension of opposites." Nowhere is this statement more appropriate than in relation to cyber-security hazards in society. Human, societal, and organizational resilience may well have been reduced by cyberspace, not increased. Furthermore, nowhere is this statement more appropriate than in relation to disaster and development studies.

Hazards, Risks, and Disasters in Society. http://dx.doi.org/10.1016/B978-0-12-396451-9.00010-X

10.2 THE LESSONS OF HISTORY: PEELERS AND ARMIES, ENIGMA

Much of what is discussed in this chapter is not absolutely new. The reaction to crime in the UK in the nineteenth century led to the establishment of the Peelers, who were to eventually become the UK police forces we know today (Gaunt, 2010). Armies evolve, some more competently than others and usually behind the pace (Dixon and Dixon, 2011), and the rules of war change (Hyslop, 2013). The Enigma code machine used by the Germans in World War II was, to all intents and purposes, a cyber-tool. It was the subject of its own battle in World War II to break the codes that enabled the Germans to pass orders through a prototype cyberspace to the disadvantage of the Allied forces (Kruh and Deavours, 2002).

Thus, much of the civil reaction to crime, the slowness of the military to adapt to new threats, and the use of new technology to try and defeat enemies is nothing new. What is new is the pervasiveness of the new technology; the lack of traditional political control; the enablement of new types of social networks; and the fact that very, very few people (relatively) actually understand the ramifications of what happens in cyberspace, and even if they do, including, of course, the author, they have only a limited purview.

The tension between the old and the new is clear.

10.3 WHAT WE KNOW, WHAT WE KNOW WE DO NOT KNOW, WHAT WE DO NOT KNOW

In Rumsfeld's (2002) famous phrase, though familiar to most risk and hazard managers, there are "known knowns, known unknowns, and unknown unknowns." To repeat a phrase, nowhere is this more appropriate than in cyberspace.

It is known roughly what the World Wide Web looks like, and who is on and where on the Internet. But in truth, remarkably little is known about cyberspace. No one can be clear as to where it begins and where it ends. Large parts of cyberspace remain unmapped; even if it was mapped, it is so dynamic that the map would be immediately out of date. Cyber activities in Russia, Eastern Europe, the Far East, South America, and the Middle East are such that very few, if any, know about them and are essentially blank spaces on the map (The Economist, 2010b; Dodge, 2008).

British Telecom (BT) has admitted variously that it did not know the limits of its own network, so intertwined with other networks was their technology platform (Roberts, 2005; Ghanea-Hercock, 2011, 2012). British Petroleum (2007) admitted some years ago that far from 5 percent of its business being linked to the Internet (as it had thought), it was closer to 85 percent.

The extent of the major networks is known and the main data highways are known, but the absolute details of all networks and all data highways are the known unknowns. The Dark Net is still not fully mapped, nor are some of the dark highways; these are also part of the known unknowns. What is not known is the question: What else? It is likely that there is something else (Sherman, 2002; Chen, 2013).

The tension between the known and the unknown is part of cyberspace.

10.4 DEFINITION OF TERMS—CYBER-SECURITY, HAZARDS, AND SOCIETY

There is a question of definition: What is cyber-security? The US Department of Homeland Security (2013) states, "Our daily life, economic vitality, and national security depend on a stable, safe, and resilient cyberspace." Cyberspace is defined, according to Oxford English Dictionary (2013), as the notional environment in which communication over computer networks occurs. As Goodger (2010) and Warwick (2013) have commented, cyberspace can now include individual humans in a real "Matrix"-style cyber-funk environment (The Matrix, 1999), and both they and Ghanea-Hercock (2011) comment on cyberspace as a complex adaptive system. Therefore, cyberspace becomes the notional environment in which communication over, and human presence, within computer networks occurs. Thus, cyber-security becomes "the protection of human existence through a stable, safe, and resilient cyberspace."

A hazard in this case is a situation that poses a level of threat to cyber-security.

Society in this environment is simply either a political or sociological grouping.

The subject of this chapter is therefore the protection of individual, sociological, and political human existence from threats to a stable, safe, and resilient cyberspace in life, political, military, critical infrastructure, organizational, economic, social, technical, environmental, legal, and criminal terms in what is a new Baroque age in an essentially unknown environment accessible by all and from a disaster and development perspective. It is more of a challenge because of a lack of widely understood symbology, another of Alexander's (2010) concerns, to describe and understand it all.

A tension exists between the symbology normally used to describe life and artifacts and the lack of a new symbology to help describe the new cyber world. Definition is a problem in such an environment.

10.5 POLITICAL THOUGHT SINCE THE GREEKS

Political thought is concerned with the state, its structure, its nature, and its purpose. It seeks to determine the morality of the state to justify its continuance in terms of the individual. Traditionally, three approaches to political thought exist: the rational-natural tradition, the will and artifice tradition, and the tradition of historical coherence. A fourth approach also now exists, the state as permanent revolution: China (Wayper, 1973).

Six things have conspired to radically change this traditional approach: globalization of economies, multinational companies, organized crime, the financial crisis of 2007 onward, "the war on terror" with the associated rise of stateless groupings, and the development of cyberspace. Politicians across the globe have not fully come to terms with the alternative scenarios that now undermine the very concept of the state as developed since the Greeks.

This is not to say that there is no attempt to bring cyberspace under the control of states. In one sense, because it controls much of the infrastructure, many of the operating companies that supply hardware and software to it, and many of the marketplaces within it, the USA has already established "de facto" rights over most of it. Of course, the European Union does not agree with this—and neither do Russia, Iran, and countless other states who rail against what they see as USA domination of cyberspace (Hyslop, 2010). States are increasingly looking to control cyberspace (The Economist, 2010b). So far they have not quite worked out how to do it. In one sense, this is a threat to cyberspace—in the sense that states will carve up cyberspace in a way that has not happened to Antarctica or space just yet. In another, it may help to bring order to an anarchic medium.

A tension exists between the traditional understanding of political thought and the cyber-challenge to traditional political thought. It is therefore important both from philosophical and practical point of view not to allow cyberspace to develop into a dystopia, rather than the utopia originally envisaged by its creators.

10.6 PHYSICAL VERSUS VIRTUAL SOCIETY (VERSUS SPIRITUAL)

Allied to this undermining of the traditional state is the development of two completely separate social, and increasingly political, societies. This is the physical society well known to "baby boomers" and their ancestors, and the physical and virtual societies, which are much more well-known to their offspring. One had developed mature structures and good infrastructure; the other has uncontrolled structures and uncontrolled infrastructure. The structure paradigm of the physical society is represented by the rules and regulations that define the way in which petroleum resources and their associated hazardous environments are governed against the uncontrolled nature and the lack of any real governance of cyberspace (Hyslop, 2010). The unstructured nature of cyberspace and the arguments between open and closed approaches, control and freedom, can be seen in the articles, blogs, and comments of I3P (2013), Anderson (2013), Anderson and Moore (2006), Schneier (2013), Gibson (2013), and Gibson and Anderson (2013) to name but six examples of interest.

This gives a number of professional psychological problems too. The best example is the traditional policeman versus the "computer technologist." The former is used to solving physical crime in the physical world. The policeman does not understand electronic crime and does not have the background, education, training, or attitude to deal with it. He or she is also at the other end of personality spectrum to the "computer technologist." Yet, the policeman needs the "technologist" to solve electronic crime. Getting the two to work together is extremely difficult. This is then compounded by police forces using different recording procedures for cybercrime (Rarden, 2014). Is it any wonder that physical crime has reduced and electronic crime increased from a zero base (Bank of Canada, 2000)?

Another is the Special Forces trooper, who fights hand to hand in appalling conditions against the Taliban versus the drone operator sitting in his office in California. Another is the disconnect between society and fighting troops. For a revealing insight into this problem from some interesting angles, see Michael Peck's Forbes magazine article on recruiting drone pilots (Peck, 2013).

Another is the disconnect between cybercrime and physical world impact. This is often seen as victimless. The film industry's campaign against cyberpiracy seen in every cinema "You Wouldn't Steal a Car Would You" is a case in point (MPAA, 2007). Another is between the members of society who are information technology literate and those who are not. The latter will be returned to later. The list goes on and on; and much of it is related to the differences between physical and virtual, or cyber, society.

Over millennia, there has been a transition of human thought from the esoteric to the scientific (Black, 2008). Now there is a potential transition from the scientific to the cyberkinetic and the cyborg; this not necessarily helpful to the development of individual imagination and may be the next retrograde step in the loss of human spirituality—a further tension.

The tensions are clear and opposite again.

10.7 BACKDROP

This section is the backdrop to a look at the cyber-security hazards in society. It is not a simple problem. It is an area dominated by Alexander's Baroque analogy and also by our joint view that a new symbology is required, perhaps making Jung an important contributor in this area.

It is an area that challenges every single part of society. It is not generally well understood. To put this in perspective, and this is the author's qualitative personal opinion from 25 years in the information technology arena, 60 million people exist in the UK, and less than 1,000 of them understand this subject really well. The author would not necessarily count himself as one. Even if this is wrong by 100-fold or even 1,000-fold, it demonstrates the gap between those who know and those who know what they do not know and those who do not even know there is anything to know (the majority of mankind). To return to the Greeks, this raises the possibility of a cyber "democracy" and "citizenry" and a non-cybercitizen and non-cyberslave population at the bidding of the cybercitizens. This is perhaps the biggest hazard to society: the development of a two-tier alternative global society. To a large extent, it already exists. This is one of the greatest tensions of all time. An argument can be made that it was ever thus, e.g., Newton and the rest of society at the time. A change of scale and quantity of information exists, however. As Stalin said, "Quantity has a quality all of its own."

The practical hazards are equally challenging.

10.8 THE LIFE HAZARD

The suicides and alleged suicides of individuals linked to cyber-grooming or cyber social networks are well documented (BBC, 2013a).

The father of a 14-year-old-girl, who was found dead in her bedroom on Friday, says she was driven to suicide by cyber bullying. Hannah Smith, from Lutterworth in Leicestershire, UK, was apparently taunted about her weight and appearance by anonymous users on the social networking site ask.fm. Her father, Dave Smith, has called for stricter controls on all internet sites used by children.

BBC (2013a)

The BBC also reported on the cyber vigilante group 'Letzgo Hunting'. (BBC, 2013b). TruTV reported on a serial killer's use of the Internet. (TruTV, 2013). All of these reports demonstrate cyberspace as a life hazard.

Over and above these specific examples, the use of the cyberspace to radicalize terrorists to kill and maim and the use of the Internet as a command and control mechanism for terrorist cells are well documented (UNODC, 2012). Stuxnet and similar viruses are deployed and triggered through cyberspace. Although not directly killers, some anecdotal comment exists that the destruction of critical infrastructures by viruses has had a collateral effect on life (Computer World, 2013).

These are the disasters. The development opportunity is both social and political. Countermeasures to these acts need to be implemented. In the UK, organizations such as the National Crime Agency can have an impact, but more needs to be done to educate and enhance individual and organizational understanding of events such as these. It is a characteristic of cyberspace that it is individual, personal, private, anarchic, group, gregarious, open, political, and social, all at the same time. It takes time for "traditional" society to catch up.

10.9 THE POLITICAL HAZARD

The more philosophical aspects of political thought have been commented upon. On a practical level, there is much of concern.

The activities of Julian Assange (Wikileaks, 2013) and Edward Snowden (exposer of the USA–British surveillance capability) (The Guardian, 2013) have threatened political stability. Andrew Parker, head of the UK's internal security service (BBC, 2013c), says that the Snowden leaks are the biggest single catastrophe to British security since Burgess and Maclean (The Sunday Times, 2013a). David Davis, a former UK home secretary (The Sunday Times, 2013b), fell into the trap that bedevils many looking at this problem: that cyber-security and the virtual world play to the same political rules as the rest of society. They do not. This misunderstanding is at the heart of many cyber-challenges to traditional political thought about the state.

Political thought is about the state. Cyberspace is not about the state. States "play" in cyberspace in much the same way as they do in space. Just as states have their own peaceful and surveillance satellites, weapons, and personnel in space, so they have their own peaceful presence, surveillance teams, weapons, and personnel in cyberspace. The key difference is that unlike space, in general at moment, cyberspace is populated by millions of individuals grouped into organizational and social networks that are themselves a potential threat to the concept of the state, as noted, in way not seen since before the Greeks.

The first disaster, and the first tension, is the inability of the current political framework to be able to accommodate the potential political power of cyberspace. It is a disaster because as the tension increases, the political fallout from Assange and Snowden types of events will destabilize the traditional political world before a replacement model is fully in place. Those in power in the traditional political infrastructure will be continually challenged by the new cyber-based political infrastructure with unknown and unintended consequences. The Assanges and Snowdens will see this as positive, and the states will see it as negative and try and stop it. Both are likely to be wrong. Something unknown is likely to emerge.

The development opportunity is huge and equally unknown. The networking opportunities will give the chance for political ideology to be determined at a global level for the first time. If you are a social democrat, it may be possible for the Swedish model to take over the world via social networking sites. Equally, Maoist China may be able to persuade the world of its own merits. More likely, a completely different type of political model is likely to emerge, and the skirmishes with Assange and Snowden are the first signs of this coming political upheaval.

The politicians over 50 years of age who dominate traditional political structures have no idea how the young can organize themselves effectively to counter them. The young themselves, despite organizing riots in various locales, have not yet bent the cyber world to a new revolutionary political ideology. No reason exists to suppose that the Linux revolution (the networked cyber-challenge to Microsoft's, and others, programming dominance) cannot have a political counterpart. The people who have organized themselves effectively in cyberspace are terrorists, criminals, and the traders of the Dark Net. These represent a political, social, and military threat. These are now organized into cyber armies, and traditional structures are creating defense forces to counter them. The Dark Net organizations may even be the new political force.

10.10 THE MILITARY HAZARD

Cyberspace is used in many ways by the military arms of many states and many non-state organizations (non-state organizations with military or pseudomilitary arms can be Chinese/USA/Russian/UK development companies and resource companies, multinational companies, Al Qaeda, Al Shabaab, Boko Haram, and Somali Pirates, to name but a few).

The Economist's (2010a) lead article noted that cyberspace had become the fifth domain of warfare. It listed a variety of cyber-attacks on Estonia (2007), Georgia (2008), and elsewhere. The Stuxnet and similar virus attacks on Iran and other countries followed. It suggested that a cyber-arms treaty would be to the advantage of the USA and suggested that the USA was wise to welcome Russia's calls for a treaty. That said, a nuclear arms type of treaty may be difficult to negotiate. Nuclear arms can be counted; cyber weapons cannot be, but can be made almost anywhere. More recently, the USA designated six cyber-tools as weapons (Shalal-Esa, 2013). There has been the impact of cyber warfare on the Syrian campaign:

The Syrian Electronic Army—supporters of President Bashar al-Assad—have warned that further attacks are on their way and media companies should 'expect us.' The SEA's spokesman stated last week, after the UK Parliament's vote against military action, that is was 'the right thing.' In addition, he stated that military intervention in Syria has 'many consequences and will affect the whole world.' In conclusion the SEA stated that their main mission was 'to spread truth about Syria and what is really happening.' In recent time the SEA has targeted some large media companies, including the BBC, CNN, New York Times, Twitter and the Guardian.

Lee (2013)

Independent cyber vigilantes such as "The Jester" react to these events in ways that rival some states' ability to respond and with no apparent legal redress (The Jester, 2013).

PLA Unit 61398 is the Military Unit Cover Designator of a Chinese People's Liberation Army advanced-persistent-threat unit that has been alleged to be the source of Chinese computer-hacking attacks. The group is also known by various other names, including "Advanced Persistent Threat 1" ("APT1") and "Byzantine Candor," a code name given by US intelligence agencies since 2002 (The Economist, 2013).

A report by the computer security firm Mandiant (2013) states that PLA Unit 61398 is believed to operate under the Second Bureau of the People's Liberation Army General Staff Department's Third Department Second Board and that there is evidence that it contains, or is itself, an entity Mandiant calls APT1, part of the advanced persistent threat that has attacked a broad range of corporations and government entities around the world since at least 2006. However, attribution is difficult, and as a consequence, physical world repercussions occur rarely for acts in cyberspace.

The military hazard has a further dimension that is a vignette of the tension that arises around this subject. A US and Allied force 2-star general, who has commanded special forces and troops in both Afghanistan and Iraq, has his picture on the relevant unit's Web site. Two clicks from his picture and his accompanying biography is his child playing tennis on a school tennis court. He does not get the hazard. Not only he does not get the hazard, the unit and the hierarchy do not get the hazard as well. The Government Communications Headquarters has to tell them after the author noticed it. Such is the tension between the old military thinking and the new, and, in passing a comment on the disengagement of both the military and society from the asymmetric war being fought at the start of, potentially, one of the most interesting realignments of global power in nearly half a millennium.

The disaster is that just as the Poles went to the War with horses in 1939, so the UK 70 years later is ordering aircraft carriers to fight a global war, which is now essentially defined in cyberspace and asymmetric in nature, but not the navy, as it needs to defend sea lanes and deter drug smugglers. The tension is between fighting the wars of yesterday and the need to anticipate the wars of tomorrow.

The disaster is that the generals running the armies do not understand the impact that cyberspace can have on their personal as well as professional lives. The disaster is that they fail in their key role to protect the citizens of the current political states in this "new Baroque age" that they do not understand. The tension is between the generals of the Cold War and the generals of cyberspace—they are not the same type of individual.

The development opportunity is for state militaries to realize it is their job to protect citizens from the hazards and threats presented by such as stateless organizations (including commercial organizations) and fanatical jihadists. They must work hard to protect essential freedoms and co-operate to install the necessary protective measures. Unlikely, but well worth pursuing.

10.11 THE ORGANIZATIONAL HAZARD

The author (Hyslop, 2013) has characterized the threats to organizations in terms of the six C's: capital, casual, competitive, cultural, criminal, and critical infrastructure.

The disasters waiting to happen are concerned with the difficulty of accessing or alternatively mislaying capital, the disgruntled employee who tries to destabilize the organization in some way, the competitor challenging the organization fairly or otherwise, the difficulty of entering or employing different cultures, the criminal threat and the dependency on critical infrastructure that includes cyberspace as a vector of both threats and engagement with the outside world. Cyberspace is fraught with dangers for the modern organization. Weinberger comments as follows (Levin et al., 2000):

The idea that we manage our world is uniquely 20th Century and chiefly American. There are tremendous advantages to believing one lives in a managed world: risk avoidance, smoothness, fairness, discretionary attention... it wasn't always thus. For millenia, we assumed that being in control was the exception and living in a wildly risk-filled world was the norm... the truth is businesses cannot be managed. They can be run but they exist in a world that is so far beyond the control of the executives and shareholders. (Cyberspace) not only gives individuals a voice it gives organisations a voice too. We are so desperate to have our voices back that we are willing to leap into the void. We embrace (cyberspace) not knowing what it is. And when the thrill of hearing ourselves speak again wears off, we will begin to build a new world. That is what cyberspace is for.

The disasters are therefore the general threats to organizations over cyberspace, the use of cyberspace, and the "gurus" who advise on its use (interesting though they are) and encourage potentially ill-advised usage. What is true, particularly in the "Twitter" age, is that organizational reputations can be damaged by anyone's voice very quickly. The tensions are again between the traditional organizations and those that can embrace the opportunities, and protect themselves from the hazards, of cyberspace. The tensions are also between the traditional marketplaces and the electronic marketplaces.

The development opportunity is huge. Organizations can appear instantly bigger than they are, overheads can be reduced, tax bills reduced, transnational markets are available to all—not just the major multinational companies—and employment opportunities now span the globe. It is of course true that many in the poverty-stricken areas of the world work for minimal wages in poor conditions and that child employment is rife. However, it should be remembered that in the context of their societies, this has often been the case, and transnational trade driven by the opportunities presented by cyberspace has helped millions get out of poverty over the last quarter century. A completely new market for hardware and software, and a new market place for all goods have been created by the development of cyberspace, e.g., Amazon.com, eBay.com, "The Silk Road" (and equivalents).

10.12 THE HAZARD TO CRITICAL INFRASTRUCTURE

Every critical infrastructure depends on information infrastructure, and thereby cyberspace, in one way, shape, or form. Critical infrastructures are defined as energy, finance, food supply, health, information, infrastructure, government services, law and order, manufacturing, national icons, transport, water, and wastewater (Dunn and Wigert, 2004; Hyslop, 2010).

The cyber-security issues surrounding these have been well documented by I3P (2013) of which the author was a charter member. Cyber-security can be broken down into the key areas of connectivity, hosting, hardware, and software. The G20 countries also have official bodies looking at the performance of different related industries (Council on Foreign Relations, 2013). In addition, a number of national and international mechanisms for developing public—private partnerships and the sharing of information have been established. No international body specifically responsible for cyber-security exists, although the UN, OECD, and NATO have made attempts (NATO, 2013).

The Stuxnet attack on Iran's nuclear capability has been well documented by the author (Hyslop, 2010) and others. This virus is remarkable because it represents an epiphany moment in warfare like the invention of the gun, machine gun, tank, airplane, and nuclear bomb. It changes the rules of the game. It has certainly changed the rules of the game in relation to the cyber-security issues surrounding the protection of energy sites and associated Supervisory Control and Data Acquisition systems (SCADA).

Die Hard 4 (2007) is a film about an attack on critical electricity infrastructure. Such attacks are not particularly well documented, but these events have happened in the UK, the US, and parts of Europe. The Maroochy Shire incident of 2000 in Australia is a related example (Abrams and Weiss, 2008). Here, a disgruntled employee used a laptop to dump wastewater/sewage into drinking water.

The disasters are the attacks on infrastructure by terrorists, disgruntled employees, and foreign governments intent on destabilizing others that provoke disruption and worse. The tensions are between the economies that can be created by the adoption of cyberspace-related techniques to manage infrastructures and the new types of hazard that come with that opportunity.

The development opportunity is therefore often misunderstood. It is linking together a number of different infrastructures together into a more resilient whole. This enables, for example, electricity networks to operate more easily across country boundaries, and as a consequence, load balancing in these networks becomes easier when supply is tight.

10.13 THE ECONOMY HAZARD

A difference exists between the cyber hazard and the cyber-security-related hazard to the economy. The former is about the threat that Amazon.com and other online retailers pose to the traditional physical shopping centers in many countries and the associated drop in tax take. The latter is about infrastructure, the difficulties banks have in managing large amounts of cash and businesses, the ease with which online banking has enabled black economy gains to be passed through the banking system, and the systemic use of online frauds to damage individuals and businesses within an economy. This is creating huge tension.

All economic players can reduce costs by using cyberspace. This is clear from, for example, the encouragement airlines give to all passengers to check in online. That said, the banks are now making life difficult for anyone in the economy to use cyberspace effectively. The reason for this is that they have now been compromised so many times that they no longer trust cyberspace to deliver the true economic gains promised. So businesses now have to go in person to banks to draw money and have to write letters to change bank details (HSBC, 2013).

The disaster is that the full economic gains for society from cyberspace will not be realized. This is because the banks in particular have not taken a proper and systemic approach to security. An example would be the Barnaby Jack-exposed vulnerabilities of ATM machines (The Register, 2010). As a consequence, their systems have been regularly compromised. This means that everyone else within their system cannot reap the rewards of an integrated electronic approach to the economy.

In the nontraditional-banking economy, the disaster is that many alternative forms of banking have grown within cyberspace that are not formally linked to the formal system, for example, the Bitcoin threat to traditional banking (swiftly shut down), Silk Road-type sites for underground activities and money laundering, money laundering via gaming sites, credit unions, crowd funding, new online currencies have all appeared—but equally, they are not always what they appear to be. The tensions between the formal and informal economies have been exacerbated by cyberspace, not alleviated (Channel 4, 2013).

The development opportunity was and remains huge. Cyberspace should give access to all markets to all—this should, theoretically, be a catalyst for the development of many impoverished areas of the world. That it is not has much to do with the hegemony of the United States, and to a lesser extent Europe, in regard to control of the cyberspace economy and the developing walls between nations that are appearing in cyberspace (The Economist, 2010a).

10.14 THE SOCIAL GROUP HAZARD

The social hazards of cyberspace have in many ways already been alluded to. A series of issues exist that have to do with social group hazard: sexting, grooming, chat rooms and impostors, burglaries and impromptu parties as a result of network-notified absences from home, pedophile rings, vigilante groups, terrorist cells, diaspora groups, stateless organizations, etc. Interestingly, really no great new political party has come into existence as a consequence of cyberspace. A major potential hazard that has been given nowhere near enough attention by the media is the potential link between the unemployment of young persons in Europe and the potential to link to action through cyberspace.

The disasters are there for all to see in the destroyed lives, property, and sometimes organizations—in any number of references to the dangers posed by social networking organizations. The author is nearly 60 years old, so he asked for younger persons' perspective on the issues of this chapter. They commented on the unseen, and unknown, dangers of social networking sites, BYOD (bring your own devices), and the lack of environmental awareness from the use of "Smart" phones. The latter has also been commented upon by Gibson (2010).

The tensions are evident between cyberspace and non-cyberspace social groups. The tension is evident between parents and children. The tension is evident between those who wish to communicate through traditional means, the political masters, and those who embrace the new ways of communicating over cyberspace.

The development opportunity is to give people a voice. The Arab Spring, whose origins probably lie in the food shortages faced by many 5 years ago (PBS, 2011), was partly a result of the social groupings brought together over cyberspace, potentially contradicting an earlier point on political grouping. Note, however, that although people organized themselves, no new political grouping has appeared yet. In fact, one of the hazards of cyberspace in regard to politics may be the development of many single-issue polities rather than the generic ones with which all are familiar.

10.15 THE TECHNOLOGY HAZARD

The technology hazard is the cyber-security hazard with which most people are familiar. This relates to the hacking stories and the takeover of networks, the botnets, and associated technocrime. Once again it is important to emphasize that very few people actually understand these issues in great depth. Anderson at Cambridge University (Anderson and Moore, 2006; Anderson, 2013) and others suggest that the technology hazard is overplayed. They conclude that it is much better to track down the criminals involved, who are fewer in number than many believe. The key issue is the applications that sit on top of the hardware and how these applications come to relate to how individuals and organizations operate their lives and human issues around the use of technology (Black, 2008; Goodger, 2012).

The disasters are well cataloged and often grab the headlines. It is difficult to get a real perspective on how much of a hazard technology is. The UK's select committee on security having been warned, by the author and others (Hyslop, 2010), of the potential hazard in using Chinese Huawei servers in major BT infrastructure hardware because of the risk that there may be a shadow "chip" sending sensitive information back to China concluded recently that this risk was real (UK Government, 2013). Technology clearly can be a hazard.

10.16 THE ENVIRONMENTAL HAZARD

The environmental hazards posed by cyberspace are by no means obvious. There are some. The wide availability of data on everything from trees to fish means that the location and use of resources is no longer limited to those who can afford to go out and look for them. The barriers to entry into the exploitation of resources market have been significantly reduced, and just as there is an upsurge in the demand for resources, so the cost and ability to look for them comes down. Cyberspace has contributed to a free-for-all in this respect. The use of rare earths in the manufacture of the equipment used in cyberspace has had an environmental impact. The location of these rare earths may be a source of power for some states, a development opportunity. The disposal of used and out-of-date resources has had an impact. Cyberspace has contributed to the development of urban areas and to the encroachment of previously wild territories for agriculture (it is safer to manage your livestock in a wilderness area with a mobile phone than without one). Geolocation technology has enhanced the search for oil but also found it in environmentally sensitive areas.

The disasters are both here already and waiting to happen. The heaps of discarded electronic devices being taken apart by low-paid workers in India and China pose health and environmental threats to individuals and locales. The stripping of resources in an unsustainable manner to fuel the demand for rare earths to make cyberspace equipment has a clear environmental impact. Lack of rare earths may be a disaster for some developed economies. The dependence of Russian gas rigs on cyberspace to locate and maintain a presence in an Arctic oil field has a potential environmental impact. These are all examples. The biggest environmental hazard of all is the collapse of cyberspace, potentially as a result of a solar storm; then most things in the world will simply close down, and there will be 3 days to Armageddon (as it will take just 3 days for the cyberspace-controlled food supply chain to run out) and the world as known will collapse. This is known from war games and exercises. Our world is much more fragile in all respects than it was at the end of World War II and nowhere near as resilient as most people think. The lack of resilience extends to people too (Hyslop, 2013). The dependence on cyberspace has enhanced this fragility, not reduced it.

10.17 THE LEGAL HAZARD

Anyone or any business (more specifically) that operates within cyberspace operates within a totally different legal context than those which just operate within the physical world.

For an individual, the hazards became plain when Sally Bercow, the wife of the Speaker of the UK House of Commons, tweeted (as in used Twitter) in 2012 to allege the involvement of another politician in a sex scandal. She was successfully sued using the normal legal system. On the other hand, it is often difficult to prosecute those who abuse others online, as the various failed attempts to extradite those abusing children from other countries and the difficulties dealing with social network sites, such as Ask.fm in the Hannah Smith case noted under The Life Hazard above, have shown.

The European Union has made various attempts to introduce eJustice in one form or another. The author was the director of the EU's acclaimed eJustice program that ran from 2002 to 2006. The program spanned 23 countries and was successful in introducing forms of cross-border arrest warrants and other electronic aids to relaxing the tensions of cross-border bureaucracy. This and subsequent attempts have failed to gain EU wide acceptability for legal documentation on a common scale. The further failure is to come to a full common agreement on how breaches of law in cyberspace are to be dealt with and indeed what the law of cyberspace is (eJustice Project, 2006).

10.18 THE CRIMINAL HAZARD

Williams (2002), an editor of "Transnational Organised Crime," shows how synergies, trends, and responses to cybercrime have been developed by organized crime. He shows how much crime is disorganized but how increasing evidence exists for organized crime exploiting the opportunities of cyberspace.

The first attempts to attribute a cost to cybercrime were made by the Ponemon Institute in 2012 (Ponemon Institute, 2012). In dealing with this issue, the Institute stated that cyber-attacks generally refer to criminal activity conducted over or through cyberspace. The attacks included stealing an organization's intellectual property, confiscating online bank accounts, creating and distributing viruses on other computers, posting confidential business information on the Internet, and disrupting infrastructure. Well-publicized cyberattacks, according to this survey, have affected private and public sector organizations such as Google, TJX Companies, TD Ameritrade, Heartland Payment Systems.

The major issue with the studies, and not a criticism of the attempt to benchmark the cost of crime, is that every cost in it is necessarily an estimate. Additionally, some of the information is skewed by particular large values, and no statistically significant difference exists between some of the costs of attack type, particularly as the sample sizes are only 45–56. So the study tells us the obvious: that cybercrime costs money. It is less good at identifying, even in general terms, how much it costs. The end result is a bit like the frequently quoted statistics that 20 percent of the money going through the City of London is drug money as it is through the rest of the country and that 50 percent of the world's GDP goes through tax havens. These are clearly estimates at best because no one actually knows. A conversation between Gibson and Anderson (Gibson and Anderson, 2013) stated the following:

The citation to your 'cost of cyber crime' was pretty good; but again the numbers are but extrapolations of extrapolated numbers. There is no country on this planet that has any kind of central reporting requirement thus making calculations difficult at best.

10.19 THE MORAL HAZARD

Moral hazard is a term normally applied to economics, where it is known that a deal is being settled knowingly to the disadvantage of one or other of the parties involved in the bargain.

In terms of the cyberspace hazard to society, the moral hazard is that the perpetrators and managers of cyberspace are knowingly and recklessly pursuing an agenda that meets the needs of the commercial entities involved to the ultimate detriment of society as a whole. This is especially so as humans are introduced into cyberspace.

So there is little point in having cyberspace at all if it is not ultimately resilient, and it is suspected so. Therefore, the tsunamis and earthquakes will be as nothing to the effect a cataclysmic failure of cyberspace would now have on the world. This is obviously not to be underestimated. Once again refer to how the care with which the regulations for the management of electricity and electronics in petroleum hazardous areas (worldwide) compare with the ungoverned approach of cyberspace to governance and regulation (US Department of Homeland Security, 2004).

10.20 SUMMARY

All sorts of cyber-security hazards in society exist. However, the biggest one is the moral hazard. Cyberspace is an accident waiting to happen. Society is no longer resilient enough to sustain a major cyberspace disaster, certainly in the developed economies.

10.21 CONCLUSION

It is a very reluctant conclusion of one who believes in the access for all principles behind much of cyberspace that such an approach is ultimately unhelpful, dangerous, and lacks the resilience required to remove it as a fundamental hazard to our way of life unless better governed, regulated, and protected.

ACKNOWLEDGMENTS

Andrew Waite (Onyx Group); John Walsh (Vanderbilt University); Joe Pennell (University of Cambridge).

REFERENCES

Abrams, M., Weiss, J., July 23, 2008. Malicious Control System Cyber Security Attack Case Study – Maroochy Water Services, Australia. Available at: http://csrc.nist.gov/groups/SMA/fisma/ics/documents/Maroochy-Water-Services-Case-Study_report.pdf (accessed 27.10.13.).

Alexander, D.E., 2010. Available at: http://d-alexander.blogspot.com (accessed 27.10.13.).

Anderson, R., 2013. Available at: http://www.cl.cam.ac.uk/~rja14/ (accessed 27.10.13.).
Anderson, R., Moore, T., 2006. The Economics of Information Security. Available at: www.cl.cam.ac.uk/~rja14/Papers/econ_science.pdf (accessed 27.10.13.).
Bank of Canada, 2000. Available at: http://www.pwgsc.gc.ca/recgen/colloquium2007/pdf/panel-discussion-jose-navarro-e.pdf (accessed 27.10.13.).
BBC, 2013a. Hannah Smith Death: Father Says Daughter Was a Victim of Cyberbullies. Available at: http://www.bbc.co.uk/news/uk-england-leicestershire-23584769 (accessed 27.10.13.).
BBC, 2013b. Dark Web Will Evolve Warns UK Cybercrime Chief Andy Archibald. Available at: http://www.bbc.co.uk/news/technology-24495029 (accessed 27.10.13.).
BBC, 2013c. Available at: www.bbc.com/news/uk-politics-24847399 (accessed 07.11.13.).
Black, J., 2008. The Secret History of the World. Quercus, London.
British Petroleum, September 19, 2007. C-Suite Officer Interview with Maitland P Hyslop.
Channel4, 2013. FBI Seizes $28 M in Bitcoin. But May Not Be Able to Use It. Available at: http://www.channel4.com/news/fbi-bitcoin-seizure-silk-road-ross-ulbricht-drugs-market (accessed 27.10.13.).
Chen, A., 2013. Silk Road's Downfall Killed the Dream of the Dark Net. Available at: http://gawker.com/tag/darknet (accessed 27.10.13.).
Computer World, 2013. Murder by Malware, Can Computer Viruses Kill?. Available at: http://blogs.computerworld.com/16801/murder_by_malware_can_computer_viruses_kill (accessed 27.10.13.).
Council on Foreign Relations, 2013. G20 Leaders Declaration September 2013. Available at: http://www.cfr.org/international-organizations-and-alliances/g20-leaders-declaration-september-2013/p31355 (accessed 27.10.13.).
Dodge, M., 2008. Understanding Cyberspace Cartographies (Ph.D. thesis). UCL.
Dixon, N.F., Dixon, M., 2011. On the Psychology of Military Incompetence. Pimlico, London.
Dunn, M., Wigert, I., 2004. Critical Information Infrastructure Protection. The Swiss Federal Institute of Technology, Zurich, Switzerland. Available at: http://www.isn.ethz.ch/crn (accessed 27.10.13.).
Die Hard 4, 2007. Directed by Len Wiseman. 20th Century Fox.
eJustice Project, 2006. Available at: www.onenortheast.co.uk/page/business/ejustice.cfm (accessed 27.10.13.).
Gaunt, R.A., 2010. Sir Robert Peel: The Life and Legacy. Tauris, London, I.B.
Gibson, E.P., 2013. Available at: http://www.infosecurityproject.com/2013/the14th_conference_speakers.html (accessed 27.10.13.).
Gibson, E.P., Anderson, R., August 28, 2013. eMail Correspondence via Hyslop, M.P.
Gibson, E.P., 2010. Available at: http://www.securityg33k.com/blog/?p=300 (accessed: 27.10.13.).
Ghanea-Hercock, R., 2010. Personal Communication, 2nd CIIP Workshop. Birkbeck College, University of London, UK. March 29, 2010 and April 28, 2010.
Ghanea-Hercock, R., 2011. Cyber-security Is about More than Physical Science. Available at: http://letstalk.globalservices.bt.com/en/tag/robert-ghanea-hercock (accessed 27.10.13.).
Goodger, A., 2010. Personal Communication, 2nd CIIP Workshop. Birkbeck College, University of London, UK. March 29, 2010 and April 28, 2010.
Goodger, A., 2012. Personal Communication, 15 November 2012.
HSBC, July 1, 2013. Fiona Swann, Senior HSBC Manager, in Conversation with Maitland Hyslop.
Hyslop, M.P., 2013. Hardening Organisations (Ph.D. thesis). University of Northumbria.
Hyslop, M.P., 2010. Critical Information Infrastructure, Resilience and Protection. Springer, Boston, USA.
I3P, 2013. Available at: http://www.thei3p.org/ (accessed 27.10.13.).
Kruh, L., Deavours, C., 2002. The commercial enigma: beginnings of machine cryptography. Cryptologia 26, 1.

Lee, D., 2013. New York Times and Twitter Struggle after Syrian Attack. Available at: http://www.bbc.co.uk/news/technology-23862105 (accessed 27.10.13.).

Levin, R., et al., 2000. The Cluetrain Manifesto. Perseus, Cambridge, MA.

Mandiant, 2013. Mandiant Exposes APT1. https://www.mandiant.com/blog/mandiant-exposes-apt1-chinas-cyber-espionage-units-releases-3000-indicators (accessed 27.10.13.).

MPAA, 2007. You Wouldn't Steal a Car Would You. Available at: http://www.youtube.com/watch?v=HmZm8vNHBSU (accessed 27.10.13.).

The Matrix, 1999. Directed by the Wachowski Brothers. Warner Brothers.

NATO, 2013. Defending against Cyber Attacks. Available at: http://www.nato.int/cps/ar/natolive/75747.htm (accessed: 27.10.13.).

Oxford English Dictionary, 2013. Oxford University Press, Oxford, UK.

PBS, 2011. Available at: http://www.pbs.org/newshour/updates/world/july-dec11/food_09-07.html (accessed 27.10.13.).

Peck, M., 2013. Why Can't the U.S. Air Force Find Enough Pilots to Fly Its Drones?. Available at: http://www.forbes.com/sites/michaelpeck/2013/08/22/why-cant-the-u-s-air-force-find-enough-pilots-to-fly-its-drones (accessed 27.10.13.).

Ponemon Institute, 2012. US Cost of Cyber Crime Study. Available at: http://www.ponemon.org/local/upload/file/2012_US_Cost_of_Cyber_Crime_Study_FINAL6%20.pdf (accessed 27.10.13.).

Rarden, P., April 8, 2014. City of London Police Comment to Hyslop, M.P.

Roberts, M., 2005. Resilience for a top FTSE 100 company — an insight from a BT perspective. In: 3rd Annual European Telecommunications Resilience and Recovery Association Conference, Sophia Antipolis, France, 29–30 June.

Rumsfeld, D., 2002. Available at: http://www.nato.int/docu/speech/2002/s020606g.htm (accessed 27.10.13.).

Schneier, B., 2013. Available at: https://www.schneier.com (accessed 27.10.13.).

Shalal-Esa, A., 2013. Six US Cyber Tools Designated as Weapons. Available at: http://www.nbcnews.com/technology/six-u-s-air-force-cyber-tools-designated-weapons-1C9277187 (accessed 27.10.13.).

Sherman, C., 2002. Mapping the Dark Net. Available at: http://searchenginewatch.com/article/2066959/Mapping-the-Dark-Net (accessed 27.10.13.).

The Economist, 2010a. Cyberwar. July 3, p. 9.

The Economist, 2010b. A Virtual Counter Revolution. September 4, p. 79.

The Economist, 2013. Chinese Cyber-Attacks. Available at: http://www.economist.com/blogs/analects/2013/02/chinese-cyber-attacks (accessed 27.10.13.).

The Guardian, 2013. Available at: http://www.theguardian.com/world/edward-snowden (accessed 27.10.13.).

The Jester, 2013. Available at: http://jesterscourt.cc/ (accessed 27.10.13.).

The Register, 2010. Armed with Exploits, ATM Hacker Hits the Jackpot. Available at: http://www.theregister.co.uk/2010/07/28/atm_hacking_demo/ (accessed 27.10.13.).

The Sunday Times, 2013a. Cracking the Secrets of the Dark Net. Available at: http://www.thesundaytimes.co.uk/sto/news/focus/article1326870.ece (accessed 27.10.13.).

The Sunday Times, 2013b. We Need an Enquiry into the Surveillance Scandal. Available at: http://www.thesundaytimes.co.uk/sto/news/focus/article1326870.ece (accessed 27.10.13.).

TruTV, 2013. John E Robinson, Sr: The Slave Master. Available at: http://www.trutv.com/library/crime/serial_killers/predators/john_robinson/index.html (accessed 27.10.13.).

UK Government, 2013. Foreign Involvement in the Critical National Infrastructure: The Implications for National Security. Available at: https://www.gov.uk/government/uploads/system/

uploads/attachment_data/file/205680/ISC-Report-Foreign-Investment-in-the-Critical-National-Infrastructure.pdf (accessed 27.10.13.).
UNODC, 2012. The Use of the Internet for Terrorist Purposes. United Nations, New York.
US Department of Homeland Security, 2013. Cybersecurity. Available at: https://www.dhs.gov/topic/cybersecurity (accessed 27.10.13.).
US Department of Homeland Security, 2004. A Comparison of Oil and Gas Segment Cyber Security Standards. Idaho National Engineering and Environmental Laboratory.
Warwick, K., 2013. Available at: http://www.kevinwarwick.com/ (accessed 27.10.13.).
Wayper, C.L., 1973. Political Thought. EUP, UK.
Wikileaks, 2013. Available at: http://wikileaks.org/ (accessed 27.10.13.).
Williams, P., 2002. Organised Crime and Cyber Crime. Available at: www.cert.org/archive/pdf/cybercrime-business.pdf (accessed 27.10.13.).

Chapter 11

Natural Disasters and Violent Conflicts[1]

Elisabeth King[1] and John C. Mutter[2]
[1] *Department of Humanities and Social Sciences, New York University, New York, USA,*
[2] *Department of Earth and Environmental Sciences and School of International and Public Affairs, Columbia University, New York, USA*

ABSTRACT
Comparisons between disasters and violent conflicts are often noted by political figures and in the news media, and those responding to conflicts and disasters witness similarities on the ground. In contrast, the academic fields studying violent conflicts and so-called natural disasters have developed separately, and practitioners usually separate the two phenomena as soon as the emergency response is over. This chapter, based on interviews with practitioners and a review of scholarly literature, makes a case for increased cross-disciplinary dialogue. We identify common consequences, responses, and even causes of conflicts and disasters. We argue that more and better partnerships between those who work on conflicts and those who work on disasters can lead to advances in understanding and responding to conflicts and disasters.

11.1 INTRODUCTION

Comparisons between disasters and violent conflicts—especially the immediate consequences—are often noted by political figures and in the news media. Earthquakes are said to resemble war zones,[2] and wars are often described with disaster language.[3] In contrast, among scholars, there has been relatively little discussion across the conflict–disaster divide. Historically, the question of culpability has been considered different in conflicts and disasters,

1. This chapter is an abbreviated version of King, Elisabeth, and John C. Mutter (2014). "Violent Conflicts and Natural Disasters: The Case for Cross-Disciplinary Dialogue" *Third World Quarterly*, 35(7): 123–139. copyright © Third World Quarterly, reprinted by permission of Taylor & Francis Ltd, www.tandfonline.com on behalf of Third World Quarterly. An earlier version of the chapter was presented at the "Dealing with Disasters" conference, Northumbria University at Newcastle, UK in November 2010.
2. Cawthorne and Delva, "Quake Destruction 'like in a war': Haiti President."
3. Wax, "Like a tsunami that won't end."

Hazards, Risks, and Disasters in Society. http://dx.doi.org/10.1016/B978-0-12-396451-9.00011-1

with conflicts thought of as "human-made" and disasters considered to be "natural."[4] These differing understandings of the locus of causation have affected the ways in which each phenomenon has been studied. With natural scientists first dominating the study of disasters, a prime focus was on predicting the time, location, and magnitude of the natural extreme event. In contrast, social scientists lead the study of conflicts, focusing on uncovering and explaining the multiple social, economic, and institutional causes of violent conflict. While social scientists studying disasters focus on these same social-structural factors, they do not typically interact with conflict scholars studying them or with natural scientists studying the same incidents. Scholars of humanitarianism often subsume conflicts and disasters into a single "emergency" category, but usually take similarities for granted rather than exploring them explicitly and do not typically consider causes.

For their part, nongovernmental organizations (NGOs) tasked with responding to conflicts and/or disasters usually do not distinguish between the two in terms of relief. According to their mission statements, Médecins Sans Frontières (MSF), for instance, "delivers emergency aid to people affected by armed conflict... and natural or human-made disasters," and Catholic Relief Services (CRS) responds to "major emergencies" of all varieties. NGOs that work on longer-term responses and prevention, however, often separate conflicts and disasters. One interviewee described International Rescue Committee (IRC) as a "conflict organization," focused on issues such as displacement and oppression, working until recently on disasters only when they arose in contexts in which they were already working. In addition, within a number of international organizations, the staff members working on conflicts and those working on disasters have little opportunity to interact; their programs, departments, deployments, and background specializations are often different. The NGO representatives with whom we spoke noted that they have had insufficient opportunity to pause and reflect upon the opportunities and challenges presented by the similarities of conflicts and disasters.

The recognition of similarities between conflicts and disasters, paired with relatively little discussion across the conflict–disaster divide, prompts us—a political scientist and a seismologist working in interdisciplinary environments—to ask: Is there a case for increased dialogue between those studying conflicts and those focused on disasters? What can such bridge building contribute to scholarship and practice? To answer these questions, we reviewed literature on the causes, consequences, and responses to conflicts and disasters. We also conducted over a dozen interviews with representatives from organizations working at the conflict–disaster nexus, including individuals based in headquarters and the field. We discussed how and why their organizations work to address conflicts and/or disasters, the programmatic and on-the-ground similarities and differences, as well as the potential causal

4. See, for instance, Albala-Bertrand, "Complex Emergencies versus Natural Disasters."

intersections between conflicts and disasters. We chose individuals who work with large, well-known, and respected international NGOs—such as CRS, MSF, IRC, and Mercy Corps—with varied approaches to conflicts and/or disasters. We also spoke with a representative from InterAction, an umbrella organization with more than 200 international relief and development member organizations. While there are many local and national organizations working to address conflicts and disasters, we focus here on international NGOs and their work in conflicts and disasters in the Global South.[5]

We argue that increased dialogue between those studying and working in conflicts and disasters is important for at least three reasons. First, in the opening section, we show that while there remain important differences, conflicts and disasters often have similar consequences. In the next section, we show how the two phenomena invoke similar responses and present similar challenges. An explicit catalog of these similarities is in itself a step forward—they are typically either presumed or overlooked—and less obvious similarities in both consequences and responses present opportunities for each field to learn from the other. Second, we argue that the still common tendency to suggest that disasters are natural and that conflicts are human-made is limiting in understanding or responding to either phenomenon. In the causes section, we show that both may have natural and human-made roots and, moreover, that there are increasingly common drivers of, and interactions between, conflicts and disasters that require interaction and dialogue across the natural and social sciences and those studying conflicts and disasters. In the end, while we do not wish to overplay the similarities, we, and many of our interviewees, opine that better acknowledging the areas of overlap can lead to advances in explaining conflicts and disasters and especially to the practical refinement of responses. The conclusion argues that stronger mitigation strategies for conflicts or disasters will, in some cases, help reduce the risk of both.

11.1.1 Comparable Consequences

There are a number of stark similarities between the immediate consequences of conflicts, by which we mean violent conflicts, and disasters, by which we mean human and capital losses and social consequences that result from societies experiencing natural extremes such as earthquakes, storms, droughts, and floods.[6] NGO representatives with whom we spoke, and especially

5. A few interviews were conducted in 2011 with the majority taking place in 2012. Whenever possible, the authors transcribed interviews verbatim. Note that all interviewees spoke in an individual rather than organizational capacity. All interviewees were given the opportunity to review an earlier draft of this chapter.

6. We do not contend, of course, that all of the consequences of conflicts and disasters are similar. The involvement of individuals as perpetrators of conflict, for instance, opens a range of additional consequences of conflict. Here, in making a case for cross-disciplinary dialogue, we highlight similarities.

responders based in the field, view the consequences of conflicts and disasters, and the initial response required, as indistinguishable. Indeed, many interviewees had an overarching term for the early phase of the two phenomena such as "emergency," "crisis," or "catastrophe." The same is true of the humanitarianism literature in which the most widely used term, "emergency," denotes a situation of urgency, where lives are at risk and people are suffering, and where their own governments are unable to meet their needs, generally without reference to specific context or causes.[7]

Indeed, both violent conflicts and natural disasters are marked by loss of life, and the scale of each is often measured in terms of casualties. Civilians are obviously the dominant casualties in disasters, and it is now a commonly accepted figure that upward of 80 percent of casualties in wars are civilians.[8] Moreover, direct deaths represent only a fraction of total deaths, with common indirect causes including infectious diseases[9] and disinvestments in child health care.[10]

Another clear parallel is in the displacement of people within countries and across borders. It is estimated that 32.4 million people were displaced due to natural disasters in 2012 and that 98 percent of those displaced people were forced to move because of weather-related disasters.[11] Likewise, in conflict, displacement can be massive; the worldwide number of refugees in 2012 was 10.5 million, alongside 6.5 million internally displaced.[12]

Both conflicts and disasters may also result in major livelihood consequences. The average cost of civil war equals upward of 30 years GDP growth for a medium-sized developing country.[13] At the individual level, over the course of the Sierra Leonean conflict, real per capita income declined by 50 percent, and in Liberia, it declined by about 80 percent from 1980 to 1997.[14] However, the costs of conflict are uneven within countries and between individuals, with the poorest usually being hardest hit. In disasters too, the poor have the least coping capacity; the poor have no insurance, little savings, very limited access to credit, and often live in countries where government institutions are weak and thus provide limited assistance.[15] The 1976

7. Barnett and Weiss, "Humanitarianism: A Brief History of the Present."; Calhoun, "The Idea of Emergency."; Orbinski, *An Imperfect Offering*.
8. Kaldor, New and Old Wars.
9. Ghobarath, Huth and Russett, "Civil Wars Maim and Kill People".
10. Antilla-Hughes and Hsiang, "Destruction, Disinvestment, and Death."
11. Yonetani, "Global Estimates 2012."
12. UNHCR, Facts and Figures about Refugees from the 2011 Global Report.
13. World Bank World Development Report 2011, 5–6.
14. Humphreys & Richards, Prospects and Opportunities for Achieving the MDGs in Post-Conflict Countries, 12.
15. Wisner et al. At Risk: Natural Hazards, People's Vulnerability and Disasters, 13.

earthquake in Guatemala is often called a "class quake" or the "Indian earthquake" because of its disproportionate impact on poor and indigenous Guatemalans.[16]

In both disasters and conflicts, the breakdown of already overextended health care is common and can lead to similar outcomes. Both disasters and conflicts can destroy schools, kill or displace teachers, and make students fearful of returning to school.[17]

Sexual violence and gendered impacts are also similar, although there are differences in motivation and scale. Sexual violence is now a distressing, but commonly acknowledged, part of violent conflict. Perhaps less-known, reports from post-earthquake Haiti present an alarming increase in opportunistic sexual violence in camps.[18] Sexual violence after disasters, such as for people housed in Federal Emergency Management Agency trailer camps after Hurricane Katrina, is so common that the National Criminal Justice Reference Service publishes a guide for the prevention and response to sexual violence in disasters.[19] In cases of both conflicts and disasters, there are also very clear gendered impacts that reach beyond sexual violence.

Finally, the environment suffers in both conflicts and disasters in somewhat similar ways. In a sense, a natural disaster is an environmental impact, almost by definition. Extreme events like floods, hurricanes, and droughts can destroy, degrade, or fundamentally and permanently alter natural and crop ecosystems, changing biodiversity and permitting invading species to flourish. Perhaps less obviously, conflict too often degrades the environment, e.g., the use of landmines pollutes soil, there is frequently illegal trade and unsustainable use of natural resources, and peacetime environmental protection agencies may be unable to operate.[20]

11.1.2 Similar Responses and Challenges

It is not surprising, given the similarities in the consequences of disasters and conflicts, that there are also underlying parallels in immediate responses.

---

16. Quoted in Gawronski & Olson, "Disasters as Crisis Triggers for Critical Junctures?"; While war has been described as "development in reverse," some alternatively "do well out of war" (Collier, "Doing Well Out of War: An Economic Perspective."), as do some in disasters, such as those in the construction and aid industries. The longer-term economic outcomes of disasters are difficult to decipher and quite variable (Mutter, "Perceiving the Social and Economic Consequences of Natural Disaster Shocks").
17. Buckland, *Reshaping the Future*; King, "The Multiple Relationships between Education and Conflict"; Nusrat and Jamil, "Earthquake and Education in Emergencies."
18. Amnesty International, *Haiti Aftershocks*: Center for Human Rights and Global Justice, "Sexual Violence in Haitian IDP camps."
19. Klein, Sexual Violence in Disasters.
20. UNEP and Saundry, "Environmental and Socioeconomic Impacts of Armed Conflict in Africa."

At the core of the humanitarian enterprise is the goal of "reliev(ing)... suffering," whatever the cause.[21] For better or worse, some use the term "the natural disaster model" to describe the apolitical needs-based framework for humanitarian assistance during and immediately after wars.[22] As one interviewee stated, "NGOs will apply the same modus operandi, the same tool kits, and come with the same solutions. A water problem is a water problem within a conflict or natural catastrophe. It's the same with malnutrition. They will address the consequence, and they don't care what causes the problem." According to several interviewees, specific context, local abilities to respond, community preparedness, and scale of the international response are the key factors in differentiating responses, usually more so than the nature of the phenomena.

Some less commonly acknowledged similarities again highlight particular value in cross-disciplinary dialogue. Many interviewees noted that there is generally less risk to relief personnel in disaster than in conflict settings. Others noted that the reality is more complicated: When conflict lines are clear, when emergency relief personnel are not in the front line, and when there is a military presence, the delivery of aid is often operationally easier than somewhere like post-earthquake Haiti, where at least one interviewee deemed insecurity to be constant in the days after the earthquake, since robberies and gangs hindered the distribution of aid.

Similarly, our interviewees suggested that it is usually thought that international relief workers need a stronger understanding of context in a conflict setting as opposed to a disaster setting because responses in conflicts are more political.[23] In conflict settings, responders are also said to face more challenges in negotiating access, working with whomever is in control of the population, and in securing materials and equipment. They must pay particular attention to context so as to "do no harm," for example by erroneously distributing aid to only one side of a conflict.[24] While such quandaries may seem less relevant to more "straightforward" responses to disasters, responses to the 2005 northern Pakistan earthquakes, for instance, were far from apolitical, when humanitarian workers faced challenges of working with a military government

21. Orbinski, *An Imperfect Offering*.
22. Woodward, "Economic Priorities for Successful Peace Implementation," 189–90; see Duffield, *Global Governance and the New Wars*.
23. This recognition contrasts with the "minimalist" humanitarian approach in which all humanitarian work is considered to be apolitical (Weiss, "Principles, Politics, and Humanitarian Action.").
24. See Anderson, *Do No Harm*.

and jihadist groups.[25] Ethnic or racial groups can be also marginalized in the event of a disaster in the same way as in a conflict.[26]

Furthermore, in both the cases, responders to conflicts and disasters were concerned about the misuse of aid and the potentially negative consequences of relief. In Ethiopia in the famine of 1984, aid was used to force migration from rebel to other areas, prompting possibly more deaths in migration than in the famine. For MSF, this experience was "seminal" and led to profound thinking about "the impact, both positive and negative, of their presence and their work which can reach the extreme of aiding and abetting crime."[27]

A related area of emerging convergence between conflict and disaster responses is securitization, which frequently includes private security forces. There is growing concern that peacebuilding and humanitarianism are being securitized and privatized, since donor governments appear concerned about conflicts not out of humanitarianism but because fragile states pose dangers to developed countries. Less commonly, the securitization paradigm is also apparent in the field of disasters, where helping states build effective response capacities may be more about preventing wider negative security and political repercussions than humanitarianism. For example, the initial stance of the government of Myanmar to refuse all outside relief for Cyclone Nargis stemmed from their belief that it would be used by foreign governments, who had made no secret of their opposition to the military rulers, to bring about regime change (Selth, 2008).

Moving from emergency relief to longer-term response widens the chasms between the responses to conflicts and those to disasters; whereas avoiding future conflicts may require peacebuilding, disaster prevention may necessitate disaster risk reduction (DRR). Nonetheless, there are similarities in these objectives as well.

In the fields studying disasters, there is a growing literature on building "adaptive capacity," fostering "resilience," or "coping capacity."[28] The notion is that there are social attributes that can make some societies more able to sustain the physical shock of natural disasters and recover fairly quickly. IRC defines DRR as a "process by which natural disaster risks are identified, analyzed, and minimized in advance in order to avoid or limit adverse impacts

25. Qureshi and Bamforth, "Political Complexities of Humanitarian Intervention in the Pakistan Earthquake."
26. Raleigh, "Political Marginalization, Climate Change, and Conflict in African Sahel States."; De Silva, "Ethnicity, Politics and Inequality."
27. Phelan, "From an Idea to Action." p.32.
28. Dayton-Johnson, "Natural Disasters and Adaptive Capacity," 6; Cutter, *Hazards, vulnerability and environmental justice*; Klein et al. "Resilience to Natural Hazards"; Buhaug, Gleditsch and Theisen, "Implications of Climate Change for Armed Conflict," 11.

on communities."[29] The goals of DRR include the ability "to absorb shocks and stresses, either by adaptation or resistance."[30]

In the peacebuilding literature, resilience is akin to concepts of building a sustainable peace.[31] Peacebuilding is "action to identify and support structures which will tend to strengthen and solidify peace in order to avoid a relapse into conflict."[32] Peacebuilding aims to empower communities to deal with conflicts in nonviolent manners. Just as hazards recur, conflicts are a constant in society, but the goal of peacebuilding is for societies to be able to absorb shocks and stresses such that conflict does not turn violent. Both DRR and peacebuilding also aim to foster development.

Interviewees raised the common challenge across the conflict–disaster spectrum of moving from relief to mitigation, a challenge that has been echoed in policy and academic literature since the 1990s. As Calhoun writes, emergency response is "at the opposite end of a continuum from development assistance that tries to address long-term issues of poverty or disempowerment."[33] MSF, for example, faces the challenge that it has defined itself as an "emergency aid organization" and currently plans on single-year cycles. Yet, postconflict and conflict lines are unclear, and disaster and postdisaster lines are similarly blurring.[34] MSF wants to make a "committed response" and is thus considering development-oriented interventions such as creating a full hospital from the beginning of an emergency response. Several interviewees spoke of the compartmentalization within organizations that hinders the relief to development (or peacebuilding and DRR) transition. One interviewee discussed how DRR work is currently housed within, and most valued by, her organization's relief/emergency team, but it may be best situated with the development team. Another interviewee, a human resources specialist, explained that at least in some organizations, the skills and training of an emergency responder are often different from those of a longer-term development or mitigation practitioner, making the transition from one phase to the next more difficult. There is a similar compartmentalization in funding: Emergency donors often provide funds that are available quickly for a short time. When emergency donors leave, responders are sometimes still on the ground without the means to continue their work. Another obstacle we heard is that while relief work is generally relatively well coordinated on the ground, thanks to the UN cluster system and the acceptance of minimum standards, development, peacebuilding, or DRR work does not have similarly preexisting coordination structures.

29. IRC, Internal Document.
30. Ibid.
31. See World Bank, World Development Report 2011, 103.
32. Boutros-Ghali, An Agenda for Peace.
33. Calhoun, "The Idea of Emergency," 52.
34. See also Duffield, "Complex Emergencies and the Crisis of Developmentalism."

11.1.3 Parallel and Interacting Causes

We still hear of "human-made" conflicts in contrast to "natural disasters" where the role of agency is limited or denied. While this distinction is outdated, as an interviewee noted, "conflicts and disasters haven't generally been married up and this is a problem," noting the need to think about "the holistic nature of conflict and disaster." We illustrate by considering the ways in which social vulnerabilities contribute to both conflicts and disasters,[35] how climate change impacts both phenomena, and the causal interconnections between conflicts and disasters.

11.1.3.1 Social Vulnerabilities

A variety of social vulnerabilities are long-recognized underliers of conflict.[36] At the macrolevel, Collier et al. find that less-developed countries face a conflict risk four times higher than countries in the OECD, and the least-developed countries face a risk 15 times as high.[37] Weak states are important root causes of war.[38] Within states, horizontal inequalities, or severe inequalities between groups, also appear to increase the propensity for conflict.[39] And the most vulnerable within a society are often the hardest hit from conflict.[40]

While natural extremes have a global distribution, hazards are likely to turn into the worst disasters in the poorest and most fragile states.[41] The widely cited "pressure and release model" identifies political/governance, economic, and social/cultural issues as "root causes" affecting if and how hazards become disasters.[42] Mortality in disasters, for instance, is much higher in poor than richer countries: The ratio of mortality in rich to poor countries is 100:1.[43] The earthquake that struck Haiti in January 2010 was far from the top of the Richter scale, but the deaths it caused were well off scale: 200,000 deaths are vastly more than that is reasonable to expect for an earthquake of magnitude 7.0. In California, earthquakes of similar magnitude result in few deaths. Haiti's death toll expresses the economic, social, and political vulnerability of many residents of Port-au-Prince more than the strength of the quake. Haiti has frequently been listed among the most failing states in the world,[44] and its long

35. While we focus on this single illustration, it does not, of course, represent the entire range of overlap.
36. Brown, *Ethnic Conflict and International Security*.
37. Collier et al. Breaking the Conflict Trap, 5.
38. Jackson, "The State and Internal Conflict."
39. Stewart, "Horizontal Inequalities and Conflict."
40. Collier et al. Breaking the Conflict Trap.
41. World Bank, World Development Report 2011, 103.
42. Wisner et al. *At Risk*.
43. Brainard, Jones & Purvis, *Climate Change and Global Poverty*.
44. Foreign Policy and the Fund for Peace, "The 2010 Failed States Index."

history of violent conflict arguably contributed to its high vulnerability to natural hazards.

The concept of vulnerability describes the likelihood that a natural hazard will lead to harm and includes such factors as poverty and inequality, class, occupation, ethnicity, gender, and social capital.[45] The death toll from Hurricane Katrina in New Orleans is 100 times higher than American norms. Louisiana ranks in the range of developing countries using an "American Human Development Index."[46] It is also one of America's most racially and economically divided cities and wealthier whites generally fared better.[47]

11.1.3.2 Climate Change

Climate change is what most interviewees presumed we meant when we asked about the possibility of common drivers of conflicts and disasters. Although surrounded by a suite of uncertainties, there is little doubt that climate is shifting and is destined to do so into the future. While there are expected to be significant variations in the impacts of climate change, the general trend affects both disasters and conflicts.[48] Climate change can be thought of as originating in the actions of rich countries of the Global North, where conflicts are rare and coping capacity high, but with the greatest consequences in the South, where conflicts and disasters are more common and have their greatest consequences.[49]

It is widely predicted that a warmer climate will lead to more weather extremes and consequently more disasters.[50] Climate change predictions suggest that the phenomena of multiyear drought, intense flooding, and severe storms will be more prevalent and widespread while gradually worsening base conditions will make those events more consequential. Tropical storms are predicted to extend their range to higher latitudes as the tropics expand into places that have never experienced hurricanes and are thus unprepared. Drought-prone regions in the African Sahel may expand southward. Sea levels are also predicted to rise, but irregularly, rising more in some places than in others and even dropping in some places. Climate change is also expected to increase social vulnerability in a way that makes disasters likelier. For

45. Wisner et al. *At Risk.*; Cutter & Finch, "Temporal and Spatial Changes in Social Vulnerability to Natural Hazards," 2303–2305; Cutter, *Hazards, vulnerability and environmental justice*; Pelling, *The Vulnerability of Cities*; Peluso & Watts, *Violent Environments*.
46. Burd-Sharps et al. *The Measure of America.*
47. Snyder, "It Didn't Begin With Katrina."
48. Parry et al. "Climate Change 2007."; Burke et al. "Warming Increases Risk of Civil War in Africa."; Raleigh, "Political Marginalization, Climate Change, and Conflict in African Sahel States."; Salehyan, "From Climate Change to Conflict?"
49. Ironically, the temperate zones, where the greatest quantity of emissions originates may be relatively robust to temperature increases.
50. Field et al. Managing the Risks of Extreme Events and Disasters to Advance Climate Change Adaptation.

instance, using historic records, Dell et al., Jones and Olken, and Hsiang all deduce significant sensitivity to temperature change on economic growth rates in the tropics.[51]

Experts also believe that climate change will have an important impact on conflict. The UN Security Council, for instance, held a first debate in 2007 on climate change as a global security issue, and the US security community commissioned a 2013 report by the National Academy of Sciences/National Research Council on the implications for security analysis of climate and social stresses.[52] Hsiang et al. used a study of conflict onset in countries where the El Nino brings harsh conditions to many poor regions of Africa. They used data from 1950 to 2004 and found robust evidence that El Nino conditions double the chances of an outbreak of conflict.[53] Others have found similar correlations.[54] The idea is that climate change will bring drought, and if drought leads to large-scale migrations, an increase in conflicts is plausible. One interviewee spoke about climate change adaptation (CCA) work, especially in East Africa and Latin America that is driving people to plant at different elevations. She noted that this CCA programming itself is moving people, sometimes onto or nearer to other people's land, and bringing different communities into competition—all of which they are monitoring for conflict potential. The summer 2013 issue of *The Aspen Idea* contended that "the events that came to be known as the Arab Spring were caused, to a large extent by climate change and the resulting effects on food prices."[55] The assertion is that climate change has increased drought severity that reduced crop yields and brought about scarcity that led to high prices.[56] Given the forecasted impact on conflicts and disasters, climate change presents a direct area of dialogue for conflict and disaster scholars and practitioners.

11.1.3.3 Causal Links between Conflicts and Disasters

There are also direct, potentially causal, connections between conflicts and disasters that point to the need for cross-disciplinary dialogue. Some protracted conflicts have quite directly resulted in disasters such as famines in

---

51. Del et al. *Climate Change and Economic Growth*; Jones and Olken, *Climate Shocks and Exports*; Hsiang, "Temperatures and Cyclones Strongly Associated with Economic Production."
52. Steinbruner et al. Climate and Social Stress.
53. Hsiang et al. "Civil Conflicts are Associated with the Global Climate."
54. See Burke et al. "Warming Increases Risk of Civil War in Africa."; Miguel et al., "Economic Shocks and Civil Conflict."; Sandholt et al., "Rain, Growth and Civil War."
55. Clark et al. "What is Food Security?" p.68
56. See also several discussion papers at http://www.iisd.org/ecp/es/climate/of the International Institute for Sustainable Development.

Ethiopia and Sudan.[57] More commonly, though, violent conflicts can render societies more *vulnerable* to hazards. Wisner et al. list war as "one of the main causes of social vulnerability": Conflict may drive marginalized people into new and less safe areas, as occurred in the civil war and genocide in Guatemala. Conflict may hinder relief efforts and divert resources that could otherwise have been used to help mitigate disasters. Conflict can ruin infrastructure such as a dam or levee that could have helped prevent a hazard from turning into a disaster. Conflict is often destructive of the environment, again increasing vulnerability to hazards.[58]

In the other direction, several recent studies have found significant relationships between various forms of disasters and the risk of conflict and marginal support for certain types of disasters affecting political instability.[59] Nel and Righarts suggest that natural disasters help provide a motive (such as widespread suffering), incentive (competition for resources), and opportunity (weak state capacity) for conflict.[60] Gawronski and Olson argue that the 1976 earthquake in Guatemala served as a nation-changing critical juncture, intensifying community self-organizing and violent responses to it.[61] Beardsley and McQuinn argue that the tsunami in 2004 served as a catalyst to rebel groups in Aceh and Sri Lanka to examine their incentive structures, the changes to them based on the tsunami, and to adjust their strategies accordingly, the former to a peace agreement and the latter to continued fighting.[62] Since disasters can displace people and displacement can generate conflicts, disasters could increase conflict.[63]

In contrast, Slettebak and de Soysa, and Slettebak find that climate-related natural disasters may *decrease* the risk of conflict.[64] Bergholt and Lujala as well as Bernauer et al. study the relationship between climate-related disasters, economic shocks, and conflicts and find that while disasters may negatively affect economic growth, and thereby increase vulnerability, they do not raise the likelihood of conflict.[65] Ciccone strongly questions the analysis of Miguel et al. that concluded that reduced rainfall triggered conflict and argues that the available data do not support a link between rainfall anomalies and outbreaks

57. Lacina & Gleditsch, "Monitoring Trends in Global Combat."
58. Wisner et al. "*At Risk*," 27–8.
59. Brancati, "Political Aftershocks"; Nel and Righarts, "Natural Disasters and the Risk of Violent Civil Conflict."; Omelicheva, "Natural Disasters."
60. Nel and Righarts, "Natural Disasters and the Risk of Violent Civil Conflict."
61. Gawronski and Olsen, "Disasters as Crisis Triggers for Critical Junctures?"
62. Beardsely and McQuinn, "Rebel Groups as Predatory Organizations."
63. Buhaug, Gleditsch, and Theisen, "Implications of Climate Change for Armed Conflict."; Reuveny, "Ecomigration and Violent Conflict"; Stern, *The Economics of Climate Change*.
64. Slettebak and de Soysa "High Temps, High Tempers?"; Sletteback, "Don't Blame the Weather!"
65. Bergholt and Lujala, "Climate Related Natural Disasters."; Bernauer et al. "Climate Variability, Economic Growth, and Civil Conflict."

of conflict.[66] In general, scholars who study conflict itself are less persuaded by the importance of climate as a factor in outbreaks of conflict than those who study climate and its effects.[67]

A recent study by the Overseas Development Institute outlines the numerous potential pathways through which disasters can make people more vulnerable to conflicts and vice versa.[68] The authors cite seven examples where disasters increased the chances of conflicts and six where conflicts increased the chance of disasters, each example being from a poor country and all but two involving meteorological phenomena. They cite only three examples where disasters may have reduced the chances of conflict, most notably the end of hostilities in Aceh. In both disasters and conflicts, cause and consequence are not simply disentangled. The nominally exogenous factor—the environment—becomes more or less consequential due to the existing endogenous factors—political, social, economic, governance—while those same endogenous factors can impact the condition of the environment. More research is certainly called for, presenting an area of direct dialogue for conflict and disaster scholars.

11.2 CONCLUSION

We suggest that the boundary between conflicts as endogenously driven and disasters as exogenously triggered is porous and limiting. It may, in fact, have led to an artificial separation between those studying each phenomenon. We illustrated that not only consequences and responses to conflict and disaster are so similar as to warrant increased cross-disciplinary dialogue, but many of the causes are sufficiently overlapping and interacting to necessitate such conversations. Understanding and explaining conflicts and disasters as separate phenomena can be improved, in some ways, by talking about them in concert.

It is perhaps on a practical level, however, that dialogue between conflict and disaster scholars and practitioners can make the strongest contribution. Since social vulnerability and environmental pressures, including climate change, may be root causes of both conflicts and disasters, well thought-out efforts to address these common root causes should help mitigate both the phenomena. Our interviewees largely agreed that emergency response programming was currently much stronger than mitigation, yet spoke of their commitment of moving from emergency response to peacebuilding and of working both "before" and after disasters.

Prevention of an extreme event turning into a disaster or of a dispute turning into violent conflict is more cost-effective than responding to either

66. Ciccone, "Economic Shocks and Civil Conflict" strongly questions the analysis of Miguel et al., "Economic Shocks and Civil Conflict."
67. Gleditsch, "Whither the Weather?"
68. Harris et al. When Disasters and Conflicts Collide.

one. The UK's Department for International Development estimates that "for every dollar invested in disaster risk reduction between two and four dollars are returned in terms of avoided or reduced disaster impacts."[69] A similar argument is made for peacebuilding: It is cheaper to prevent conflicts than to respond to them.[70] Most importantly though, peacebuilding and DRR can both save lives and improve human security.

In acknowledging the crossover between conflicts and disasters, we could think about "DRR for peacebuilding" and "conflict prevention for DRR," leveraging interventions to get at "two birds with one stone." Since at least some of the underliers are the same, addressing them should help mitigate both conflicts and disasters. This is certainly an area where increased collaboration among disaster and conflict scholars and practitioners would be fruitful.

FURTHER READINGS

Albala-Bertrand, J.M., 2000. Complex emergencies versus natural disasters: an analytical comparison of causes and effects. Oxford Dev. Stud. 28 (2), 187–204.

Amnesty International, 2011. Haiti Aftershocks: Women Speak Out against Sexual Violence in Haiti's Camps. Amnesty International Publications, London.

Anderson, Mary B., 1999. Do No Harm: How Aid Can Support Peace or War. Lynne Riener, Boulder.

Antilla-Hughes, J.K., and Solomon Hsiang. "Destruction, Disinvestment, and Death: Economic and Human Losses Following Environmental Disaster". Working paper (Forthcoming). http://papers.ssrn.com/sol3/papers.cfm?abstract_id=2220501.

Barnett, Michael, Weiss, Thomas, 2008. Humanitarianism: a brief history of the present. In: Barnett, M., Weiss, T. (Eds.), Humanitarianism in Question: Politics, Power, Ethics. Cornell University Press, Ithaca, pp. 1–48.

Beardsley, Kyle, McQuinn, Brian, 2009. Rebel groups as predatory organizations: the political effects of the 2004 tsunami in Indonesia and Sri Lanka. J. Confl. Resolut. 53 (4), 624–645.

Bergholt, Drago, Lujala, Paivi, 2012. Climate-related natural disasters, economic growth, and armed civil conflict. J. Peace Res. 49 (1), 147–162.

Bernauer, Thomas, Koubi, Vally, Kalbhenn, Anna, Spilker, Gabrielle, 2012. Climate variability, economic growth, and civil conflict. J. Peace Res. 49 (1), 113–127.

Boutros-Ghali, Boutros, 1992. An Agenda for Peace: Preventative Diplomacy, Peacemaking and Peace-keeping. United Nations, New York.

Brainard, Lael, Jones, Abigail, Purvis, Nigel, 2009. Climate Change and Global Poverty: A Billion Lives in the Balance? The Brookings Institute, Washington, DC.

Brancati, Dawn, 2007. Political aftershocks: the impact of earthquakes on intrastate conflict. J. Confl. Resolut. 51 (5), 715–743.

Brown, Michael E., 1993. Ethnic Conflict and International Security. Princeton University Press, Princeton.

Brown, Oli, McLeman, Robert, 2009. A recurring anarchy? The emergence of climate change as a threat to international peace and security. Confl. Security Dev. 9 (3), 289–305.

69. UK Department for International Development (in IRC, *Internal Document.*).
70. Reychler, "Proactive Conflict Prevention."

Buhaug, Halvard, Petter Gleditsch, Nils, Theisen, Ole Magnus, 2010. Implications of climate change for armed conflict. In: Mearns, Robin, Norton, Andy (Eds.), Social Dimensions of Climate Change: Equity and Vulnerability. World Bank, Washington, DC, pp. 75–101.

Burd-Sharps, Sarah, Lewis, Kristen, Martins, Eduardo Borges, 2008. The Measure of America: American Human Development Report 2008–2009. SSRC/Columbia University Press, New York.

Burke, Marshall, Miguel, Edward, Satyanath, Shanker, Dykeme, John A., Lobell, David B., 2009. Warming increases risk of civil war in Africa. Proc. Natl. Aca. Sci. 106 (49), 20670–20674.

Buckland, Peter, 2005. Reshaping the Future. The World Bank, Washington DC.

Calhoun, Craig, 2010. The idea of emergency: humanitarian action and global (dis)Order. In: Fassin, D., Pandolfi, M. (Eds.), Contemporary States of Emergency: The Politics of Military and Humanitarian Interventions. Zone Books, New York, pp. 29–58.

Cawthorne, Andrew, Delva, Joseph Guyler, January 15, 2010. Quake destruction "like in a war": Haiti President. Reuters. http://mobile.reuters.com/article/topNews/idUSTRE60E60T20100115.

Center for Human Rights and Global Justice, 2011. Sexual Violence in Haitian IDP Camps: Results of a Household Survey, NYU School of Law. http://www.chrgj.org/press/docs/Haiti%20Sexual%20Violence%20March%202011.pdf.

Ciccone, Antonio, 2011. Economic shocks and civil conflict: a comment. Am. Econ. J. Appl. Econ. 3 (4), 215–227.

Clark, Peggy, Glickman, Dan, Monsma, David, Verstandig, Toni, Buckley, Nic, 2013. What is food security? The Aspen Idea 78–81. Summer.

Collier, Paul, 2000. Doing well out of war: an economic perspective. In: Berdal, Mats, Malone, David (Eds.), Greed and Grievance: Economic Agendas in Civil Wars. Lynne Rienner, Boulder, pp. 91–111.

Collier, Paul, 2007. The Bottom Billion: Why the Poorest Countries Are Failing and what Can Be Done about it? Oxford University Press, London.

Collier, Paul, Elliott, Lani, Hegre, Håvard, Hoeffler, Anke, Reynal-Querol, Marta, Sambanis, Nicholas, 2003. Breaking the Conflict Trap: Civil War and Development Policy. World Bank, Washington, D.C.

Corty, Jean-Francois, October 20, 2005. It reminds me of images from a war zone. Doctor's Without Borders. http://www.doctorswithoutborders.org/news/article.cfm?id=1613.

Cutter, Susan, 2006. Hazards, Vulnerability and Environmental Justice. Earthscan, London.

Cutter, Susan L., Finch, Christina, 2008. Temporal and spatial changes in social vulnerability to natural hazards. Proc. Natl. Aca. Sci. U.S.A. 105 (7), 2301–2306.

Dayton-Johnson, Jeff, 2004. Natural Disasters and Adaptive Capacity. OECD Development Center Working Paper No. 237. OECD, Paris.

Dell, Melissa, Jones, Benjamin F., Olken, Benjamin A., 2008. Climate Change and Economic Growth: Evidence from the Last Half Century. National Bureau of Economic Research, Cambridge MA.

De Silva, Amarasiri, 2009. Ethnicity, politics and inequality: post-tsunami humanitarian aid delivery in ampara district, Sri Lanka. Disasters 33 (2), 253–273.

De Waal, Alex, 1994. Famine Crimes: Politics and the Disaster Relief Industry. Indiana University Press, Bloomington.

Duffield, Mark, 2001. Global Governance and the New Wars: Merging of Development and Security. Zed Books, London.

Duffield, Mark, 1994. Complex emergencies and the crisis of developmentalism. IDS Bull. 25 (4), 1–15.

Field, C.B., Barros, V., Stocker, T.F., Qin, D., Dokken, D.J., Ebi, K.L., Mastrandrea, M.D., Mach, K.J., Plattner, G.K., Allen, S.K., Tignor, M., Midgley, P.M. (Eds.), 2012. Managing the Risks of Extreme Events and Disasters to Advance Climate Change Adaptation: Special Report of the Intergovernmental Panel on Climate Change. Cambridge University Press, Cambridge and New York.

Foreign Policy and the Fund for Peace. "The 2010 Failed States Index." http://www.foreignpolicy.com

Flint, Julie, Waal, Alex de, 2008. A Short History of a Long War. Zed Books, New York.

Gawronski, Vincent T., Olsen, Richard Stuart, 2013. Disasters as crisis triggers for critical junctures? the 1976 Guatemala case. Lat. Am. Polit. Soc. 55 (2), 133–149.

Ghobarath, Hazem, Huth, Paul, Russett, Bruce, 2003. Civil wars kill and maim people — long after shooting stops. Am. Polit. Sci. Rev. 97 (2), 189–202.

Gleditsch, Nils Petter, 2012. Whither the weather? Climate change and conflict. J. Peace Res. 49 (3), 3–9.

Harris, Katie, Keen, David, Mitchell, Tom, 2013. When Disasters and Conflicts Collide: Improving Links between Disaster Resilience and Conflict Prevention. Overseas Development Institute, Brighton.

Homer-Dixon, Thomas, 2001. Environment, Scarcity, and Violence. Princeton University Press, Princeton.

Hsiang, Solomon M., 2010. Temperatures and cyclones strongly associated with economic production in the Caribbean and Central America. Proc. Natl. Acad. Sci. 107 (35), 15367–15372.

Hsiang, Solomon M., Meng, Kyle C., Cane, Mark A., 2011. Civil conflicts are associated with the global climate. Nature 476 (7361), 438–441.

Humphreys, Macartan, Richards, Paul, 2005. Prospects and Opportunities for Achieving the MDGs in Post-Conflict Countries: A Case Study of Sierra Leone and Liberia. The Earth Institute and Center on Globalization and Sustainable Development, New York.

International Rescue Committee. Internal Document. New York, Undated.

Jackson, Richard, 2001. The state and internal conflict. Aust. J. Int. Affairs 55 (1), 65–81.

Jones, Benjamin F., Olken, Benjamin A., 2010. Climate Shocks and Exports. National Bureau of Economic Research, Cambridge MA.

Kaldor, Mary, 2006. New and Old Wars: Organised Violence in a Global Era, second ed. Polity Press, Cambridge.

King, Elisabeth, 2011. The multiple relationships between education and conflict: reflections of rwandan teachers and students. In: Mundy, Karen, Dryden-Peterson, Sarah (Eds.), Educating Children in Conflict Zones: A Tribute to Jackie Kirk. Teachers College Press, New York, pp. 137–151.

Klein, Alisa, 2008. Sexual Violence in Disasters: A Planning Guide for Prevention and Response. Center for Disease Control and Prevention and the US Department of Justice, Office on Violence Against Women, Washington, DC.

Klein, Richard J.T., Nicholls, Robert J., Thomalla, Frank, 2003. Resilience to natural hazards: how useful is this concept? Environ. Hazards 5 (1), 35–45.

Lacina, Bethany, Gleditsch, Nils Petter, 2005. Monitoring trends in global combat: a new dataset of battle deaths. Eur. J. Popul. 21, 145–166.

Miguel, Edward, Satyanath, Shanker, Sergenti, Ernest, 2004. Economic shocks and civil conflict: an instrumental variables approach. J. Polit. Econ. 112, 725–753.

Mutter, John C., 2012. Perceiving the social and economic consequences of natural disaster shocks: getting ready for climate change. In: Sclar, Elliott, Volavka-Close, Nicole,

Brown, Peter (Eds.), The Urban Transformation Health, Shelter and Climate Change. Routledge, London, pp. 173–196.

Nel, Philip, Righarts, Marjolein, 2008. Natural disasters and the risk of violent civil conflict. Int. Stud. Quart. 52, 159–185.

Nusrat, Rabia, Jamil, Baela Raza, September 11–13, 2007. Earthquake and education in emergencies: an opportunity and a Challenge. In: Paper Presented at the 9th UKFIET International Conference on Education and Development, Oxford.

O'Brien, Karen, Sygna, Linda, Leichenko, Robin, Neil Adger, W., Barnett, Jon, Mitchell, Tom, Schipper, Lisa, Tanner, Thomas, Vogel, Coleen, Mortreux, Colette, 2008. Disaster Risk Reduction, Climate Change Adaptation and Human Security. University of Oslo, Global Environmental Change and Human Security (GECHS), Oslo.

Omelicheva, Mariya Y., 2011. Natural disasters: triggers of political instability? Int. Interact. Empirical Theoretical Res. Int. Relations 37 (4), 441–465.

Orbinski, James, 2008. An Imperfect Offering: Humanitarian Action in the Twenty-first Century. Doubleday Canada, Toronto.

Parry, Martin, Canziani, Osvaldo, Palutikof, Jean, van der Linden, Paul, Hanson, Clair, 2007. Climate Change 2007: Impacts, Adaptation and Vulnerability. Contribution of Working Group 2 to the Fourth Assessment Report of the Intergovernmental Panel on Climate change. Cambridge University Press, Cambridge.

Pelling, Mark, 2003. The Vulnerability of Cities: Social Resilience and Natural Disaster. Earthscan, London.

Peluso, Nancy Lee, Watts, Michael (Eds.), 2001. Violent Environments. Cornell University Press, Ithaca.

Phelan, Kevin, 2008. From an idea to action: the evolution of Médecins sans Frontières. In: Stout, Chris E. (Ed.), The New Humanitarians: Inspiration, Innovations and Blueprints for Visionaries. Praeger, Westport CT, pp. 1–30.

Qureshi, Jawad Hussain, Bamforth, Thomas, 2007. Political complexities of humanitarian intervention in the Pakistan earthquake. J. Humanitarian Assis. http://sites.tufts.edu/jha/archives/12.

Raleigh, Clionadh, 2010. Political marginalization, climate change, and conflict in African Sahel states. Int. Studies Rev. 12 (1), 69–86.

Reuveny, Rafael A., 2008. Ecomigration and violent conflict: case studies and public policy implications. Hum. Ecol. 36, 1–13.

Reychler, Luc, 1998. Proactive conflict Prevention: impact assessment? Int. J. Peace Stud. 3 (2).

Salehyan, Idean, 2008. From climate change to conflict? No consensus yet. J. Peace Res. 45 (3), 315–326.

Sandholt-Jensen, Peter, Gleditsch, Kristian Skrede, 2009. Rain, growth and civil war: the importance of location. Defence Peace Econ. 20 (5), 359–372.

Schubert, Renate, Schellnhuber, Hans Joachim, Buchmann, Nina, Epiney, Astrid, Griebhammer, Rainer, Kulessa, Margareta E., Messner, Dirk, Rahmstorf, Stefan, Schmid, Jürgen, 2008. Climate Change as a Security Risk/German Advisory Council on Global Change. Earthscan, London.

Selth, Andrew, 2008. Even paranoids have their enemies. Contemporary Southeast Asia 30, 379–402.

Slettebak, Rune T., Soysa, Indra de, June 21–24, 2010. High temps, high tempers? Weather-related natural disasters & civil conflict. In: Draft Paper for the Conference on Climate Change and Security. Trondheim, Norway.

Slettebak, Rune T., 2012. Don't blame the weather! Climate-related natural disasters and civil conflict. J. Peace Res. 49 (1), 163–176.

Snyder, Mary Gail, 2005. It didn't begin with katrina, national housing Institute. Shelterforce 143. http://www.nhi.org/online/issues/143/beforekatrina.html.

Steinbruner John, D., Stern, Paul C., Husbands, Jo L., 2013. Climate and Social Stress: Implications for Security Analysis. National Academies Press, Washington DC.

Stern, Nicholas, 2006. The Economics of Climate Change: Stern Review. HM Treasury, London.

Stewart, Frances, July 9–10, 2007. Horizontal inequalities and conflict: understanding group violence in multiethnic societies. In: Conflict Prevention and Peaceful Development: Policies to Reduce Inequality and Exclusion, a CRISE Policy Conference, Oxford.

United Nations Environment Programme (UNEP), 2007. Sudan Post-Conflict Environmental Assessment. http://www.unep.org/sudan/post-conflict/.

United Nations Environment Programme (UNEP), 2008. Environmental and socioeconomic impacts of armed conflict in Africa. In: Cleveland, Cutler J. (Ed.), The Encyclopedia of Earth. Environmental Information Coalition, National Council for Science and the Environment, Washington, DC. Available at: http://www.eoearth.org/view/article/152607/.

United Nations High Commissioner for Refugees (UNHCR), 2011. Facts and Figures about Refugees from the 2011 Global Report. UNHCR, Geneva.

Weiss, Thomas G., 1999. Principles, politics, and humanitarian action. Ethics Int. Aff. 13 (1), 1–22.

Wisner, Ben, Blaikie, Piers, Cannon, Terry, Davis, Ian, 2003. At Risk: Natural Hazards, People's Vulnerability and Disasters, second ed. Routledge, London.

Woodward, Susan L., 2002. Economic priorities for successful peace implementation. In: Stedman, Steven, Rothchild, Donald, Cousens, Elizabeth M. (Eds.), Ending Civil Wars: The Implementation of Peace Agreements. Lynne Rienner, Boulder, pp. 183–214.

World Bank, 2011. World Development Report 2011: Conflict, Security and Development. World Bank, Washington DC.

Yonetani, Michelle, 2013. Global Estimates 2012: People Displaced by Disasters. Internal Displacement Monitoring Centre and Norwegian Refugee Council, Oslo.

Chapter 12

Everyday Practices and Symbolic Forms of Resistance: Adapting to Environmental Change in Coastal Louisiana

Julie Koppel Maldonado
American University, Washington DC, USA

ABSTRACT

Environmental and technological disasters, extractive industries, river mismanagement, and climate change are drastically transforming coastal Louisiana's sea- and landscape. The environmental degradation and related social, cultural, health, and economic impacts are particularly significant for Louisiana's tribal communities that have called the region home for centuries. Despite being faced with an uncertain future of whether the diminishing land they live on will stay above the rising tide, the tribes still persist and continue to adapt to a changing environment and a transforming subsistence-based livelihood. This chapter focuses on the everyday strategies the tribes and residents employ to be included in hurricane protection systems, maintain their communities and culture, be included as citizens and acknowledged as Natives. As communities around the world face increasing impacts and challenges from global environmental change and dominant political, economic, and social structures and ideologies, it is important to look not just at the forms of resistance that make headlines but also the everyday forms. It is through both the everyday practices and locally formed partnerships that a democratization of the adaptation process could emerge, one that includes decision-making based on more equitable distribution of knowledge sharing and integration, public participation, local input, and science and multiple perspectives.

Living here is a commitment. You have to do it in spite of the challenges of storms, flooding, distance to everything. But the good outweighs the bad. When there are no storms, no flooding, raising the kids on the Island, that is the good. It's not everywhere that you can be outside your house, have a nice breeze, stay outside in comfort and safety knowing everyone around you. We have got to slow the rising tide. What it was, what it is, that's what keeps me here.

Tribal Member, Isle de Jean Charles Band of Biloxi-Chitimacha-Choctaw Indians, Louisiana. 2012

Hazards, Risks, and Disasters in Society. http://dx.doi.org/10.1016/B978-0-12-396451-9.00012-3

Environmental and technological disasters, extractive industries, river mismanagement, and climate change are drastically transforming coastal Louisiana's sea- and landscape. The environmental degradation and related social, cultural, health, and economic impacts are particularly significant for Louisiana's tribal communities.[1] The Isle de Jean Charles Band and Grand Caillou/Dulac Band of Biloxi-Chitimacha-Choctaw Indians and the Pointe-au-Chien Indian Tribe, located approximately 80 miles southwest of New Orleans in coastal Louisiana, are historically subsistence-based fishers, trappers, farmers, and hunters who have experienced a long history of displacement, discrimination, and environmental degradation. Oil and river management development projects, oil disasters, increased exposure to hurricanes, and relative sea level rise threaten the tribes' culture and water-based settlements and livelihoods, severely degrading their surrounding landscape, what many of them describe as their freedom, their paradise.

Despite being faced with an uncertain future of whether the diminishing land they live on will stay above the rising tide, many still persist and continue to adapt to a changing environment and a transforming subsistence-based livelihood. Saltwater intrusion, land loss, expansion of large-scale commercial fisheries and import industry, and oil spills and chemicals from dispersants have depleted and contaminated their once abundant seafood catch, trapping grounds and gardens.

This chapter focuses on the everyday strategies the tribes and residents employ to be included in hurricane protection systems, maintain their communities and culture, be included as citizens and acknowledged as Natives. Their strategies are not the way resistance is typically perceived, through forms like protesting. Much like the continued flow of water up and down the bayous they live along, their resistance is subtle, quiet, but persistent. It is not about seeing themselves as activists or resisters, but more about daily actions and interactions.

Not to over-romanticize the type of resistance that Scott described as "weapons of the weak" (1985), people from the three communities employed certain everyday practices and locally organized forms of symbolic resistance, such as the ideological struggle against the public discourse that they no longer exist as Native peoples. Despite experiencing oppression since French colonialists displaced their ancestors in the 1700s, and thus far being denied federal recognition, they continue to adapt to environmental change and resist against state and private industry actors discounting them in development and restoration plans.

While their agency is constrained by structural violence (Farmer, 2004), people from the communities often exercised their agency through examples

1. While this chapter focuses on three tribal communities in coastal Louisiana, there are many other communities in the region that are also facing similar impacts and to which the issues and arguments in this chapter could also be applied.

of adaptation. The communities' leaders approached adaptation as a process, not an outcome, one in which adaptation is connected to social agency, power relations, and issues of environmental justice (Peets and Watts, 1996). When used in a social scientific way, adaptation "refers mainly to changes in belief and/or behavior in response to altered circumstances to improve the conditions of life (or survival)" (Oliver-Smith, 2009, p. 12). Adapting to environmental change involves not just the natural stresses, but also the connection between natural stresses and other sources of system stress (Nelson et al., 2009, p. 272).

To fully understand the process of adaptation to change, it is important to consider the ways that culture is linked to power, which enables us to see the social flows of patterns within complex and linked elements and creates understanding of how humans adapt to change (Wolf, 1999). It is also significant to consider the intersection between particular structural forces and agents of power, as well as the sociohistorical context under which oppressed groups and individuals in society came to inhabit specific environments. This chapter highlights the proactive measures individuals took to adapt to environmental change, and, for some tribes, steps toward community-led relocation.

12.1 METHODOLOGY

The quotes and observations throughout this chapter come from anthropological dissertation fieldwork I conducted with the three communities between October 2011 and September 2012. I utilized a variety of methods, including participant observation; intentional conversations, in which I wove specific topics into longer one-on-one conversations; digital storytelling (short narratives created in video format that bring together participants' words, voices, and pictures); and story circles. Fifty-six people participated in intentional conversations and/or story circles and five people created complete digital stories. These methods helped me better understand the tribes' history of environmental degradation and oppression and drew out people's reflections in nuanced ways. The methodological approach informed my understanding of the social, political, economic, and ecological processes that have taken place to form, shape, and reshape the sea- and landscape the tribes have called home for centuries.

12.2 LAYERED DISASTERS

Coastal Louisiana is an elaborate network of diverse habitats and landforms, including wetlands, barrier islands, natural levees, land ridges, and marshes connected together in an ecosystem formed over thousands of years from deposits of sediment, soil, and clay from the Mississippi River (Barry, 1997; Coastal Louisiana Ecosystem Assessment and Restoration/CLEAR, 2006). However, that ecosystem has been severely degraded over the past century. Since 1932, 1,880 square miles of land has been lost from Louisiana's coast

(Couvillion et al., 2011; CPRA, 2012; NOAA, 2013). While the organic sediments deposited along the coast undergo a natural process of compacting, consolidation, and oxidation, which results in subsidence, other compounding factors have greatly contributed to the intense amount of land loss.

This picture demonstrates how fast the land is eroding away. The dead tree, from saltwater intrusion, has land in front of it, October 2011. *Source: Author.*

Dredging thousands of miles of marsh and wetlands along Louisiana's coast for pipelines caused considerable land loss and erosion and extensive saltwater intrusion. The construction of dikes and levees and damming of the Mississippi River by the Corps of Engineers and local Levee Districts, other flood control measures and development of the Mississippi River Gulf Outlet prevent sediment and silt from reaching the delta (Button and Peterson, 2009; Freudenberg et al., 2009; Laska et al., 2005). These processes have resulted in extreme loss of wetlands and reduced natural protection against hurricanes and storms. Since 2005, the three tribal communities have survived the impacts of six major storms. The dredged canals become substantially wider as the current and hurricanes pull more marshland away. With increased impacts from hurricanes and storms, rates of coastal land loss have accelerated in recent years (Couvillion et al., 2011). Whereas before they rarely experienced flooding from hurricanes, now flooding can occur with just high tide.

Climate change-induced sea level rise compounds the effects of erosion and subsidence (Burkett and Davidson, 2012). Southeast coastal Louisiana has experienced slightly faster than twice the global rate of relative sea level rise, with over an 8-inch rise in the last 50 years (Karl et al., 2009; NCADAC, 2013; NOAA, 2012). Furthermore, runoff containing highly contaminated nitrogen and synthetic chemicals from large-scale agricultural development flows down the Mississippi River into the Gulf of Mexico, creating the largest hypoxic "dead zone" in US coastal waters and negatively impacting fishing resources (CLEAR, 2006; CPRA, 2012; Morris, 2012; Rabalais et al., 1999).

Same tree with the land now eroded away behind it, July 2012. Local residents repeatedly requested the Parish Council to put rocks in front to protect the land from erosion, but no action was taken by the Council. *Source: Author.*

12.3 THE IMPACTS OF RAPID ENVIRONMENTAL CHANGE

With severe erosion and intense saltwater intrusion, the majority of the communities' trees are dead and their traditional and medicinal plants, gardens, and trapping grounds, along with the animals, are gone. Encroaching toxic industries, chemicals from dispersants, oil spills, and poststorm debris contaminate the air and the bayous, creating severe health and livelihood impacts (Button and Peterson, 2009; Louisiana Workshop, 2012; Maldonado et al., 2013).

The land upon which they planted gardens that provided a self-sustaining way of life is now either underwater, or the soil is so saturated with saltwater that it is too difficult for plants to grow. There is no more trapping nearby, as the trapping grounds for muskrats and possums are no longer there.

The British Petroleum Deepwater Horizon Oil disaster greatly exacerbated the impacts on an already declining fishing industry, including shrimp, crabs, and oysters. State regulations put in place for the large-scale commercial fishing industry and to protect oil interests have also contributed to local people's economic and subsistence displacement (Harrison, 2012). For example, instead of being able to freely harvest oysters, they now have to lease oyster beds from the oil companies, major land developers, and US Fish and Wildlife Service. When they ate from the land and waters, they shared with each other. But now they have the economic burden of having to buy more food and the sense of sharing has diminished.

Some community members still maintain a subsistence livelihood. Crab traps, Pointe-au-Chien, August 2012. *Source: Author.*

The communities' have faced severe subsistence and livelihood impacts. Isle de Jean Charles, April 2012. *Source: Author.*

Losing their subsistence-based livelihoods, many tribal members now work on oil rigs or tugboats, as that is the only work available, causing them to be gone from their families for long periods of time. Or they had to move to find work in other places, which also breaks apart families and changes community dynamics. However, some people still maintain a subsistence livelihood and depend on their catch to feed their families.

A few shrimping boats remain in Grand/Caillou Dulac, July 2012. *Source: Author.*

For generations, residents have navigated and embedded their way of life in coastal Louisiana's sea- and landscape. Thus, the land loss and environmental changes are not just physical but impact peoples' sense of place, belonging, cultural identity, and local knowledge. For example, an elder from Isle de Jean Charles described where he used to travel by pirogue (small, dugout boats), explaining that "(n)ow it's hard to distinguish where those places are." As opposed to the environmental determinism within which such issues are often framed in public discourse, these problems are not isolated occurrences. Rather, the environmental issues and the so-called "natural" disasters are embedded within forms of structural power and shaped by the social, political, and economic system that increases people's vulnerability to such changes through the everyday social order and systems of domination and inequality (Hewitt, 1983; Hilhorst and Bankoff, 2004; Oliver-Smith, 2004; Quarantelli, 1998; Wisner, 2004; Wolf, 1999).

12.4 THE SOCIAL, POLITICAL, AND ECONOMIC CONTEXT OF ENVIRONMENTAL CHANGE

The state of Louisiana has historically asserted that land loss is due to sediment deprivation (Turner, 1997). However, the disappearance of Louisiana's coastal wetlands has been concurrent with the ongoing extensive oil and gas development and extraction activities. Louisiana is the country's top crude oil producer and the second largest natural gas producer when including the Outer Continental Shelf (US EIA, 2009). Since the 1930s, the oil and gas companies, with the backing of the state, have cut through Louisiana's coastal marshes and left the once fertile lands barren from saltwater intrusion (Austin, 2006).

Emphasizing the protection of oil interests and private property, along with the drive for industrialization and economic gain, the long-standing partnership between the oil industry and the state helped turn coastal Louisiana into an energy sacrifice zone (Colten, 2012).

A sacrifice zone is "a place where human lives are valued less than the natural resources that can be extracted from the region" (Buckley and Allen, 2011, p. 171). Neoliberal policies focused on free trade and privatization, that guide our current economic, political, and social systems, encourage the creation of sacrifice zones, and have resulted in "frontier dispossession and reckless accumulation" (Watts, 2012, p. 458; also Harvey, 2005; McNeil, 2012).

During the first half of the twentieth century, representatives from the oil companies made residents sign leases they could not read or understand to lease or sell their land for oil extraction. Similar practices occur today through the process of, and policies related to, land loss. For example, once land disappears under water, the state takes it over and the submerged land can be leased or sold to private corporations. The tribes are blocked from restoring the remaining land, such as their sacred, ancestral mounds, because private corporations now control the water surrounding the land. Thus, sacrifice zones such as coastal Louisiana have become "the new geographies of domination" (Reid and Taylor, 2010, p. 11), within which local populations become further marginalized. One form of domination is seen, for example, in the political decision-making process that bases restoration decisions on cost–benefit analyses (CBA).

12.5 THE POLITICS OF COASTAL RESTORATION

Thus far, the communities have been mostly left out of cost–benefit-based coastal restoration efforts, including the US Army Corps of Engineers', Morganza to the Gulf of Mexico Hurricane Protection Project, and are given minimal attention in Louisiana's 50-year Master Plan for a Sustainable Coast. Without restoration and flood protection measures, all of the land in and around the tribal communities is predicted to be gone by 2050 (CPRA, 2012). And the land could be gone even sooner, as recent data from the National Oceanic and Atmospheric Administration predict the region will experience the highest rate of relative sea level rise worldwide, with an additional 4.3 feet of water rising by the end of this century (Marshall, 2013).

While the Gulf Coast Ecosystem Restoration Council notes the importance of including tribal input in restoration activities (Gulf Coast Ecosystem Restoration Council, 2013, p. 1), the plan to restore the Gulf Coast acknowledges affected tribes as those that are federally recognized (White House, 2010). This excludes the three tribes because they have thus far been denied tribal federal recognition, despite the Bureau of Indian

Affairs recognizing their Indian ancestry (Bureau of Indian Affairs, 2008). Although the tribes were officially recognized as Indian tribes by the state of Louisiana in June 2004, without federal recognition they lack government support offered to federally recognized tribes.

When French settlers pushed the tribes' ancestors out of their homelands in the 1700s, their ancestors escaped to the dense forested swamps down the bayous to survive persecution. Historical injustices, such as being forced into isolation and lack of formal treaties with colonial settlers, made it extremely difficult for the tribes to prove the seven criteria required for recognition (Miller, 2004).[2] The tribes are once again being excluded through political decisions made based on CBA.

Furthermore, the CBA used to make coastal restoration decisions does not account for the distribution of costs and benefits or important social and cultural factors, such as people's identity, beliefs, traditions, livelihoods, and sacred places (Cernea, 2000, 2008; Mayo, 2010; Oliver-Smith, 2010). By basing restoration activities on CBA, at "the root of these discussions lie differentials in power: power to tell the story of the future and then to enact it" (Westman, 2013, p. 112). As the tribal members' livelihoods, cultural identity, and sense of place are at risk, they carry out various forms of resistance and adaptation.

12.6 RESISTANCE AND ADAPTATION

The government doesn't do anything to save our land and neither do the oil companies who are responsible for digging the damaging canals many years ago. These companies reap the rewards while we are left to sink into the Gulf of Mexico... They say we can't be saved. They say that we have to relocate to preserve life. Once again they have proven us to be expendable in their eyes. But we have proven that we never back down from the impossible. We will continue to fight for our way of life, for ours is unlike any other. We will fight until the last tree has died, until the last bit of land has washed away. We will always fight for our 'Home'.

Tribal member, Grand Caillou/Dulac Band of Biloxi-Chitimacha-Choctaw Indians, Louisiana, 2012

The communities' experiences are more than about the current political and economic structures, but are part of a multicentury story of resisting destruction of culture and identity. While they resist against tangible objects like oil pipelines, widening channels, land loss, occupation of their homelands, and placements of storm protection systems, the forces being resisted run much deeper. It is a resistance against continued colonial practices and

2. For a listing of the seven mandatory criteria, see Bureau of Indian Affairs (2013). Procedures for Establishing that an American Indian Group Exists as an Indian Tribe. http://www.bia.gov/cs/groups/public/documents/text/idc-001115.pdf.

neoliberal policies that threaten their livelihoods and very existence. It is both a conscious and unconscious resistance.

One of the major challenges they face is the very acknowledgment of their presence in the present. For example, in a meeting in New Orleans, I listened to Senator Landrieu discuss the importance of restoring the coast and stated how Native Americans had been there many centuries before. One of the local residents sitting next to me whispered that there is no recognition they are still here. Thus, their struggle is not just over property rights and livelihoods, but is also "a struggle over the appropriation of symbols, a struggle over how the past and present shall be understood and labeled, a struggle to identify causes and assess blame, a contentious effort to give partisan meaning to local history" (Scott, 1985, p. xvii).

12.6.1 Symbolic Forms of Resistance

People from the tribal communities struggle against the public discourse that they no longer exist as Native peoples, as discussed in the example above. They attend public forums and meetings to show that there are still Native communities living down the bayous. For example, at a public meeting for the draft Master Plan, one resident stood up and said, "How do we know that our comments will be considered for real and not just to fit into your guidelines?" Standing there in front of the panel, in the middle of the crowded auditorium, she pushed back her long black hair, revealing the words written across her t-shirt, "Sure you can trust the government, just ask an Indian."

They also resist having their history erased by seeking federal recognition. As one leader explained, "We are trying to get federally recognized so we can maintain our community, our elders can be supported and we can have our own schools back and our own education for our kids. I would give my life to see us federally recognized to show who we really are." The leaders of the communities are working to obtain federal recognition and maintain their culture through the very recognition process itself, such as storytelling by elders to preserve the tribes' history. In being recognized, they are working to reclaim their ancestral lands that were taken by the government and private industry. Continuing to pursue federal recognition for the past 20 years, they resist a network of intersecting dominant ideologies about belonging that constrains, dominates, and denies them cultural citizenship (Ong, 2004).

Community members are also actively reinvigorating their cultural traditions, such as creating a drumming group, naming ceremonies, building a traditional palmetto hut, and preserving their food and community traditions. Cultural reinvigoration is not only happening among coastal Louisiana's tribal communities, but also within other populations with long historical ties to the region, such as the Cajuns and Isleños (Canary Islanders).

Palmetto hut built by community members, Pointe-au-Chien, August 2012. *Source: Author.*

Some tribal members who have relocated continue to claim the land they once lived on. For example, in the annual Christmas parade for one of the communities, as the parade goes down the road and Santa passes out the donated gifts, many families stand where their houses used to be, now empty spaces grown over with weeds, remnants of what recent hurricanes left behind.

As their identity is intimately tied to place, the drive to reclaim the land is interwoven in both present actions and memories of what their ancestors did to protect their land that fed them, tied them together, and served as a safe haven. For example, one community member's grandfather was given a $500 check after his land was taken from him, but he never cashed the check because he had not agreed to sell the land. While oil companies continue to take over their waters, tribal members get together to lease some of the waters back so their people can continue to access the water for fishing.

Sign on Isle de Jean Charles, July 2010. *Source: Author.*

Talking about reclaiming land, one resident noted, "If we get back land, it'll be for everybody. Hopefully one day it will happen. This land won't be there long, it's washing away quick." This comment and the earlier excerpt from the digital story evoke the notion of fighting to take back land, while at the same time recognizing the land that is left is disappearing. So what keeps them resisting with so much uncertainty of whether the land will remain above the rising tide? Their intention, much like their ancestors, is survival and persistence (Scott, 1985). Their ancestors resisted genocide by going to the ends of the bayous. However, it is about more than just survival; it is about their cultural perseverance and identity that is deeply tied to the land. As one resident said, "this land has fed our people, it may not be much but it's ours."

12.6.2 Adaptation Strategies

The tribes also employ many adaptation strategies, such as emphasizing the importance of traditional, local knowledge. Their multigenerational knowledge of the surrounding waterways, habitats, and landforms enables them to see the changes happening and what needs to be done to mitigate and adapt to the impacts. As one local fishermen explained, "I say we've been here all our lives. We know how the water works."

They adapt practices to maintain some of their subsistence and traditions, like still planting despite the saltwater intrusion. For example, as one person said, "I'm going to be planting as long as I live." She said this even though she acknowledged a moment later that nothing is being done to protect her community and authorities will not listen to them. Despite the uncertainty, a few days later we chatted together as she planted in a makeshift raised-bed garden she created in an old toilet bowl she had cleaned out and was using to keep her plants out of the encroaching saltwater.

Other strategies for adapting to increased flooding and impacts from hurricanes and storms include elevating their houses above the floodwaters. This has its consequences however, as people, especially elderly, have injured themselves going down the steps and the houses are more vulnerable to the wind during storms. Yet, many people have been able to stay in-place because they raised their houses.

Some people adapt to the land loss and erosion by putting oyster shells, rocks, and dirt around their houses and trying to rebuild important places, such as ceremonial mounds. As one resident explained, they want to save the mounds because "If we let that wash away, your heritage wash away, so we try to stop some of it." But they have run into obstacles with permitting and land ownership. Yet, some of them still continue to press local authorities to take action to protect the remaining land.

A house elevated above the floodwaters from Hurricane Isaac. Isle de Jean Charles, September 2012. *Source: Author.*

Piles of dirt and oyster shells to help protect the cemetery on Isle de Jean Charles from flooding during Hurricane Isaac, September 2012. *Source: Author.*

12.6.3 "Accidental Activists"

While community members do not directly confront authority in a coordinated manner (e.g., organized protests), several of them have become "accidental activists." For example, one community elder from Isle de Jean Charles sent letters to local parish representatives about needing restoration efforts. And while the community is still cutout of the Morganza Hurricane Protection Project, the local parish built a small ring levee that has somewhat helped mitigate flooding during high tide. The tribal leaders interact with organizations and agencies from the local to the international level, such as the United Nations, US Environmental Protection Agency, local levee districts, and numerous organizations along the Gulf Coast. Some other community members continue to speak out to save their land, as one Island resident described, to "keep us on the map," while others maintain silent demonstrations, like putting up signs to announce that they are not going anywhere.

Many community members continue to practice everyday forms of resistance, like the night before a hurricane having a crab boil together and through the preservation of their food and community traditions. However, some people also create formal organizations. For example, through the US Department of Agriculture's Natural Resources Conservation Service, the three tribes, along with Grand Bayou, another nearby tribal community, have created the First People's Conservation Council to address natural resource issues occurring on the tribal and coastal lands. The organization was formed through the tribal leaders and the support of the Lowlander Center and the Wisconsin Tribal Conservation Advisory Council.

More informally, but still coordinated, through the work of researchers at the Lowlander Center and a nearby religious congregation, the tribes hold events with community representatives from Alaska and Minnesota, sharing knowledge and cultural traditions, such as basket-making, ethnobotany, and experiences of extractive industries impacting their communities. Reflecting on her time with the communities, a Native leader from Minnesota explained, "Through everything these communities have been through, they're not whiny, they're looking to solve their problems, they're looking to have people not bail them out but just to help them, to understand who they are."

While many community members continue to stay in-place, some of the tribes are facing the need to relocate, as their homelands are sinking into the Gulf of Mexico and they are no longer protected from the increasing intensity and impacts of more frequent hurricanes and storms.

12.7 "RESTORATION" INSTEAD OF "RELOCATION"

Some people from the three communities have already relocated to higher ground further inland because of increased flooding and hurricane impacts and loss of subsistence-based livelihoods. While the Pointe-au-Chien tribe, with some storm protection and land left does not yet need to consider relocation, Grand Caillou/Dulac is starting to face the need to consider community-led relocation plans. Most immediate though, leaders from Isle de Jean Charles have been looking to relocate the community, which has seen the most drastic changes of the three communities. For example, Isle de Jean Charles had 78 houses in 2002; 10 years later in 2012 there were about 25 houses left. People from the Island described it being about 5 miles wide and 12 miles long in the 1950s; by 2012 it was approximately one-quarter mile wide and 2 miles long.

To bring their people back together, the leaders of Isle de Jean Charles are working to relocate the community to higher ground, while still working to save their ancestral land. The Tribal Council's relocation plan includes maintaining community and cultural integrity and promoting traditional livelihood, economic development, and sustainability. However, the Tribal Council has run into countless regulatory challenges, policy obstacles, and practical challenges. For example, there are no funds for the lengthy and sensitive process of moving an entire community and there is no federal government agency delegated to support communities' proactive relocation efforts (Bronen, 2011; Maldonado et al., 2013; NCADAC, 2013). Despite the need to relocate, the Council frames their efforts in a positive context. As one leader explained, he prefers to say "restoration" instead of "relocation."

12.8 CONCLUSION

Through their adaptive strategies, the tribal members engage in everyday and symbolic forms of resistance against the visible, hidden, and invisible power structures, influences, and social relations involved in creating justice in adaptation. They resist the visible power observed in the processes and procedures of decision-making, such as by speaking out at public forums, advocating being included in restoration activities, and pursuing their own restoration efforts, despite obstacles faced. They resist the hidden power of those who frame what is included in public discourse, such as by talking about the environmental changes being caused by oil and gas corporations, as opposed to the political discourse promoting the change in waterways as the sole cause of land loss. And they resist the invisible power found within the ideological foundations of societal structures by demonstrating that they are still here. Their resistance can be both self-indulgent and revolutionary, with the intent for survival and immediate and long-term gains. However, their resistance takes place within a specific social, political, and economic context. As Marx stated, "Men make their own history, but they do not make it just as they please" (Marx, 1994/1888, p. 188). Thus, the tribes both resist the dominant system and find ways to survive within it.

As communities around the world face increasing impacts and challenges from global environmental change and dominant political, economic, and social structures and ideologies, it is important to look not just at the forms of resistance that make headlines but also the everyday forms that are "the stubborn bedrock upon which other forms of resistance may grow, and they are likely to persist after such other forms have failed or produced, in turn, a new pattern of inequity" (Scott, 1985, p. 273). It is through both the everyday practices and locally formed partnerships that a democratization of the adaptation process could emerge, one that includes decision-making based on more equitable knowledge sharing, public participation, local input, and inclusion of diverse knowledges. Such an adaptation process could help create greater justice in an era of rapid global environmental change.

REFERENCES

Austin, D., 2006. Cultural exploitation, land loss and hurricanes: a recipe for disaster. Am. Anthropol. 108 (4), 671–691.

Barry, J.M., 1997. Rising Tide: The Great Mississippi Flood of 1927 and How It Changed America. Simon and Schuster Paperbacks, New York.

Bronen, R., 2011. Climate-induced community relocations: creating an adaptive governance framework based in human rights doctrine. N.Y. Univ. Rev. Law Soc. Change 35, 356–406.

Buckley, G.L., Allen, L., 2011. Stories about mountaintop removal in the Appalachian coalfields. In: Morrone, M., Buckley, G.L. (Eds.), Mountains of Injustice: Social and Environmental Justice in Appalachia. Ohio University Press, Athens, OH, pp. 161–180.

Bureau of Indian Affairs, 2013. Procedures for Establishing that an American Indian Group Exists as an Indian Tribe (accessed 26.02.14.). http://www.bia.gov/cs/groups/public/documents/text/idc-001115.pdf.

Bureau of Indian Affairs, 2008. Summary under the Criteria and Evidence for Amended Proposed Finding against Federal Acknowledgment of the Biloxi. Chitimacha Confederation of Muskogees, Inc., Washington, DC.

Burkett, V., Davidson, M. (Eds.), 2012. Coastal Impacts, Adaptation and Vulnerability: A Technical Input to the 2012 National Climate Assessment. Cooperative Report to the 2013 National Climate Assessment.

Button, G.V., Peterson, K., 2009. Participatory action research: community partnership with social and physical scientists. In: Crate, S.A., Nuttal, M. (Eds.), Anthropology and Climate Change: From Encounters to Actions. Left Coast Press, Walnut Creek, pp. 327–340.

Cernea, M.M., 2008. The Impoverishment Risks and Reconstruction Model: Resettlement and Benefit-Sharing. Asian Development Bank, Manila.

Cernea, M.M., 2000. Risks, safeguards and reconstruction: a model for population displacement and resettlement. Econ. Political Wkly. 35 (41), 3659–3678.

Coastal Louisiana Ecosystem Assessment and Restoration (CLEAR), 2006. Reducing Flood Damage in Coastal Louisiana: Communities, Culture and Commerce (accessed 27.11.13.). http://www.clear.lsu.edu.

Coastal Protection and Restoration Authority of Louisiana (CPRA), 2012. Louisiana's Comprehensive Master Plan for a Sustainable Coast. Baton Rouge, LA. (accessed 02.12.13.). http://www.lacpra.org/assets/docs/2012%20Master%20Plan/Final%20Plan/2012%20Coastal%20Master%20Plan.pdf.

Colten, C.E., June 2012. An incomplete solution: oil and water in Louisiana. J. Am. Hist. 91–99. http://dx.doi.org/10.1093/jahist/jas023.

Couvillion, B.A., Barras, J.A., Steyer, G.D., Sleavin, W., Fischer, M., Beck, H., Trahan, N., Griffin, B., Heckman, D., 2011. Land Area Change in Coastal Louisiana from 1932 to 2010. US Geological Survey Scientific Investigations Map. 3164.

Farmer, P., 2004. An anthropology of structural violence. Curr. Anthropol. 45 (3), 305–325.

Freudenberg, W.R., Gramling, R., Laska, S., Erikson, K.T., 2009. Catastrophe in the Making: The Engineering of Katrina and the Disasters of Tomorrow. Island Press, Washington DC.

Gulf Coast Ecosystem Restoration Council, 2013. Draft Initial Comprehensive Plan: Restoring the Gulf Coast's Ecosystem and Economy (accessed 19.11.13.). http://www.restorethegulf.gov/sites/default/files/Gulf%20Restoration%20Council%20Draft%20Initial%20Comprehensive%20Plan%205.23.15.pdf.

Harrison, J.A., 2012. Buoyancy on the Bayou: Shrimpers Face the Rising Tide of Globalization. Cornell University Press, Ithaca, NY.

Harvey, D., 2005. A Brief History of Neoliberalism. Oxford University Press, Oxford.
Hewitt, K., 1983. Interpretations of Calamity. Allen and Unwin Inc., Winchester, MA.
Hilhorst, D., Bankoff, G., 2004. Introduction: mapping vulnerability. In: Bankoff, G., Franks, G., Hilhorst, D. (Eds.), Mapping Vulnerability: Disasters, Development and People. Earthscan, London, pp. 1–9.
Karl, T.R., Melillo, J.M., Peterson, T.C. (Eds.), 2009. Global Climate Change Impacts in the United States: A State of Knowledge Report from the US Global Change Research Program. Washington, DC.
Laska, S., Wooddell, G., Hagelman, R., Grambling, R., Teets-Farris, M., 2005. At risk: the human, community, and infrastructure resources of Coastal Louisiana. J. Coastal Res. 44, 90–111.
Louisiana Workshop, 2012. Stories of change: coastal Louisiana tribal communities' experiences of a transforming environment. Input to the National Climate Assessment. In: Maldonado, J.K. (Ed.), Participating Tribes: Grand Bayou Village, Grand Caillou/Dulac Band of the Biloxi-Chitimacha Confederation of Muskogees, Isle de Jean Charles Band of the Biloxi-Chitimacha Confederation of Muskogees, Pointe-au-Chien Indian Tribe.
Maldonado, J.K., Shearer, C., Bronen, R., Peterson, K., Lazrus, H., 2013. The impact of climate change on tribal communities in the US: displacement, relocation, and human rights. Clim. Change 120 (3), 601–614. http://dx.doi.org/10.1007/s10584-013-0746-z.
Marshall, B., 2013. New Research: Louisiana Coast Faces Highest Rate of Sea-Level Rise Worldwide. The Lens. February 21.
Marx, K., 1994(1888). The communist manifesto; preface to a contribution to the critique of political economy; capital, volume one. In: Simon, L.H. (Ed.), Selected Writings. Hackett Publishing, Indianapolis, IN, pp. 209–300.
Mayo III, R.J., 2010. Democracy and Environmental Decision-Making: A Review and Analysis of the Theoretical and Empirical Relationships. Master's Thesis. State University of New York, Syracuse, USA.
McNeil, B.T., 2012. Combating Mountaintop Removal: New Directions in the Fight against Big Coal. University of Illinois Press, Urbana, Illinois.
Miller, M., 2004. Forgotten Tribes: Unrecognized Indians and the Federal Acknowledgment Process. University of Nebraska Press, Lincoln.
Morris, C., 2012. The Big Muddy: An Environmental History of the Mississippi and Its Peoples from Hernando de Soto to Hurricane Katrina. Oxford University Press, New York.
NCADAC (National Climate Assessment and Development Advisory Committee), 2013. Draft National Climate Assessment (accessed 02.12.13.). http://ncadac.globalchange.gov.
Nelson, D.R., West, C.T., Finan, T.J., 2009. Introduction to "In focus: global change and adaptation in local places". Am. Anthropol. 111 (3), 271–274.
NOAA (National Oceanic and Atmospheric Administration), 2013. Underwater: Land Loss in Coastal Louisiana since 1932 (accessed 22.11.13.). http://www.climate.gov/news-features/featured-images/underwater-land-loss-coastal-louisiana-1932.
NOAA (National Oceanic and Atmospheric Administration), 2012. Global Sea Level Rise Scenarios for the United States National Climate Assessment. NOAA Technical Report OAR CPO-1.
Oliver-Smith, A., 2010. Defying Displacement: Grassroots Resistance and the Critique of Development. University of Texas Press, Austin.
Oliver-Smith, A., 2009. Sea Level Rise and the Vulnerability of Coastal Peoples: Responding to the Local Challenges of Global Climate Change in the 21st Century. UNU Institute for Environment and Human Security, Bonn, Germany.

Oliver-Smith, A., 2004. Theorizing vulnerability in a globalized world: a political ecological perspective. In: Bankoff, G., Franks, G., Hilhorst, D. (Eds.), Mapping Vulnerability: Disasters, Development and People. Earthscan, London, pp. 10–25.

Ong, A., 2004. Cultural citizenship as subject-making: immigrants negotiate racial and cultural boundaries in the United States. In: Baker, L.D. (Ed.), Life in America: Identity and Everyday Experiences. Wiley-Blackwell, Malden, MA, pp. 156–175.

Peet, R., Watts, M. (Eds.), 1996. Liberation Ecologies: Environment, Development, Social Movements. Routledge, London.

Quarantelli, E.L. (Ed.), 1998. What Is a Disaster? A Dozen Perspectives on the Question. Routledge, New York.

Rabalais, N.R., Turner, R.E., Justic, D., Dortch, Q., Wiseman Jr., W.J., 1999. Characterization of Hypoxia: Topic 1 Report for the Integrated Assessment of Hypoxia in the Gulf of Mexico. US Department of Commerce, National Oceanic and Atmospheric Administration. NOAA Coastal Ocean Program Decision Analysis Series No. 15.

Reid, H., Taylor, B., 2010. Recovering the Commons: Democracy, Place, and Global Justice. University of Chicago Press, Urbana, IL.

Scott, J., 1985. Weapons of the Weak: Everyday Forms of Peasant Resistance. Yale University Press, New Haven.

Turner, R.E., 1997. Wetland loss in the northern Gulf of Mexico: multiple working hypotheses. Estuaries 20 (1), 1–13.

The White House, 2010. Executive Order 13554 – Gulf Coast Ecosystem Restoration Task Force. October 5 (accessed 30.11.13.). http://www.whitehouse.gov/the-press-office/2010/10/05/executive-order-13554-gulf-coast-ecosystem-restoration-task-force.

US Energy Information Administration, 2009. Louisiana Energy Fact Sheet – October 2009 (accessed 29.11.13.). http://www.eia.gov/state/state-energy-profiles-print.cfm?sid=LA.

Watts, M., 2012. A tale of two gulfs: life, death, and dispossession along two oil frontiers. Am. Q. 64 (3), 437–467.

Westman, C.N., 2013. Social impact assessment and the anthropology of the future in Canada's tar sands. Hum. Organ. 72 (2), 111–120.

Wisner, B., 2004. Assessment of capability and vulnerability. In: Bankoff, G., Franks, G., Hilhorst, D. (Eds.), Mapping Vulnerability: Disasters, Development and People. Earthscan, London, pp. 183–193.

Wolf, E., 1999. Envisioning Power: Ideologies of Dominance and Crisis. University of California Press, Berkeley.

Chapter 13

Political Responses to Emergencies

David Alexander
Institute for Risk and Disaster Reduction (IRDR), University College London, UK

ABSTRACT
This chapter defines politics in relation to disaster risk reduction and resilience. It considers the political organization of society in the context of the architecture and effectiveness of emergency management systems. Ethical questions are dealt with in relation to denial of human rights, corruption, lack of regulation, and use of relief funds for "buying" votes. It is concluded that the ethical framework within which politics are conducted has a strong bearing on the ability to manage disaster risk. The political roots of strategies to improve resilience are examined, and the widespread lack of concern for women's issues is marked as one of the most serious political failings that need to be addressed. Global political processes of disaster reduction are surveyed critically, and finally trends and future prospects are assessed. One of these is the increasing use of the term *governance* to denote political processes that require transparency and a participatory approach. In this sense, we may be seeing the beginnings of a swing away from total reliance on representative democracy to a resurgence of the participatory kind.

13.1 INTRODUCTION

The term politics came into the English language from Middle French in the Renaissance, although the concept was, of course, used by Aristotle to describe the government of a city state, and the phenomenon is as old as human communities can be. Politics are the principles, processes, ethics, and decisions involved in governing the state, where this is the organ that comprises and represents the people (Minogue, 2000). As politics embodies the concept of power, the form that it takes depends on its distribution in any given society. Some concepts that help define politics are the nation-state (the body politic), in association with its presumed sovereignty (i.e., its inviolate right to self-determination), representative democracy (usually achieved by voting for representatives), and participatory, or direct, democracy. The last of these is

Hazards, Risks, and Disasters in Society. http://dx.doi.org/10.1016/B978-0-12-396451-9.00013-5

difficult to implement at the broad scale of a highly populated society, as it relies heavily on person-to-person contact, but it has value at the local scale (Mutz, 2006).

This chapter first defines the political context of disasters and then considers the relations between different levels of government, and their implications for resilience against disasters. The effects of authoritarianism and denial of human rights are also examined in relation to their impacts upon disaster risk reduction. This leads to a broader consideration of political ethics in disaster management, and some of the negative connotations, such as the impact of corruption on safety. The relationship between politics and resilience is considered in detail, and finally the analysis is broadened to the global scale, with consideration of the prospects for consensus and concerted action on worldwide disaster risk reduction.

13.2 THE POLITICAL CONTEXT OF DISASTERS

Social scientists have long argued that disaster is a social phenomenon more than it is a physical one. The decisions made in society, the collective will of groups of people, the dominant interests in society, and the way people perceive hazards are all vital factors that contribute to the making of disaster. They act primarily on vulnerability, which is simultaneously created and reduced as social activities put people and assets at risk and at the same time strive to reduce those risks (Birkmann et al., 2013).

The basis of modern politics is the nation-state, with its concept of sovereignty and autonomy over its internal affairs (Opello and Rosow, 2004). In turn, the nation-state relies on the concept of personal identity, which, from the sixteenth century to modern times, has steadily evolved and become more sophisticated. Personal identity carries with it the notion of citizenship, which carries rights and responsibilities, including those pertaining to the election of representatives to national assemblies. As a concept, personal identity is seldom questioned these days, although it is sometimes misappropriated. It has considerable importance in the fields of security and counterterrorism, in which it is linked to questions of intelligence gathering in relation to illegal activities (Kephart, 2007). In contrast, the concept of the nation-state looks less secure than it did, for example, a century ago. Multinational commerce, global initiatives, and increasing confederation of states all conspire to reduce the power and scope of nationality. Moreover, in times of recession, economic stringency increasingly dictates that policy on certain key issues be made at the level of international finance, not national governments.

Many aspects of politics influence disaster risk reduction. These include the dichotomy between centrism and devolution, which is discussed in the next section, the roles of authoritarianism and democracy, corruption, vote buying, and ethics. The next section deals with the political organization of society in terms of how it affects a country's ability to manage civil contingencies.

13.3 CENTRISM AND DEVOLUTION

The form that a nation-state takes will have a considerable impact on its arrangements for dealing with emergencies, crises, and disasters. Broadly, these arrangements take two forms: civil protection against natural hazards, industrial accidents, and localized mass casualty emergencies and civil defense against armed aggression and terrorism (Alexander, 2002). It should be noted, however, that the distinction is often poorly drawn: for example, in New Zealand, the national Ministry of Civil Defence and Emergency Management is largely occupied with civil protection rather than civil defense, *sensu stricto*. Although civil defense needs to be governed nationally, and is therefore organized "top-down," whatever the role of national civil protection, it is unlikely to function well unless it has adequate local bases, and hence to a greater or lesser extent is organized on a "bottom-up" basis. During the latter part of the twentieth century, civil protection essentially grew out of civil defense, and hence in many countries it can be considered as a form of extension, outgrowth, or reorganization of previous emergency arrangements (Alexander, 2007).

In vertically stratified emergency management systems, such as those that have a military origin, the hierarchy is generally based on the following levels: policies (and ethics—see later sections), strategic, tactical, and operational. Essentially, the policy level determines what resources are made available to manage a major emergency (usually at the national level), the strategic command finds them, the tactical one allocates or deploys them to local users, and the operational command utilizes them. Political decision making will have predetermined the architecture of the system. In this respect, most countries have enacted a basic law that governs the ways in which major emergency situations are managed, and particularly how state funds are capable of being utilized (Handmer and Dovers, 2007).

Emergency management is invoked when a situation crosses the threshold from something that can be handled using ordinary resources to a contingency that requires the suspension of regular activities and the implementation of special measures, which, one trusts, have been planned and organized in advance. As the theatre of operations is inevitably the local area, most emergencies require a local response, even though for large events this needs to be replicated in many different places. However, a purely "grassroots" approach to emergency management would tend to fail unless it were coordinated, supported, and harmonized by higher levels of government, preferably at the national level. Moreover, for the many countries that lack sufficient resources to be autonomous when major disasters occur, the national level acts as the interface between the local application of resources and the international community that donates many of them (Katoch, 2006).

Coordination of emergency intervention, and, of course, disaster mitigation measures, gives the nation-state two possible roles. The first is to support and harmonize efforts at lower levels of government, and the second is to ensure that national interests are protected in cases where disaster threatens them (e.g.,

UK Cabinet Office, 2010). The latter instance begs the question of the extent to which national forces can and should take over local efforts. It is a valid general principle that local autonomy should be upheld, as there is always the likelihood that, sooner or later, outside forces will leave an area and local ones will have to fend for themselves. However, a poor local response may need to be substituted by regional or national efforts if it is fundamentally incapable of resolving the problems arising from disaster (Aloyo, 2006).

An essential distinction needs to be made between federal and nonfederal nations. The division of competencies and responsibilities is different between these two entities. Even a nation with strong federalist tendencies does not have the same division of powers as one that is federated or confederated. In the case of major emergencies in a federal country, the central government is generally relegated to support roles with respect to the constituent states and their local governments. The one exception to this is usually that of events in which the health, integrity, or survival of the nation might be threatened unless the central government takes over the management of the event.

Under all of these circumstances, disaster management can become a barter market. One of the functions of government, and politics, is to make decisions about the apportionment of scarce resources, in other words, to apply economics. In a major disaster, local government will want more resources, whereas national government will want to ration those resources that it has to distribute among local authorities. Intermediate levels of government (counties, provinces, regions, departments, etc.) will find themselves exercising both roles. In the context of specific major disasters, countries that are not endowed with sufficient resources may launch an international appeal through the United Nations (UN) Office for the Coordination of Humanitarian Affairs, at which point the national government becomes an interface with UN agencies and donors, whether these be international nongovernmental organizations or foreign governments (Coppola, 2011).

After disaster has struck, governments that are adept at disaster management will exert strong leadership at the national level, specify exactly what is needed to augment their efforts, and fend off unwanted aid. Both nationally and internationally, the last of these actions can be politically difficult, as it creates a negative image, but in reality aid that is not needed can be debilitating. For example, the arrival of 500 tons of medical supplies at Yerevan Airport after the 1988 Armenian earthquake used up resources in warehousing, classifying and eventually disposing of them, whereas only a very small proportion of them was useful (Filson et al., 1990, Chapter 12).

More serious imbalances in disaster mitigation can be created by lack of democracy, as the following section shows.

13.4 DICTATORSHIP, DEMOCRACY, AND DISASTERS

Politically, civil protection has been unpopular with countries in which there have been latent instability and fear of *coups d'état*. To a lesser extent, so has civil defense. To a greater or lesser extent, emergency management systems require the creation of a civilian force that can be deployed around the country to manage exceptional contingencies. The more exceptional the circumstances are, the more out of the ordinary the measures employed to deal with them. Fears have been expressed in national parliaments that such a force could be subverted to support the overthrow of legitimate government. However, in countries directly involved in the Cold War, such as those in Europe, such fears tended to disperse with the end of that epoch and the subsequent relaxation of international tensions and dissolution of blocs (Alexander, 2006).

Basic human rights are codified in various documents, principally the UN's Universal Declaration on Human Rights (United Nations, 1948). Respect for human rights varies considerably around the world and is an important factor in disaster risk reduction. To a greater or lesser extent, lack of rights means that people and their communities are deprived of the information they need to manage hazards, as well as the necessary resources and the opportunities to act (Alexander and Davis, 2012). It could be argued that the lack of means inherent in deprivation of human rights also signifies an inability to assume responsibility for hazard abatement, vulnerability reduction, and risk management (Hurst, 2010).

The cases in which human rights are restricted vary considerably from small communities to the populations of entire countries. At the community level, if not on a broader scale, one is dealing with marginalization, which has long been correlated with disaster vulnerability (Polack, 2008). Marginalization can have political, economic, or sociocultural causes, but it boils down to the inability of groups to achieve full participation in a nation's life, either through repression or through deprivation. Studies have shown that it greatly increases vulnerability to disaster (e.g., Wisner and Luce, 1993). Marginalization is thus the opposite of democratic participation and inclusion.

Human rights are part of a code of ethics, which will now be considered more broadly within the context of disaster risk reduction.

13.5 DISASTERS, POLITICS, AND ETHICS

One question is the extent to which political decision making influences the system in its operation. In essence, the politics of disaster response (and, indeed, disaster risk reduction) should be governed by ethical considerations. One of these is that vulnerability reduction, disaster response, and recovery processes should be guided by principles of fairness and equity (Beatley, 1989). Another is that decisions on the allocation and use of resources should be made transparently and in the interests of the beneficiaries and general population. Moreover, no community, group, or person should be adversely exploited as a result of disaster.

In many countries, the concept of "welfare" is rich in political overtones. One possible definition of welfare (my own) is as follows: "the provision of care to a minimum acceptable standard to people who are unable adequately to look after themselves."[1] This requires the construction of a safety net for those members of society who are vulnerable and disadvantaged and are thus unable adequately to provide for themselves. The existence, coverage, and performance of such safety nets varies considerably from one nation to another. In some countries, the only significant source of welfare is the family or extended family. In some communities, part of the burden of welfare is taken up by organized crime. Alternatively, the only fallback mechanism is to resort to the black economy. Its role is viewed equivocally by observers of disaster: it provides employment where none would otherwise exist, but it provides no tax revenue for disaster mitigation and can undermine attempts to reduce vulnerability or recover from disaster.

The extent to which disaster response is treated ethically in the political decision-making process is highly variable. Disaster has long been viewed as a "window of opportunity" for positive change (Solecki and Michaels, 1994). The aftermath is a period in which public opinion will be highly sensitized and may strongly support efforts to make life safer by reducing disaster risk. However, an alternative view of disaster involves the "Pandora's box" model, in which the window of opportunity is a negative one. Disaster is an opportunity to be exploited by people and groups whose agenda is ethically dubious, including organized crime groups (see next section) and those who would impose unwelcome political ideologies on the public. Thus, the 2004 tsunami in Aceh Province, Indonesia, was followed by the imposition of Sharia law. Restrictions on social liberties were also imposed in Sumatra after the Padang earthquake of 2009 (Alexander, 2012).

13.6 CORRUPTION AND DISASTERS

In politics, corruption is a widespread and endemic phenomenon, one that touches all societies. No direct measures of corruption exist, as it is by nature occult. Moreover, attempts to gauge or measure it can be hazardous, as powerful and unscrupulous interests may be behind it. Hence, all we have is an index of how much corruption is perceived in a country's affairs (Transparency International, 2012), which puts Denmark as the country that is perceived to be least corrupt and Afghanistan, North Korea, and Somalia in joint 175th position as the most corrupt. As corruption may occur at all levels from that of the most minor local official to that of a national leader of government, it can infect a system from top to bottom.

Corruption weakens the decision-making process and reduces opportunities for democratic participation (Escaleras et al., 2007). It bypasses due

1. The academic literature on disasters and welfare is diffuse and inadequate. Most of it relates to economic well-being in a general sense, and the residual amount to family survival. There is thus an unfulfilled need for a general examination of the question of what welfare means in the context of disasters.

process and favors the lowering of standards. It subverts the money that is earmarked for disaster response or reduction, and it can paralyze operations. It renders situations less amenable to clear interpretation and thus hinders the diffusion of useful or essential knowledge about risk and disaster. Finally, corruption favors the influence of organized crime syndicates over processes of government, which can thus be subverted to their own interests (Schultz and Søreide, 2008).

One example of corruption at work after disaster concerns the money and resources that flowed into Nicaragua after the earthquake of December 1972. The ruling Somoza family expropriated a large proportion of the funds and apparently sent them abroad to foreign bank accounts. The net effect was to slow down recovery from the disaster, and in the case of the lowest, least influential classes of society, to stop it altogether. The end product of this rank exploitation of a postdisaster situation was the overthrow of the Somoza dictatorship by the Sandinista revolution of 1979, but that did not mean that the relief money reappeared (Brown and Cloke, 2005). In another example, the November 1980 earthquake in Southern Italy required reconstruction and development work in more than 600 towns and villages. The process is reputed to have seriously enriched a number of leading political families from the disaster area (Wikipedia, 2014), although there is little indication of the extent to which this was a direct result of corruption. Nevertheless, organized crime moved into the area in a big way and dominated the reconstruction process (Alexander, 2005).

As the distinguished Italian journalist Roberto Saviano noted, cement is a great basis for organized crime (Saviano, 2011). The construction industry is easy to dominate and manipulate and difficult to subject to legal and administrative controls. Moreover, it can be subverted to serve political ends. Hence, it is remarkably common to find a discrepancy between published standards, for example, pertaining to antiseismic construction, and their enforcement, either because inspectors can be corrupted or because there is no inspection (Ambraseys and Bilham, 2011). Turkey is a country that lacks neither earthquakes nor antiseismic building codes, but in the 1999 İzmit earthquake, large numbers of relatively modern buildings collapsed because they were not built to code. Similar problems were experienced in Mexico during the 1985 earthquake. In Italy in 1980, the maternity wing of a large hospital collapsed, killing staff and patients. It had been built with absolute economy of materials and in flagrant violation of both its own design blueprint and the prevailing building codes.

Ambraseys calculated that five-sixths of deaths in earthquakes occur in countries with low rankings in the corruption perception index (Ambraseys, 2010). Ambraseys and Bilham (2011) found a relationship between poverty and corruption that they went on to relate to earthquake deaths, in the sense that some poorer countries are more corrupt than richer ones and also account for the greater part of seismic mortality. In pointing out that an unregulated construction industry is a killer, these authors noted that "the structural integrity of a building is no stronger than the social integrity of the builder."

This is true, but the problem with correlation exercises such as the ones conducted by Ambraseys and Bilham is that they do not verify causality or identify its direction of operation. Are countries poor because they are corrupt, or corrupt because they are poor? Is the lack of safeguards against earthquake mortality a result of poverty or one of its contributing factors, or both? The same criticism could be leveled against the more detailed study by Escaleras et al. (2007), which found a positive correlation between public sector corruption and earthquake impacts in 42 countries. Nevertheless, in these studies, the central relationships seem clear.

Ambraseys and Bilham (2011, p. 154) noted that a connection exists between the increase in global population and the rise in seismic mortality. This suggests that the more people inhabit seismic zones, the more will be killed in earthquakes, ceteris paribus. It also implies that seismic hazard mitigation is outpaced by population growth, but all such generalizations could be invalidated by corruption, if we only understood it. The fact that poorer countries appear at various levels of the corruption perceptions index shows that it is dangerous to generalize without a robust model. However, it is equally difficult to create one, as corruption tends to defy measurement.

The next section discusses a related aspect of the distortion of the political process, namely, the ways in which votes can be bought using publicly funded largesse.

13.7 FORGIVENESS MONEY

In the United States, the disbursement of Federal funds is governed by the Stafford Act (FEMA, 2007), which is one legal instrument that, during its 40-year evolution, has struggled to regulate and contain the flow of government moneys to the victims of disaster. A ruling in the US Federal Register (Vol. 63(24): 5895-7, 1997) states unequivocally that: "federal assistance will only be supplied in cases in which the appropriate written laws of protection against disasters have been formally adopted by state and local governments before the declaration of a disaster." Moreover, various provisions in the Stafford Act (e.g., Section 312 "Duplication of Benefits") endeavor to ensure that beneficiaries are not overpaid. However, it is difficult to use this, or any other, legal instrument as a means of ensuring that people, businesses, and communities prepare for disaster and reduce their vulnerabilities. Numerous instances have occurred in which victims have been denied payments under the Stafford Act but have received them from other Federal sources, for example, the Small Business Act (15 USC, Chapter 14A).

In the United States and many other countries, the victims of disaster are also voters. If not a means of buying votes, compensation after disaster is certainly a way in which voters' opinions of parties can be swayed. The paradox of disaster risk reduction is that it resembles the "no votes in sewage" syndrome, in which mitigation policies are not particularly attractive to voters, but postdisaster recovery money is more attractive, as it is evidence of compassion. Of course, this is offset by the burden of the taxation needed to pay for disaster recovery, which may be regressive, for example, by increasing fuel duty or value added tax (Delany, 2012).

Mention of the taxation dilemmas begs the examination of wider questions of how to transform disaster recovery into resilience. A gathering consensus has come to agree that this needs to be done by improving conditions, not merely restoring them (the "bounce forward rather than bounce back" concept: Manyena et al., 2011). However, there are questions as to how the political conditions to achieve this can be created.

13.8 THE POLITICS OF "BOUNCE FORWARD" IN DISASTER RISK REDUCTION

In a democratic society, however imperfect its democracy, the political consensus or majority defines the will of the people. Some policies are unpopular and some are imposed by circumstances, such as excessive national debt. However, politics are intended to be the summation and averaging of what people want. One problem with this is that disaster risk reduction is seldom a popular option. As noted above, its negative overtones do not endear it to voters. Unfortunately, even the concept of resilience is difficult to sell in electoral terms. Social surveys have shown that the public tends to have a poor understanding of probability, and therefore of the magnitude and frequency of particular hazards. Support for disaster risk reduction therefore waxes and wanes in proportion to events. It is the greatest during the short-lived "window of opportunity" for change that opens in the immediate aftermath of disaster. However, it is a rare coincidence when the cycle of disasters follows the political cycle of elections and governments (Birkmann et al., 2010).

Democracy ought to ensure that disaster risk reduction is taken seriously, and thus implemented on behalf of the people, with their support, whereas the electoral cycle tends to impede it. Commonly, little is done that does not fit into the cycle, with the duration of government set at between a few months and 5 years. Part of an administration's mandate is taken up with settling in, and part with bowing out, as there is little incentive to embark on complex and lengthy projects when a government faces the risk of losing the next election. A common reaction of electorates is to think in terms of the transfer of responsibility, by means of electing a government, from personal to official levels. Hence, as democracy has become more sophisticated as it has developed, personal responsibility to an extent has been diminished by putting the onus for public safety on governments and public administrators (Horlick-Jones, 1995).

Nevertheless, the concept of "resilience" is resonant with positive overtones. It implies prudence, safety, defense, protection, and robustness under duress. In theory, if not in practice, this should make it attractive to an electorate (Aven, 2011). Increasingly, resilience against disasters is seen as a process of not only reducing vulnerability in workaday actions, but also using recovery from disaster to "bounce forward," or "build back better," in the sense of improving physical assets, legal instruments, and social institutions as they are rebuilt (Manyena et al., 2011). In this context, disaster often involves step-wise change, in which the majority of the improvements occur within a short period of the disaster, rather than steadily over periods of years. To acquire resilience is an expensive process that requires the transfer of resources from other worthy causes. In rich societies, disasters are at best a middling priority among all forms of risk reduction, whereas in poor societies, deprivation and poverty greatly overshadow disaster as workaday problems, and fatalism may prevail in the face of catastrophe risk.

One important aspect of resilience that has frequently been neglected is gender. An increasing body of work demonstrates that, in many different countries and situations, women and girls are more adversely affected by disaster than are men and boys (e.g., Plan International, 2013). The gender subfield of disaster risk reduction is, of course, about both, but its overriding aim is to help ensure that disaster and its aftermath are not occasions for renewed or increased exploitation of women and girls. Characteristically, they lack political weight, and perhaps even representation. The consequences can be very serious, and not merely in poorer countries, as a report on women in the Great Eastern Japan Earthquake and Tsunami of 2011 forcefully demonstrated (Domoto et al., 2013). The gender question is one of a number of concerns that involve politics at the global scale.

13.9 THE GLOBAL POLITICS OF DISASTER

In 1984, Dr Frank Press, President of the US National Academy of Sciences, proposed the designation of the 1990s as a decade for reducing natural hazards. He was speaking in San Francisco at the Eighth World Conference on Earthquake Engineering, and was evidently seeking to unite disparate attempts at scientific monitoring (Press, 1986, p. 153). He was also concerned that the fruits of scientific and engineering endeavor were not being adequately utilized in disaster reduction processes. When presented for ratification in the UN General Assembly, delegates from developing countries argued that the initiative should have a greater sociopolitical component and less emphasis on science and technology. Hence, the International Decade for Natural Disaster Reduction, 1990–2000, was born. The UN coordination of this event was consolidated into the permanent International Strategy for Disaster Reduction, which at a meeting in Japan in 2005 launched the Hyogo Strategy for Disaster Reduction (Hyogo Framework for Action (HFA)), 2005–2015 (UNISDR, 2005). The HFA is due for renewal at another global meeting in Japan in 2015, this time in Sendai, scene of the devastating tsunami of March 2011.

The HFA, and the processes that created it, are political rather than scientific. This is not a criticism, as it is clear that the document is backed by scientific knowledge. Moreover, since the inauguration of the HFA, the research inputs to the process have grown strongly and steadily (see, for example, the Global Assessment Reports, e.g., UNISDR, 2013a). However, the value of the HFA lies in its function as a set of guidelines for policy on disaster reduction.

Criticisms of the HFA (e.g., BOND, 2013) have tended to focus on producing a revised or alternative instrument for the period 2015–2025. For example, HFA contains no mention of gender, yet this has proved to be a central issue in disasters (see above). It does not provide for monitoring progress, a deficiency that will also necessarily be rectified. In fact, one problem with international instruments such as the HFA is that governments can subscribe to them, but this does not guarantee that there will be any improvements in disaster risk reduction. HFA is not legally binding and so cannot be enforced. Moreover, it appears that the representatives of countries that have poor records in guaranteeing their citizens' human rights have sometimes attempted to legitimize themselves by signing international agreements for disaster reduction, for example, the UN International Strategy for Disaster Reduction (UNISDR) "Safe Cities" initiative (UNISDR, 2010).

Many other international initiatives are, of course, in favor of disaster risk reduction. I noted earlier in this chapter that national representative democracy is somewhat undermined by globalization, in which the parameters of international trade and finance can dictate important elements of a government's agenda. This begs the question of how globalization affects the politics of disaster. In a positive sense, it has facilitated the exchange of information, expertise, and other resources for mitigating and responding to disasters. In a negative manner, disaster has been seen as an opportunity for free market capitalism to exploit the affected countries and societies. Naomi Klein advanced this thesis in some detail in her book *The Shock Doctrine* (Klein, 2008), in which she suggested that predatory capitalists exploit disaster to reduce the regulatory barriers to their activities. Critics of Klein's analysis (e.g., Redburn, 2007) have argued that capitalist entrepreneurialism stops the affected societies from lapsing into stagnation, and that market behavior and predatory corporate behavior are not necessarily the same thing.

13.10 CONCLUSIONS

Several conclusions can be drawn from the analysis reported in this chapter. First, the determinants of politics are changing in an increasingly globalized and interconnected world. The sovereignty of the nation-state is qualified by globalization, which can extend exploitation and inequality to victims and states affected by disaster. Second, although the nation-state has proved to be an enduring concept in the modern age, and the concept of personal identity has remained strongly bound up with it, this picture masks considerable variations and tensions in the political organization of disaster risk reduction and emergency response measures. Hence, many nations exhibit a certain degree of tension between political forces that centralize and those that devolve power. This has strong implications for the functioning of measures designed to counter disaster.

Vulnerability to disasters is at least partly a function of the degree to which a population at risk enjoys full human rights. Moreover, it is critically affected by any processes of corruption or failure to create and maintain adequate regulatory mechanisms. Concepts of welfare can be distorted by political considerations when votes are effectively "bought" by distributing relief money without objective criteria for its allocation. Empirical analyses of corruption show that it appears to be positively related to the severity of disaster impact: if the causal mechanism (yet to be adequately verified) is robust, then reducing corruption will undoubtedly reduce the toll of casualties and damage in disasters.

Global political processes are struggling to cope with disaster, which is having an increasingly serious impact on lives, livelihoods, and assets as our crowded world struggles to develop and sustain itself. Change tends to occur unevenly, often in response to particular events, such as the Indian Ocean tsunami of 2004, or the earthquake and tsunami of March 2011 in Japan, both of which had major international implications for well-being.

Last, the term governance arrived in the English language in the late fourteenth century from Medieval French and Latin as a synonym for government. Since then it has acquired various overtones (see Oxford English Dictionary). Currently, in the field of disaster risk reduction, it implies transparent government and strong, functional institutions that encourage the public to respect and participate in the political process (UNISDR, 2005, p. 5). In recent years, many studies have shown that these goals are central to resilience and disaster risk reduction (e.g., Wilkinson, 2013). They are now an essential component of an international strategy: as UNISDR (2013b, p. 11) noted in its monitoring of implementation of the HFA, "all countries affirm that governance and accountability are key drivers of success." Some have argued that fragmentation of approaches and inadequacies of coordination are the principal barriers to disaster risk reduction (UNESCAP, 2011; UNISDR, 2013c, p. 4). Others have suggested that the international community is failing adequately to confront political shortcomings, including corruption, authoritarianism, discrimination, and indifference (Alexander and Davis, 2012). One might conclude that governance is difficult to measure, or at least that doing so is a complex and demanding process. According to UNISDR (2013d, p. 6), the parameters are effectiveness of government, quality of regulations, accountability, and control of corruption. All are vital factors that back up disaster risk reduction, but there is an element of subjectivity in measuring them. Nonetheless, practitioners in the field have been unequivocal in stating that improvements in politics, such as greater inclusiveness and transparency, are essential if disaster risk is to be reduced (GNCSODR, 2011).

REFERENCES

Alexander, D.E., 2002. From civil defence to civil protection — and back again. Dis. Prevention Manage. 11 (3), 209–213.

Alexander, D.E., 2005. The Italian Mafia's Legacy of High-rise Death Traps. Global Corruption Report 2005: Corruption in Construction and Post-Conflict Reconstruction. Transparency International, Berlin, pp. 26–27.

Alexander, D.E., 2006. Globalization of disaster: trends, problems and dilemmas. J. Int. Aff. 59 (2), 1–22.

Alexander, D.E., 2007. Disaster management: from theory to implementation. J. Seismol. Earthquake Eng. 9 (1), 39–49.

Alexander, D.E., 2012. A tale of three cities and three earthquake disasters. Tafter J. 50, 1–10.

Alexander, D.E., Davis, I., 2012. Disaster risk reduction: an alternative viewpoint. Int. J. Dis. Risk Reduction 2, 1–5.

Aloyo, O.O., 2006. Emergency preparedness is all local. J. Emerg. Manage. 4 (1), 11–12.

Ambraseys, N., 2010. A note on transparency and loss of life arising from earthquakes. J. Seismol. Earthquake Eng. 12 (3), 83–88.

Ambraseys, N., Bilham, R., 2011. Corruption kills. Nature 469, 153–155.

Aven, T., 2011. On some recent definitions and analysis frameworks for risk, vulnerability, and resilience. Risk Anal. 31 (4), 515–522.

Beatley, T., 1989. Towards a moral philosophy of natural disaster mitigation. Int. J. Mass Emerg. Dis. 7 (1), 5–32.

Birkmann, J., Buckle, P., Jaeger, J., Pelling, M., Setiadi, N., et al., 2010. Extreme events and disasters: a window of opportunity for change? Analysis of organizational, institutional and political changes, formal and informal responses after mega-disasters. Nat. Hazards 55 (3), 637–655.

Birkmann, J., Cardona, O.D., Carreño, M.L., Barbat, A.H., Pelling, M., Schneiderbauer, S., Kienberger, S., Keiler, M., Alexander, D., Zeil, P., Welle, T., 2013. Framing vulnerability, risk and societal responses: the MOVE framework. Nat. Hazards 67 (2), 193–211.

BOND, 2013. Priorities for the Post-2015 Disaster Risk Reduction International Framework. BOND Disaster Risk Reduction Working Group. BOND, London, 4 pp.

Brown, E., Cloke, J., 2005. Neoliberal reform, governance and corruption in Central America: exploring the Nicaraguan case. Political Geogr. 24 (5), 601–630.

Coppola, D., 2011. Introduction to International Disaster Management, second ed. Elsevier, Amsterdam. 696 pp.

Delany, T., 2012. Compensation for natural disasters: the tax consequences. Taxation Australia 46 (6), 248–252.

Domoto, A., Ohara, M., Rieko, A., Hara, H., Amano, K., 2013. Japanese Women's Perspective on 3–11. Japan Women's Network for Disaster Risk Reduction. Tokyo, 31 pp.

Escaleras, M., Anbarci, N., Register, C.A., 2007. Public sector corruption and major earthquakes: a potentially deadly interaction. Public Choice 132, 209–230.

FEMA, 2007. Robert T. Stafford Disaster Relief and Emergency Assistance Act, as Amended, and Related Authorities, FEMA 592, June 2007. US Federal Emergency Management Agency, Washington, DC, 113 pp.

Filson, J.R., Sharp, R.V., Borcherdt, R.D., et al., 1990. Medical and healthcare aspects. Armenia 1988 earthquake reconnaissance report. Earthquake Spectra 5 (S1), 161–169.

GNCSODR, 2011. If We Do Not Join Hands... Views from the Front Line. Global Network of Civil Society Organisations for Disaster Reduction, Twickenham, UK, 51 pp.

Handmer, J., Dovers, S., 2007. Handbook of Disaster Policies and Institutions. Earthscan, London, 187 pp.
Horlick-Jones, T., 1995. Modern disasters as outrage and betrayal. Int. J. Mass Emerge. Dis. 13 (3), 305–315.
Hurst, J.L., 2010. Establishing human rights protections in post-disaster contexts. J. Emerge. Manage. 8 (6), 7–14.
Katoch, A., 2006. The responders' cauldron: the uniqueness of international disaster response. The Globalization of Disaster J. Int. Aff. 59 (2), 153–172.
Kephart, J.L., 2007. Identity and security: moving beyond the 9/11. J. Homeland Security. Published 3/28/2007.
Klein, N., 2008. The Shock Doctrine: The Role of Disaster Capitalism. Penguin, Harmondsworth, 576 pp.
Manyena, S.B., O'Brien, G., O'Keefe, P., Rose, J., 2011. Disaster resilience: a bounce back or bounce forward ability? Local Environ. 16 (5), 417–424.
Minogue, K., 2000. Politics: A Very Short Introduction. Oxford University Press, Oxford, 128 pp.
Mutz, D.C., 2006. Hearing the Other Side: Deliberative versus Participatory Democracy. Cambridge University Press, Cambridge, 184 pp.
Opello, W.A., Rosow, S.J., 2004. The Nation-State and Global Order: An Historical Introduction to Contemporary Politics, second ed. Lynne Rienner Publishers, Boulder, Colorado. 320 pp.
Plan International, 2013. Because I am a girl. The state of the world's girls 2013. In: Double Jeopardy: Adolescent Girls and Disasters. Plan International, Woking, UK, 219 pp.
Polack, E., 2008. A right to adaptation: securing the participation of marginalised groups. IDS Bull. 39 (4), 16–23.
Press, F., 1986. The role of science and engineering in mitigating natural hazards. In: Keynote Address at the Eighth World Conference on Earthquake Engineering, San Francisco, 23 July 1984. Earthquake Engineering and Structural Dynamics, vol. 14(1), pp. 150–154.
Redburn, T., September 29, 2007. It's all a grand capitalist conspiracy. New York Times.
Saviano, R., 2011. Gomorrah: Italy's Other Mafia. Pan Books, London, 300 pp.
Schultz, J., Søreide, T., 2008. Corruption in emergency procurement. Disasters 32 (4), 516–536.
Solecki, W.D., Michaels, S., 1994. Looking through the post-disaster policy window. Environ. Manage. 18 (4), 587–595.
Transparency International, 2012. Corruption Perceptions Index 2012. Transparency International, Berlin, 5 pp.
UK Cabinet Office, 2010. Responding to Emergencies: The UK Central Government Response – Concept of Operations. United Kingdom Government, London, 80 pp.
UNESCAP, 2011. Achievements and Challenges in Implementing the Hyogo Framework for Action in Asia and the Pacific. UN Economic and Social Commission for Asia and the Pacific, Bangkok, 3 pp.
UNISDR, 2005. Hyogo Framework for Action 2005–2015: Building the Resilience of Nations and Communities to Disasters. UN International Strategy for Disaster Reduction, Geneva, 22 pp.
UNISDR, 2010. Making Cities Resilient: My City Is Getting Ready. UN International Strategy for Disaster Reduction, Geneva, 24 pp.
UNISDR, 2013a. Third Global Assessment Report on Disaster Risk Reduction 2013. From Shared Risk to Shared Value: The Business Case for Disaster Risk Reduction. UN International Strategy for Disaster Reduction, Geneva, 283 pp.
UNISDR, 2013b. Implementation of the Hyogo Framework for Action: Summary of Reports, 2007–2013. International Strategy for Disaster Reduction, Geneva, 62 pp.

UNISDR, 2013c. Towards a Post-2015 Framework for Disaster Risk Reduction. International Strategy for Disaster Reduction, Geneva, 10 pp.

UNISDR, 2013d. Towards the Post-2015 Framework for Disaster Risk Reduction: Indicators of Success: A New System of Indicators to Measure Progress in Disaster Risk Management. International Strategy for Disaster Reduction, Geneva, 11 pp.

United Nations, 1948. Universal Declaration on Human Rights. United Nations Organisation, Geneva. http://www.un.org/en/documents/udhr/ (accessed 06.01.14.).

Wikipedia, 2014. Terremoto Dell'Irpinia 1980. Wikipedia in italiano (accessed 06.01.14.).

Wilkinson, E., 2013. Disaster Risk Governance in Volcanic Areas. Concept Note. Overseas Development Institute, London, 31 pp.

Wisner, B., Luce, H.R., 1993. Disaster vulnerability: scale, power and daily life. GeoJournal 30 (2), 127–140.

Chapter 14

Double Disaster: Disaster through a Gender Lens

Sarah Bradshaw[1] and Maureen Fordham[2]

[1] *Department of Law and Politics, Middlesex University, London, UK,* [2] *Department of Geography/ Gender and Disaster Network (GDN), Northumbria University, Newcastle, UK*

ABSTRACT

This chapter explores the impact of disasters on women and girls, with particular reference to the context of the developing world. It critically explores the conceptual and theoretical basis for assuming that a differential impact exists. It highlights that disasters are gendered events and women and girls experience them differently from men, suffering longer term and more intangible impacts such as a rise in violence or greater insecurity in employment. Given women and girls are impacted more and differently than men and boys, it might be expected gender issues would be a key policy concern, yet the chapter highlights that gender is still excluded from much policy on disaster risk reduction. Drawing on the lessons learned from processes to "engender development," it suggests that, although exclusion remains an issue, how women are included in disaster risk reduction and response can also raise concerns. It concludes by highlighting that tackling gendered risk demands both a reconceptualization of "disaster" and for disasters to become a development issue.

14.1 INTRODUCTION

This chapter explores the impact of disasters on women and girls, with particular reference to the developing world context. It begins by considering why "natural" disasters should be understood as gendered events, before critically examining the existing evidence that suggests there is a differential impact on women and girls. Through consideration of the "double disaster" women face, including increased violence, poverty, and insecurity postevent, it problematizes ideas of what the disaster really is for women. Drawing on evidence from the developing world, it then explicitly considers the situation of adolescent girls to highlight how they often fall through programming cracks, being conceptualized neither as women nor children, meaning their specific needs are largely invisible

Hazards, Risks, and Disasters in Society. http://dx.doi.org/10.1016/B978-0-12-396451-9.00014-7

and remain unmet. The final sections highlight how the capacities of women and girls mean they are increasingly being seen as much as policy solutions as policy problems and suggests inclusion can bring its own problems. It concludes by highlighting that to reduce disaster risk for women demands an explicit focus on reducing gender inequalities. Without this, although a focus on women within disaster risk reduction (DRR) and response will reduce the risk of disasters, it will not reduce women and girls' disaster risk.

14.2 WHY SHOULD DISASTERS BE UNDERSTOOD AS GENDERED EVENTS?

Although "disaster" is a contested notion (Cardona, 2004; Oliver-Smith, 2002; Quarantelli, 1998), it is usually understood that one occurs when an individual or group is vulnerable to the impact of a natural or human-made hazard, that is they are unable to "anticipate, cope with, resist, and recover from" an event (Blaikie et al., 1994). The root causes of vulnerability lie with the lack of access to the resources that allow people to cope with hazardous events, such as income, education, health, and social networks. This access may be gendered, in that women and girls in general tend to have less access to, or control over, assets than men and boys. Thus, from the outset, it may be assumed women and girls will be more vulnerable to hazards than men and boys. Vulnerability may also be related to the roles women and men play in society. Typically women (and especially poor women) juggle triple roles simultaneously: a reproductive role (this is mainly related to childbearing/rearing responsibilities and domestic tasks); a productive role (this is paid work or subsistence/home production); and a community managing role (these are voluntary, unpaid roles carried out in and for the community) (Moser, 1993). These roles add to women's vulnerability through making them time poor and have implications for their health and well-being. Since reproductive and community roles do not generate an income they are little valued, and this, coupled with the fact women's jobs generally pay less than men's, means women may be poorer and/or dependent on a male income, giving them less voice than men in the home and the community. Gendered norms may impact directly on women's ability to respond to a natural hazard. Their socially prescribed roles mean they are not taught to swim or climb trees, skills that may be vital for survival during floods and tsunamis, or the fact they are slowed down by their clothing, for example (Oxfam, 2012) may mean they are more likely to die during an event than men. Studies suggest gendered social norms mean women are not only less able, but also less willing to save themselves, given they may be constrained by socially constructed roles and norms that suggest they cannot leave the home without male permission or because care of children and the elderly slows their escape (Chowdhury et al., 1993; Oxfam, 2005; WEDO, 2008).

Vulnerability to an event then is not based on sex or biological differences between men and women, but rather due to how society constructs what it means to be a man or a woman—what roles they should play and how they should behave. As gender intersects with characteristics such as class, ethnicity, and sexuality, and interacts with age/life course, different women experience different levels of vulnerability.

Although it is important to take into account the different characteristics of a woman when determining vulnerability, understanding vulnerability as a product of the intersection of gender, generation, and other characteristics such as disability or race/ethnicity should not result in checklists of characteristics used in a cumulative fashion to predict who will be "most" vulnerable. For example, the assumption that women over 65 years of age would be more vulnerable to the negative impact of hurricane Katrina was not borne out, and what was important in understanding impact was the twinning of race and gender (Willinger and Knight, 2012). One group generally highlighted in the literature as particularly vulnerable is single mothers/female heads of household since they are seen to be among those most asset-poor and "have been found to be among those most affected by natural disasters" (DFID, 2004: 3). However, little empirical evidence exists to support such an assertion, since disaggregated data are scant, and the assumption rests on the contested assertion that female heads are the "poorest of the poor" in "normal" times (see Chant, 2003, 2008 for discussion). Furthermore, as the risk of disaster is a product of both hazard and vulnerability, in many ways checklists of vulnerability are nonsensical without a hazard to which to relate them.

Women's vulnerability should not be assumed, nor can the hazards they face be known in the current "risk society" (Beck, 1998). Exogenous shocks by their nature are often "known unknowns" but increasingly the "unknown" element is being stressed as climate-induced hazards are increasingly unpredictable, whereas financial, fuel, and food crises make it more difficult to predict vulnerability. Due to existing inequalities of power, women and girls may be the most impacted by exogenous shocks that bring both environmental and financial crises. Ironically, it is their lack of access to resources that may make them most practiced at responding to crisis, through living this on a daily basis. Women's responses to economic crisis brought about by the neoliberal policies of the 1980s demonstrated their ability to adapt and cope; they also demonstrated the heavy price women pay in their time and well-being to ensure the survival of the household (Elson, 1991). Although women may have ideas about how best to adapt, they may be least able to translate ideas into practice, given their limited voice in the household and community. In the short term then we might expect to see greater gender losses. Since the Indian Ocean Tsunami, women's greater vulnerability to hazards has been assumed; however, the evidence on which to base this assumption is not as robust as might be imagined.

14.3 EVIDENCE FOR A GENDERED IMPACT OF NATURAL HAZARDS

A disaster is generally defined as such if the levels of loss of life, numbers injured, or loss of material goods goes over a specific, but subjective, threshold. Loss of material goods while devastating at a household or individual level often does not even register at a national level where loss is usually recorded in terms of economic and social infrastructure, or in terms of output and production. Holland (2008) attempted to estimate the household cost of the floods in Fiji in 2004, and suggested a loss of F$4,815 per impacted household. Given the average household income was F$3,500 a year, this clearly would have forced many families to fall below the poverty line if no external assistance was provided to counteract this. No estimate of the differential loss by sex was provided. Losses women suffer to their asset base, such as cooking utensils or chickens raised to supplement food consumption or for sale, often go unrecorded, yet loss of this output can have a large economic cost; for example, the loss of subsistence egg production to Nicaragua was calculated to be between US$90,000 and US$120,000 per month immediately after hurricane Mitch (Bradshaw, 2004).

Perhaps the most obvious evidence of women and girls being more vulnerable to hazards would be if larger numbers of them die or are injured. Biological differences between men and women may suggest more women than men will die, but not because women are born physically "weaker" than men but because women's lack of access to resources such as food may mean they are undernourished lowering their physical strength to respond or because their socially constructed roles impact on their ability to respond effectively. Ironically, concerns about women's safety may also have a role to play as many parents in Bangladesh consider cyclone shelters to be unsafe for girls, for example, and prefer to leave them at home (Plan, 2011). The key constraints then are not biological but arise from the socially constructed roles of women and men, and the social norms that govern their behavior. Given this, in some cultures and situations men may be more likely to die, especially in sudden-onset events. Gendered ideology and gendered practice give rise to systematic gender differences in the perception of risk (Gustafson, 1998) and studies suggest white males are likely to rank a variety of risks significantly lower than women or minority groups (Finucane et al., 2000; Fothergill, 1996). Men may display more risk-taking behavior than women, explaining why men accounted for 73 percent of all road traffic deaths in 2002 (Waldron et al., 2005). In Latin America cultural ideas of maleness may mean men exhibit more risky behavior (Bradshaw, 2004), which may help explain why more men than women were said to have died post–hurricane Mitch (Gomáriz, 1999). Jonkman and Kelman's (2005) study of UK and European flood fatalities also points to higher risk-taking by men. In Australian bushfires, men are most often killed outside while attempting to protect the home and other assets, whereas most female and child fatalities occur while sheltering in the house or when fleeing, usually too late (Haynes et al., 2010). Lack of information, education, and engagement with preparedness activities means women when faced by a perceived risk often do not feel they can act, do not know when to act, or how to act on warnings (Tyler and Fairbrother, 2013) but when warnings are in the hands of women casualties overall are dramatically reduced (Buvinić, 1999).

How does this translate into actual impact? First and foremost, it is important to note reliable fatalities data, disaggregated by gender and generation, is still largely missing (Mazurana et al., 2011). Even the Emergency Events Database does not hold data disaggregated by gender and generation, not because they do not choose to or are unaware of the need for such data, but because the organizations who provide data for the database do not collect it (Fordham et al., 2007). In this context the study by Neumayer and Plümper (2007) is of particular note. The researchers constructed indicators of disaster magnitude and of women's socioeconomic status and explored how these relate to the size of the gender gap in life expectancy. They concluded that in countries where a disaster had occurred, where the socioeconomic status of women is low, more women than men die or die at a younger age. In the absence of gender disaggregated date on deaths due to disasters this study is often cited as evidence to support the notion that women are more vulnerable to hazardous events, but perhaps more importantly it suggests that some women are more vulnerable than others. The studies highlight that socioeconomic status and gender matter, but more importantly highlight the need for data to be disaggregated by gender and generation in order to understand and address this.

There is then growing evidence to suggest that natural hazards may have a different impact on men and women, on women and girls, and on different groups of women; however, the impact may not just be felt in terms of loss of life and material goods, but there may be longer term "secondary" impacts.

14.4 THE DOUBLE IMPACT OF DISASTERS ON WOMEN AND GIRLS

A disaster tends to be measured in terms of loss of life and/or economic output and social infrastructure, and measured immediately after an event, yet as the discussion below suggests the real disaster for women might develop over time and be experienced differently than for men (see also Bradshaw and Fordham, 2013). This raises the old question of what a "disaster" is (Quarantelli, 1998). If, as will be argued, the "real" disaster for women and girls is the "secondary" impacts felt in intangible changes to well-being or increased time burden then the notion of "disaster" needs to be reconceptualized if gendered risks are to be addressed (see Bradshaw, 2013a).

14.4.1 Time Burden

Perhaps the biggest impact on women and girls is the escalation of hours in the working day. Studies suggest women in general are "time poor" compared to men and postevent women often find themselves having to spend even more time in reproductive activities (because that work has become more difficult postdisaster), having to continue in, or to find productive work (because of the loss of other income sources), and are also increasingly targeted as "participants" in processes of reconstruction, spending time rebuilding homes, livelihoods, and communities. The elongation of an already long working day may have a negative impact on women's health and well-being and a knock-on effect on their daughters' well-being if they have to step in to help their mothers. Activities will also be made more burdensome by the context of insecurity in which they occur.

14.4.2 Violence against Women and Girls

Violence against women and girls (VAWG) is beginning to be taken seriously at all policy levels as its inclusion in the recent illustrative goals for the post-2015 development agenda highlights (HLP, 2013; OWG 2014). Violence, here understood as physical, emotional, and economic violence, and the threat of violence, are important in policy terms, not only because of the moral issues they raise, or the public health issues, but also since the threat of "domestic" violence keeps women in the home, whereas the construction of outside the home as populated by dangerous unknowns further constrains women's movements and actions and limits their life choices. Data on the prevalence and incidence of violence are patchy and commonly unreliable, but it is estimated that over 30 percent of women suffer physical partner abuse during "normal" times (Ellsberg et al., 2000).

Theoretically, we might expect to see an increase in violence or that violence has become more visible postevent (Merton, 1970), yet from the outset authors have stressed that for violence to increase there must be other factors present, that is, the disaster in itself is not enough to "cause" violence (Barton, 1970). It could be argued that disasters, or their aftermath, reveal existing actual levels of violence or the potential for violence. Those who are displaced and forced to live communally may continue to carry on their private lives in public, including violence. It may be the case that the levels of violence increase due to the frustration felt by men unable to fulfill their socially constructed gendered roles of protector and provider. It may be the case that the nature of violence changes, with higher levels of stranger violence as social systems and structures of protection break down (see Bradshaw, 2013a; Phillips et al., 2010 for discussion). In particular, large-scale "refuges" and displaced people's camps have seen high levels of sexual violence by "unknowns" (Delaney and Shrader, 2000; Duramy, 2011; Dynes and Rodríguez, 2007) and their design has been critiqued for facilitating this. Although sexual violence and forced sex is often assumed to increase post-event, care needs to be taken as dramatic media reports may promote that perception, whereas data to support this are still lacking (GSDRC, 2013). Moreover, intimate partners and close family may remain the highest risk to women and girls: a study in internally displaced person camps in northern Uganda found women were 8–10 times more likely to experience violent assault by their husband than by a stranger (IFRC, 2012: 85).

The 2012 World Disasters Report noted that, although violence in a disaster context might be complex, it is not inevitable; rather it is predictable and preventable (IFRC, 2012: 120). VAWG is increasingly a humanitarian concern; however, survivors of violence still suffer from the continued construction of humanitarian response as short term, which means, in the medium term, their needs fall into the "gap" between when the emergency response ends and "normal" services resume, or in the developing world context, between relief and development (Bradshaw, 2013b).

14.4.3 Psychosocial Impact

Living through a violent and destructive natural event is assumed to bring further distress to those with mental health issues and to provoke emotional distress in those who have not suffered previously. The International Federation of Red Cross and Red Crescent Societies suggests addressing psychosocial needs should be based on the principle that most acute stress problems during emergencies are best managed without medication, following the principles of "psychological first aid" (IFRC, 2009). However, the interaction of long-term conflict and rapid-onset "natural" disaster can provoke severe psychosocial response. Although the applicability of post-traumatic stress disorder (PTSD) in a non-Western context has been questioned (Norris et al., 2001; PTSD Research Quarterly, 2009) as has the "medicalization of distress" (Summerfield, 1999), PTSD is increasingly recognized as an issue in the developing as well as "developed" world context postevent.

A study in Nicaragua that asked people about the "emotional impact" of hurricane Mitch found 74 percent of those who reported being affected were women (CIET/CCER, 1999a,b), and research in the occupied Palestinian territories found exposure to traumatic events was associated with more severe psychiatric disorders among women than men (Punamäki et al., 2005). Young people may display different symptoms than adults, and variation exists between the very young and adolescents (Hamblen, 2006). Gender differences also occur, with boys becoming more aggressive and antisocial, and often taking longer to recover than girls who are more verbal and show more distress. Why women suffer more or more severely is less to do with them being women than to do with their position and situation in society as women. A study from Tamil Nadu (Kumar et al., 2007) showed PTSD was higher among individuals with no household incomes and those who were illiterate. Because women tend to occupy a lower socioeconomic position than men, they may be more susceptible to psychological problems. Emotional health is also related to physical health and, once again, due to reproductive health risks and the social and economic limitations around dealing with these, women tend to be less physically well than men. Caution needs to be exercised when evaluating mental health data, however, as history demonstrates that women have been diagnosed as "ill" for displaying attitudes considered radical at the time or actions that are outside the gender norm (Showalter, 1985).

In contrast to PTSD, posttraumatic growth suggests positive change is possible after a traumatic event or recognizes that at least people often feel a mixture of positive and negative emotions (Tedeschi and Calhoun, 2004). Research suggests some people find positive changes arise from struggling with the aftermath of trauma and develop a deeper appreciation of life. One woman survivor of the Marmara earthquakes in Turkey in 1999 reported that the changes in opportunities postearthquake were so profound that she had "adopted a new identity" (Ozsoy and Sariipek, 2010: 103), whereas research post-Katrina highlights for some young women the event was a catalyst for them feeling more able to openly express their sexual identity (Overton, 2014).

14.4.4 Deterioration in Reproductive and Sexual Health

Although women and girls may be more susceptible to health threats given their lower entitlements to resources such as food and water, they also face specific health issues related to their sex. Pregnant women, and those women who give birth soon before, during, or immediately after an event, face specific health concerns. Nursing mothers, for example, are particularly susceptible to malnutrition and dehydration and this might be assumed to provoke higher mortalities and morbidity. However, lack of centralized, systematic records postevent means the extent of the problem is largely unknown.

The World Health Organization suggests a core package of minimum reproductive health interventions should be put in place in emergency settings; however, only limited awareness exists of this and limited funding. Even as late as 2005 in the Indian Ocean tsunami response, "hygiene kits" for women, although provided, were underresourced (Nobel de Silva, 2006). Often products, where available, are sourced from overseas and these may not be culturally appropriate and thus not used, and privacy issues and cramped living exacerbate the situation, leaving many women suffering from urinary tract and other infections (Ha-Redeye, 2006).

Although sexuality is seldom discussed in the emergency context, the specific sexual health needs of those living outside the heteronormative expectations of a society may not be met and the postevent context may leave them more vulnerable to assault and discrimination. An increase in sexual assaults can lead to an increase in sexually transmitted infections and, in a situation of limited access to contraception, unplanned pregnancies for women forced to have sex with known or unknown men.

14.4.5 Poverty and Insecure Employment

It is often suggested that women and girls postdisaster will be forced into sex work or into "transactional sex," where women and girls are coerced into providing sex in exchange for food and other basic goods or "protection." For example, a study by United Nations High Commissioner for Refugees (UNHCR) found that the practice of women and adolescent girls engaging in transactional sex in Port-au-Prince after the earthquake was "widespread" (UNHCR, 2011), whereas a survey in a refugee camp and surrounding host villages near Tanzania's border with Democratic Republic of the Congo found refugees were two-and-a-half times more likely to have experienced forced sex and three times more likely to have engaged in transactional sex (IFRC, 2012: 91). In Sierra Leone in 1999, 37 percent of those involved in sex work or transactional sex were found to be under the age of 15 years (IFRC, 2012: 82). However, it is not clear the extent to which levels of such sexual exchanges increase postevent, or if it is the nature of sexual exchanges that changes, in terms of with whom women engage in sex or which women engage in such transactions, since large-scale studies do not exist, especially tracking change over time. The impact on the self-perception of the women and how they are viewed by others is also an unknown.

14.4.6 Changes in Networks and Family Support

Although community and familial networks are important for ensuring survival during and immediately after the event as well as therapeutic communities, "corrosive" communities have been noted (see Brunsma et al., 2007 for discussion) and at the very least social and familial networks may be put under pressure. Although women may have better and wider social networks than men, their ability to effectively transform good networks into tangible resources may be less than a man's since her access to "household" resources may be determined by her ability to negotiate with a male partner (Bradshaw, 2013a). All people may suffer a loss of friendship and kinship networks postevent and for women loss of networks may have implications for child care and their ability to (re)engage in income-generating activities as well as increasing their vulnerability to violence (Duramy, 2011). On the other hand, calling on familial networks for support may allow the reestablishment or reenforcement of familial power relations, which further removes agency and control from women in particular (Hoffman, 1998).

14.4.7 Change in Self-Perception

A less well-recognized consequence of disasters might be women's perception of a loss of status and dignity. Being in a refuge may add to this through the "inhuman" conditions found in camps such as those in Haiti after the 2010 earthquake (Horton, 2012). The media coverage, which constructed men as violent criminals and women as prostitutes, meant the women's desire for respect and dignity was also undermined, something also found post-Katrina where African-American women had suffered from being constructed as undeserving recipients of benefits long before the event (Sterett, 2012). However, it is not only external perceptions that may have a negative impact, but also processes of reconstruction may not always have the positive outcomes assumed. In post-Mitch Nicaragua, female heads of household, despite continuing with their reproductive and productive roles and taking an active role in reconstruction, were less likely to recognize their own contribution to the household and named an adult son as the main contributor instead (Bradshaw, 2002). Differences in how disasters are understood may also mean not all disasters are "disasters" for all women. Cupples (2007) and Fordham et al. (2007) research in very different parts of the world highlight "disasters" can be seen as "windows of opportunity" in which women find their prospects have changed radically for the better; at least for a time.

How women's perception of self is changed by an event remains an issue still to be researched, and how response and reconstruction impact upon this is still largely unknown. The impact of disasters on young women is even less known, and how adolescent girls grow up under the shadow of disasters is only now starting to be explored.

14.4.8 Specific Issues Faced by Adolescent Girls

During the relief period, adolescent girls are often seen only as "daughters" with the focus being on supporting those who care for them or as "orphans" when the focus lies on finding someone to care for them. Reconstruction projects similarly do not target adolescent girls as beneficiaries of aid, if they are not seen to be providers for others, as "workers," or mothers, but instead constructed as dependents. Girls then fall between the gaps in programming. This is exacerbated in the developing world by the division between disaster and development, meaning a 13-year-old girl who survives an event effectively grows up in "transition." By the time development funding returns, girls of a disaster will be young women, and the bounded choices they were able to make, or had made for them, will have determined to a large extent their "development" possibilities (Bradshaw, 2013b).

While there is limited information on women, there is even less on adolescent girls especially as there have been no systematic follow-up studies tracking the situation of girls after the reconstruction period is "over" to determine if the impact of the issues highlighted here persist in the longer term. This is compounded by the fact the concept of "girls" is a contentious one and "adolescence" may be a short or nonexistent period, and at very least is definitionally difficult to pin down. The notions of child and adolescent, young and adult woman then should be seen as fluid and not determined by age but by life course and context.

Notions of childhood and youth are associated, especially in the developed world, with education. Willingness to school girls is far more strongly determined by income and the broader costs of education, than is the case for boys. Thus, losses postevent, including the need to replace books and uniforms, may mean girls more than boys are delayed in returning to the classroom, while insecurity of roads and the fear of violence might further limit their return. The opportunity cost to leaving girls in school may increase postevent if they have to replace their mother's role in the household due to her death, disability, or migration, or as their mothers are occupied with reconstruction. Even if they remain in school, their reduced time for learning due to household chores, compounded by stress and trauma, may adversely affect their performance.

While adolescent girls juggle productive, reproductive, and community management roles, aid agencies prioritize and promote their role as "school girl," blurring their dual identity as adult/child. This blurring is important as the specific needs of adolescent girls may be ignored if they are classified as "children" rather than young women, in particular their sexual and reproductive health needs are often bottom of the checklist of relief agencies, if they feature at all (UNFPA, 2013).

Yet adolescent girls may be sexually active, sometimes by choice, but other times due to lack of choices. As noted above, a rise in transactional sex is suggested to occur postevent and this may include the trafficking of women, girls, and boys across countries. Trafficking may increase but it is important to note that the forced movement of children and adolescents by their parents for financial gain, including being sold as domestic servants or being "lent" to relatives, may be widespread in some cultures.

While some girls leave behind the school room for precarious employment, for others their identity changes as they move from being daughter to wife, from child to mother. Forced and early marriage is an existing "problem" across the world, and in the developing world one out of seven girls marries before the age of 18 years (UNFPA, 2012). An increase in early marriages postevent has been suggested to have occurred in Haiti, Pakistan, and in various countries impacted by the Indian Ocean tsunami. Early marriage may also be associated with slow-onset "disasters" and food insecurity with girls being exchanged for food, and this may increase as part of a climate change (mal) "adaptation" strategy (Deen, 2010).

Young married women may face extreme isolation and this may be something more generally shared by adolescent girls. In general, the ability to restore friendship networks may be more difficult for girls, given the increase in time helping their mothers, and as parents may be more protective of them and not encourage them to go "out" and meet new friends (IFRC, 2012). This may leave girls more isolated and without emotional support postevent with long-term implications for their mental health and self-perception. In contrast, for some young women post-Katrina the changes may have been more positive with young women establishing new friendships and social support networks (Overton, 2014).

14.5 GENDERED CAPACITIES: INCLUDING WOMEN AND GIRLS IN DRR

While women's vulnerability has become assumed, their capacity to respond not only individually but also collectively to hazards is less well documented. While often depicted as dependent on heroic men (often Western military personnel and male relief workers arriving to "save" them), women do take actions to save themselves and others. A study on post-Mitch Nicaragua documented how women helped to save people and possessions. However, women's actions were constructed in postevent narratives as "helping" men and even this help was often forgotten over time (Bradshaw, 2001, 2002). Typically, women's actions are associated with rescuing other women, children, and the infirm, not men. Yet, Fordham's (2006) study highlights that through engagement in local-level DRR women come to understand they can take a lead role at all stages, including by rescuing men. Women's engagement in early warning systems and other preparedness activities can reduce the need for them and others to be "rescued" (Buvinić, 1999).

Where they have the opportunity to do so, research suggests that women are more likely to receive and act upon early warnings (Fothergill, 1996; Fordham, 2001; Enarson, 2006). Women have their own social networks, often formed through parenting and caring roles, which make them best able to identify those at greatest risk in the community. Social networks and social capital are also important resources to help respond to a crisis in the short and medium term. Girls too can be important communicators of risks and risk management options especially in terms of communications with parents, adults, or those outside the community (Plan UK, 2013).

At the international level, DRR initiatives have also made tentative steps toward including gender into policy frameworks. The Hyogo Framework for Action (HFA 2005–2015), which provides the framework for DRR at an international level, in the opening section states that a gender perspective should be integrated into all disaster risk management policies, plans, and decision-making processes. However, as this did not even result in gender being integrated throughout the HFA itself, the commitment to gender seems to have been more rhetoric rather than reality. In a survey in 2010 by the International Strategy for Disaster Reduction the theme that scored lowest on both expertise and recognition of the need to strengthen knowledge of this area was "gender," with only 13 percent suggesting they had expertise in the area, yet only 13 percent suggesting it was an area that needs more expertise.

It seems that despite the growing evidence to support the notion of disasters as gendered events, there is still a long way to go to convince DRR professionals of the importance of adopting a gender approach. In contrast, in development programs the benefits of involving women in projects seem more accepted.

14.6 "ENGENDERING" POLICY INITIATIVES

The parallels between how gender and development and gender and disasters emerged and developed have been noted in the literature (see Bradshaw, 2015). The former is more "advanced" not only in terms of the achievements made to date to integrate gender into development, but also in terms of how this has been systematically documented. For the last 15 years the global development agenda has been framed by the Millennium Development Goals. While critiqued by gender activists from the outset (see WICEJ, 2004) the 10 goals do include a stand-alone goal focused on gender equality and women's empowerment, as well as a goal focused on maternal mortality. Despite their limitations, they suggest an acceptance of gender as an important issue, something that is still lacking from DRR policy. A consideration of why gender has become "accepted" within development suggests, however, that "inclusion" can bring its own problems.

Based on evidence that suggests that societies that discriminate by gender tend to experience less rapid economic growth and poverty reduction than societies that treat males and females more equally, it became understood by actors such as the World Bank that social gender disparities produce economically inefficient outcomes (World Bank, 2001). The World Bank notes that "gender-sensitive" development strategies "contribute significantly to economic growth, as well as equity objectives" (World Bank, 2002) leading to the assertion that reducing gender inequalities is "smart economics" (World Bank, 2006). Yet, while the evidence highlights gender equality is good for growth, it also highlights that economic growth is not necessarily good for gender equality.

Within this economic development discourse women have been recognized as efficient providers of goods and services to others (see Bradshaw et al., 2013). Women are the preferred providers of "social protection" in programs such as Conditional Cash Transfers designed to reduce poverty in the short and longer term, programs that are increasingly being utilized in postdisaster settings (Heltberg, 2007; Vakis, 2006). Women are targeted with resources because they are seen to use them more efficiently than men (Bradshaw, 2008; Molyneux, 2007), yet this targeting often leads policy makers to claim these are gendered projects bringing women's empowerment. Jackson (1998) noted in the late 1990s that poverty reduction programs targeted at women became equated to gender equality projects, yet the assumption that poverty reduction programs automatically bring gender equality can be challenged. She pointed to the need to "rescue gender from the poverty trap," highlighting if gender equality is the goal then gender equality projects are the only means to achieve this goal.

Translating the lessons learned from gender and development to the field of disasters highlights that if losses are to be avoided or lessened, and if women are more vulnerable to disaster loss as suggested by the evidence presented above, then drives to promote the role of women and girls in disaster response and risk reduction could be justified in "value for money" terms alone. Including women and girls in all the stages of the disaster "cycle" can improve the outcomes or bring "efficiency" gains. Yet, as noted from the development literature, while including women and girls in this way will improve disaster response and reduce disaster risk, unless DRR and response activities specifically address gender inequalities they will not change the situation and position of women. They will not then fundamentally change women's and girls' risk of disaster.

14.7 CONCLUDING COMMENTS

Reducing the impact of disasters on women and girls cannot be solved by disaster-based actions alone. It is inequities in the everyday, and not just in times of disaster, that create greater risk and reduce life chances for women and girls. Thus, action across the gender–disaster–development nexus is key to creating lasting change. This would imply not only reconceptualizing "disaster" to construct it as a "development" concern, but also understanding that to reduce disaster risk for women demands a focus on reducing gender inequalities. Without such fundamental change in how disasters are understood women will continue to be at greater risk of a natural hazard becoming a disaster for them. They will face the "double disaster" of material losses from the event itself being compounded by longer term losses in health and well-being, through violence, insecurity, and time poverty. While there is growing recognition of the need to focus on women in disasters, the double identity of adolescent girls as child and as adult, rather than leading to a double response, sees them falling between the gaps in programming. The disaster for them will be measured not only in the short or long term, but also in their lifetime.

ACKNOWLEDGMENTS

This paper draws on a report commissioned by the UK Department for International Development (DFID) on "Women and Girls in Disaster." We would like to thank DFID and all those involved for their support. We would also like to thank Brian Linneker, and all those who commented on earlier drafts of the paper, for their useful suggestions.

REFERENCES

Barton, A.H., 1970. Communities in Disaster: A Sociological Analysis of Collective Stress Situations. Doubleday, New York.

Beck, U., 1998. Risk Society: Towards a New Modernity. Sage, London.

Blaikie, P., Cannon, T., Davis, I., Wisner, B., 1994. At Risk: Natural Hazards, People's Vulnerability, and Disasters (also published in 2003 with these authors listed: B. Wisner, P. Blaikie, T. Cannon, I. Davis). Routledge, London, New York.

Bradshaw, S., 2001. Dangerous Liaisons: Women, Men and Hurricane Mitch/Relaciones Peligrosas: Mujeres, Hombres y el Mitch. Puntos de Encuentro, Managua, Nicaragua.

Bradshaw, S., 2002. Exploring the gender dimensions of reconstruction processes posthurricane Mitch. J. Int. Develop. 14, 871–879.

Bradshaw, S., 2004. Socio-economic Impacts of Natural Disasters: A Gender Analysis, 32. United Nations Economic Commission for Latin America and the Caribbean (ECLAC) Serie Manuales. Available at: http://www.eclac.org/publicaciones/xml/3/15433/lcl2128i.pdf.

Bradshaw, S., 2008. From structural adjustment to social adjustment: a gendered analysis of conditional cash transfer programmes in Mexico and Nicaragua. Global Soc. Policy 8 (1), 188–207.

Bradshaw, S., 2013a. Gender, Development and Disasters. Edward Elgar, Northampton.

Bradshaw, S., 2013b. Bridging the humanitarian - development divide' opinion piece in. In: Double Jeopardy: Adolescent Girls and Disasters because I Am a Girl Report: The State of the World's Girls 2013. Plan International, UK.

Bradshaw, S., 2015. Engendering Development and Disasters (forthcoming in Disasters, January 2015).

Bradshaw, S., Castellino, J., Diop, B., May 2013. Women's Role in Economic Development: Overcoming the Constraints. Background paper prepared for the High Level Panel on the post-2015 agenda. UN Sustainable Development Solutions Network. http://www.post2015hlp.org/the-report/.

Bradshaw, S., Fordham, M., August 2013. Women and Girls in Disasters. Report Produced for the Department for International Development (DFID), UK. https://www.gov.uk/government/uploads/system/uploads/attachment_data/file/236656/women-girls-disasters.pdf.

Brunsma, D., Overfelt, D., Picou, J.S., 2007. Introduction: Katrina as paradigm shift: reflections on disaster research in the twenty first century. In: Brunsma, D., Overfelt, D., Picou, J.S. (Eds.), The Sociology of Katrina: Perspectives on a Modern Catastrophe. Rowman and Littlefield, Maryland USA, pp. 1–24.

Buvinic, M., 1999. Hurricane Mitch: Women's Needs and Contributions. Inter-American Development Bank, Sustainable Development Department Technical Papers Series.

Cardona, O.D., 2004. The need for rethinking the concepts of vulnerability and risk from a holistic perspective: a necessary review and criticism for effective risk management. In: Bankoff, G., Frerks, G., Hilhorst, T. (Eds.), Mapping Vulnerability: Disasters, Development and People. Earthscan, London, pp. 37–51.

Chant, S., 2003. Female Household Headship and the Feminisation of Poverty: Facts, Fictions and Forward Strategies. London School of Economics and Political Science, London, UK. Gender Institute. Available at: http://eprints.lse.ac.uk/574/ (14.08.12).

Chant, S., 2008. Dangerous equations? how female-headed households became the poorest of the poor: causes, consequences and cautions. In: Momsen, J. (Ed.), Gender and Development: Critical Concepts in Development Studies. Routledge, London, UK.
Chowdhury, A.M.R., Bhuyia, A.U., Choudhury, A.Y., Sen, R., 1993. The Bangladesh cyclone of 1991: why so many people died. Disasters 17 (4), 291–304.
CIET-CCER, 1999a. Social Audit for the Emergency and Reconstruction: Phase 1-April 1999. Civil Coordinator for Emergency and Reconstruction/CIETinternational/Carqui Press, Managua, Nicaragua.
CIET-CCER, 1999b. Auditoría Social para la Emergencia y la Reconstrucción - Fase 2 -Noviembre. Civil Coordinator for Emergency and Reconstruction/CIETinternational/Carqui Press, Managua, Nicaragua.
Cupples, J., 2007. Gender and hurricane Mitch: reconstructing subjectivities after disaster. Disasters: J. Disaster Studies, Policy Manage. 31 (2), 155–175.
Deen, T., 2010. Famine Marriages - Just One By-Product of Climate Change. webpost 9/3/10 on. www.globalchange.me.
Delaney, P., Shrader, E., 2000. Gender and Post-Disaster Reconstruction: The Case of Hurricane Mitch in Honduras and Nicaragua. Decision review draft presented to the World Bank, January 2000. Available online: www.gdnonline.org.
DFID, 2004. Adaptation to Climate Change: Making Development Disaster-Proof. DFID Key Sheet 06. Department for International Development, UK.
Duramy, B.F., 2011. Women in the aftermath of the 2010 Haitian earthquake. Emory Int. Law Rev. 25, 1193–1215.
Dynes, R., Rodríguez, H., 2007. Finding and framing Katrina: the social construction of disaster. In: Brunsma, D., Overfelt, D., Picou, J.S. (Eds.), The Sociology of Katrina: Perspectives on a Modern Catastrophe. Rowman and Littlefield, Maryland USA, pp. 23–34.
Ellsberg, M., et al., 2000. Candies in hell: women's experiences of violence in Nicaragua. Soc. Sci. Med. 51 (11), 1595–1610.
Elson, D., 1991. Male bias in macro-economics: the case of structural adjustment. In: Elson, D. (Ed.), Male Bias in the Development Process. Manchester University Press, Manchester, pp. 164–190.
Enarson, E., 2006. SWS Fact Sheet: Women and Disaster. Available at: http://www.socwomen.org/wp-content/uploads/2010/05/fact_10-2006-disaster.pdf (accessed 22.09.14).
Finucane, M.L., Slovic, P., Mertz, C.K., Flynn, J., Satterfield, T.A., 2000. Gender, race, and perceived risk: the 'white male' effect. Health, Risk Soc. 2 (2), 159–172.
Fordham, M., 2001. Challenging boundaries: a gender perspective on early warning in disaster and environmental management. In: United Nations Division for the Advancement of Women (DAW) International Strategy for Disaster Reduction (ISDR) Expert Group Meeting on Environmental Management and the Mitigation of Natural Disasters: A Gender Perspective, Ankara.
Fordham, M., 2006. Unpublished Field Notes: El Salvador 2006, Plan International.
Fordham, M., Ariyabandu, M.M., Gopalan, P., Peterson, K., 2007. Please Don't Raise Gender Now - We're in an Emergency! 2006 World Disasters Report International Federation of Red Cross and Red Crescent Societies.
Fothergill, A., 1996. Gender, risk and disaster. Int. J. Mass Emerg. Disasters 14 (1), 33–56.
Gomáriz Moraga, E., 1999. Género y desastres: Introducción conceptual y análisis de situación (Paper prepared for the IDB technical meeting on the effects of Hurricane Mitch on women and their participation in the reconstruction of Central

America). Available from: http://www.cridlac.org/digitalizacion/pdf/spa/doc11630/doc11630.htm.
GSDRC, 2013. Helpdesk Research Report: Violence against Women and Girls in Haiti. Governance and Social Development Resource Centre.
Gustafson, E., 1998. Gender differences in risk perception: theoretical and methodological perspectives. Risk Anal. 18, 6.
Ha-Redeye, O., 2006. "Engendering" change in disaster response: increasing women in leadership roles. In: Provost, M. (Ed.), International Centre for Emergency Management Studies 2006 Gender and Disaster in Canada: New Thinking, New Directions, Record of Proceedings, October 31st to November 2nd, 2006.
Hamblen, J., 2006. PTSD in Children and Adolescents: A National Center for PTSD Fact Sheet. Available at: http://www.ptsd.va.gov/professional/treatment/children/ptsd_in_children_and_adolescents_overview_for_professionals.asp.
Haynes, K., Handmer, J., McAneney, J., Tibbits, A., Coates, L., 2010. Australian bushfire fatalities 1900–2008: exploring trends in relation to the 'prepare, stay and defend or leave early' policy. Environ. Sci. Policy 13 (3), 185–194.
HLP, 2013. A New Global Partnership: Eradicate Poverty and Transform Economies through Sustainable Development. The report of the High level Panel of Eminent Persons on the Post-2015 Development Agenda. http://www.post2015hlp.org/the-report/.
Heltberg, R., 2007. Helping South Asia cope better with natural disasters: the role of social protection. Develop. Policy Rev. 25 (6), 681–698.
Holland, P., 2008. An Economic Analysis of Flood Warning in Navua, Fiji. Pacific Islands Applied Geoscience Commission (SOPAC), Suva.
Hoffman, S., 1998. Eve and Adam among the members: gender patterns after the Oakland Berkeley firestorm. In: Enarson, E., Morrow, B. (Eds.), The Gendered Terrain of Disasters, Westport. Praeger, Connecticut and London, pp. 55–62.
Horton, L., 2012. After the earthquake: gender inequality and transformation in post-disaster Haiti. Gender Develop. 20 (2).
IFRC, 2009. Psychosocial Interventions: A Handbook. International Federation of Red Cross and Red Crescent Societies Reference Centre for Psychosocial Support, Denmark.
IFRC, 2012. World Disasters Report: Focus on Forced Migration and Displacement. International Federation of Red Cross and Red Crescent Societies.
Jackson, C., 1998. Rescuing gender from the poverty trap. In: Pearson, R., Jackson, C. (Eds.), Feminist Visions of Development: Gender Analysis and Policy. Routledge, London, pp. 39–64.
Jonkman, S.N., Kelman, I., 2005. An analysis of causes and circumstances of flood disaster deaths. Disasters 29 (1), 75–97.
Kumar, M.S., Murhekar, Manoj V., Hutin, Yvan, Subramanian, Thilakavathi, Ramachandran, Vidya, Gupte, Mohan D., 2007. Prevalence of posttraumatic stress disorder in a coastal fishing village in Tamil Nadu, India, after the December 2004 Tsunami. Am. J. Public Health 97 (1), 99–101.
Mazurana, D., Benelli, P., Gupta, H., Walker, P., 2011. Sex and Age Matter: Improving Humanitarian Response in Emergencies. CARE International, Feinstein International Center, Tufts University, UN OCHA - Office for the Coordination of Humanitarian Affairs. Available at: https://www.care.org/careswork/whatwedo/relief/docs/sex-and-age-disag-data.pdf.
Merton, R., 1970. Foreword. In: Barton, A.H. (Ed.), Communities in Disaster. Doubleday, Garden City, NY.
Molyneux, M., 2007. Two cheers for conditional cash transfers. IDS Bullet. 38 (3), 69–75.
Moser, C., 1993. Gender and Development Planning. Routledge, London.

Neumayer, E., Plümper, T., 2007. The gendered nature of natural disasters: the impact of catastrophic events on the gender gap in life expectancy, 1981–2002. Ann. Assoc. Am. Geogr. 97 (3), 551–566.

Nobel de Silva, S., 2006. The trauma of tsunami-affected women in Sri Lanka. In: Domroes, M. (Ed.), After the Tsunami: Relief and Rehabilitation in Sri Lanka. Mosaic Books, New Delhi, pp. 61–70.

Norris, F.H., Weisshaar, D.L., Conrad, M.L., Diaz, E.M., Murphy, A.D., Ibañez, G.E., 2001. A qualitative analysis of posttraumatic stress among Mexican victims of disaster. J. Trauma. Stress 14, 741–756.

Oliver-Smith, A., 2002. Theorizing disasters: nature, power and culture. In: Hoffman, S., Oliver-Smith, A. (Eds.), Catastrophe & Culture: The Anthropology of Disaster. School of American Research Press, New Mexico, pp. 23–48.

Overton, L., 2014 forthcoming. Girls interrupted: girls growing up post-Katrina, (Ph.D. thesis). Middlesex University.

Oxfam, October 2012. Post-earthquake response and reconstruction gender-sensitive advocacy in Indonesia. Gender Equality Emerg. Oxfam GB, Oxford.

Oxfam, 2005. The Tsunami's Impact on Women. Oxfam Briefing Note, Oxfam International, March 2005. Available at http://www.preventionweb.net/files/1502_bn050326tsunamiwomen.pdf.

OWG, 2014. Outcome Document - Open Working Group on Sustainable Development Goals. Open Working Group on Sustainable Development Goals. United Nations. Available at: http://sustainabledevelopment.un.org/content/documents/4518SDGs_FINAL_Proposal%20of%20OWG_19%20July%20at%201320hrsver3.pdf.

Ozsoy, A., Sariipek, D.B., 2010. Women and their socio-economic situations. In: Dasgupta, S., Siriner, I., De, P.S. (Eds.), 2010 Women's Encounter with Disaster. Frontpage, Kolkata.

Phillips, B.D., Jenkins, P., Enarson, E., 2010. Violence and disaster vulnerability. In: Phillips, B., Thomas, D., Fothergill, A., Blinn-Pike, L. (Eds.), Social Vulnerability to Disasters. CRC Press, Boca Raton, Fl, pp. 279–306.

Plan, 2011. Weathering the Storm: Adolescent Girls and Climate Change. Plan International, UK.

Plan UK, 2013. In: Double Jeopardy: Adolescent Girls and Disasters because I Am a Girl Report: The State of the World's Girls 2013. Plan International, UK.

PTSD Research Quarterly 2009. Special edition on the 'The impact of disasters and political violence on mental health in latin America,' 20(4).

Punamäki, R.L., Komproe, I.H., Qouta, S., Elmasri, M., de Jong, J.T., 2005. The role of peritraumatic dissociation and gender in the association between trauma and mental health in a Palestinian community sample. Am. J. Psychiatry 162 (3), 545–551.

Quarantelli, E.L. (Ed.), 1998. What Is a Disaster? Perspectives on the Question. Routledge, London and New York.

Showalter, E., 1985. The Female Malady: Women, Madness and English Culture, 1830–1980. Penguin Books, New York.

Sterett, S., 2012. State policy and disaster assistance: listening to women. In: David, E., Enarson, E. (Eds.), The Women of Katrina: How Gender, Race and Class Matter in an American Disaster. Vanderbilt University Press, Nashville, pp. 118–129.

Summerfield, D., 1999. A critique of seven assumptions behind psychological trauma programs in war-affected areas. Soc. Sci. Med. 48, 1449–1462.

Tedeschi, R., Calhoun, L., 2004. Posttraumatic growth: conceptual foundations and empirical evidence. Psychol. Inquiry 5 (1), 1–18.

Tyler, M., Fairbrother, P., 2013. Bushfires are "men's business": the importance of gender and rural hegemonic masculinity. J.Rural Studies 30, 110–119.

UNFPA, 2012. From Childhood to Womanhood: Meeting the Sexual and Reproductive Health Needs of Adolescent Girls. Fact Sheet: Adolescent Girls' Sexual and Reproductive Health Needs. The United Nations Population Fund, December 2012. Available at: http://www.unfpa.org/webdav/site/global/shared/factsheets/srh/EN-SRH%20fact%20sheet-Adolescent.pdf.

UNFPA, June 2013. Report Calls for Sharper Focus on Adolescent Sexual Health Needs in Humanitarian Response. UNFPA.

UNHCR, 2011. Driven by Desperation: Transactional Sex as a Survival Strategy in Port-au-Prince IDP Camps. Haiti, Port-au-Prince. http://www.urd.org/IMG/pdf/SGBV-UNHCR-report2_FINAL.pdf.

Vakis, R., 2006. Complementing Natural Disasters Management: The Role of Social Protection, Social Protection Discussion Paper 0543. The World Bank, Washington DC.

Waldron, I., et al., 2005. Trends in gender differences in accidents mortality: relationships to changing gender roles and other societal trends. Demographic Res. 2005 (13), 415–454.

WEDO, 2008. Gender, Climate Change and Human Security: Lessons from Bangladesh, Ghana and Senegal. Prepared for ELIAMEP, May 2008. WEDO, New York. Available at: http://www.wedo.org/wp-content/uploads/hsn-study-final-may-20-2008.pdf.

WICEJ (Ed.), 2004. Seeking Accountability on Women's Human Rights: Women Debate the Millennium. Women's International Coalition for Economic Justice, New York.

Willinger, B., Knight, J., 2012. Setting the stage for disaster: women in New Orleans before and after Katrina. In: David, E., Enarson, E. (Eds.), The Women of Katrina: How Gender, Race and Class Matter in an American Disaster. Vanderbilt University Press, Nashville, pp. 55–75.

World Bank, 2001. Engendering Development through Gender Equality in Rights, Resources, and Voice. Oxford University Press, New York.

World Bank, 2002. Integrating Gender into the PRS Monitoring and Evaluation. World Bank PRSP Source Book.

World Bank, September 2006. Gender Equality as Smart Economics: A World Bank Group Gender Action Plan (Fiscal Years 2007–10). World Bank Group.

Section 3

Cross-Disciplinary and Non-Mainstream Futures of Dealing with Hazards, Risks and Disasters in Society

Chapter 15

Disaster Risk Reduction in the Shadow of the Law

Michael Eburn
ANU College of Law, The Australian National University, Canberra, ACT, Australia

ABSTRACT

Law is a tool that can be used by interest groups to advance what they perceive to be their best interests, whether that is advocating for their right to build homes in hazard-prone environments or advocating against such development because of the social and economic costs of disasters. Identifying how interests are to be balanced, and responsibility for disaster risk reduction is to be shared, requires more than knowledge of local law. The allocation of responsibility must be negotiated with stakeholders, negotiation that takes place in the "shadow of the law," that is, with an expectation of what the law requires and how others may behave to comply with legal obligations.

There are, however, some (reasonably) fixed legal rules that set the outer limits of possible negotiation. With reference to international law and the laws of Australia, the United Kingdom, and the United States (nations with a shared legal history based on the English common law), this chapter will identify some legal boundaries that help define the limits of shared responsibility. Identifying these boundaries helps to show where law's shadow falls, and so empowers stakeholders by helping define what is possible in the negotiations for a safer community.

The modern view of disaster management is that governments, individuals, and communities must share responsibility for preparing for, responding to, and recovering from the impact of natural and other hazards (Victoria, 2010; COAG, 2011; Cabinet Office, 2013; FEMA, 2011, 2013). Recognizing that shared responsibility *exists* does not, however, determine *how* that responsibility is to be shared. As the Royal Commission inquiring into the 2009 Victorian (Australia) bushfires said "(s)hared responsibility does not mean equal responsibility" and "there are some areas in which the State should assume greater responsibility than the community" (Victoria, 2010, para 9.1). It is important for governments, communities, individuals, and the nongovernment sector to negotiate how responsibility is to be shared, who is to be responsible for what, and who owns the risk that comes with natural hazards.

Hazards, Risks, and Disasters in Society. http://dx.doi.org/10.1016/B978-0-12-396451-9.00015-9

It is said that bargaining or negotiation occurs in the shadow of the law (Depoorter, 2010; McAdams and Nadler, 2008; Stevenson and Wolfers, 2006; Mnookin and Kornhauser, 1979). People who have an interest in negotiating their respective rights do so with some expectation of how, if they cannot reach agreement, the law would resolve their negotiations: "the outcome that the law will impose if no agreement is reached gives each (party) certain bargaining chips-an endowment of sorts" (Mnookin and Kornhauser, 1979, p. 968). It is not only what outcome the law will impose that is relevant, but also an expectation of how others, if they comply with the law, will behave

...where people have an incentive to coordinate their behaviour, law can provide a framework for understanding and predicting what others are likely to do. The shadow cast by law in these situations helps people predict not what a court is likely to do but rather what other people who are also aware of the law are likely to do.

McAdams and Nadler (2008), p. 867.

It is assumed that governments and individuals have an incentive to coordinate their behavior in order to maximize resilience and minimize the impact of hazards, both personally and at a whole of community level. If shared responsibility requires negotiation, knowledge of the law allows the parties to understand each other's legal obligations so that they can know what to expect from the other stakeholders, how the law might deal with any disputes or claims where agreement is not reached, and what rights and interests may be compromised or traded to achieve a desired outcome. Knowledge of the law will define the shadows within which the negotiations take place.

In the absence of legal knowledge, shared responsibility could be placed anywhere on the spectrum ranging from complete regulation to a complete laissez-faire approach to emergency management. The theoretical spectrum is shown Figure 15.1.

Neither of those extremes is acceptable; responsibility for emergency management has to sit somewhere in between, in part because of the practical realities; individuals cannot fully prepare themselves as they depend on others such as the weather bureau to warn them of impending danger; and no government can afford to have sufficient resources standing by in order to respond

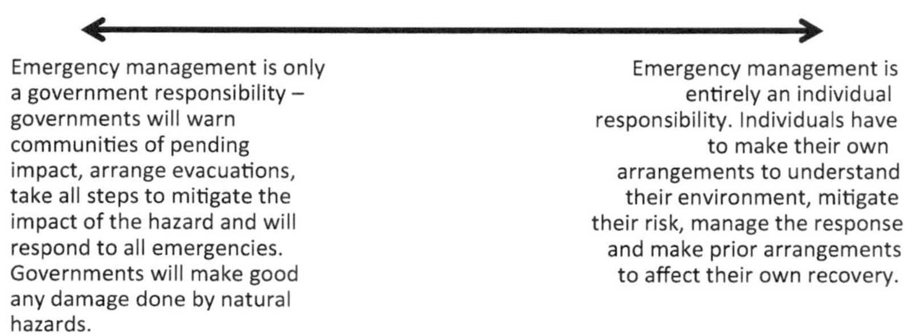

FIGURE 15.1 Possible shared responsibility spectrum.

to every home threatened by wildfire, flood, or other hazard. Further, and the subject of discussion in this chapter, the extremes are not legally permissible or required. The law (discussed below) will not allow governments to simply abandon citizens to their own resources, neither does it require governments to guarantee individual safety.

This chapter contributes to the discussion on developing shared responsibility by identifying what "the" law requires of governments and individuals. This analysis will identify some limits of shared responsibility. It will then be up to stakeholders to negotiate, within law's shadow, how responsibility for effective emergency management is to be negotiated in their particular local, regional, and national context.

Before commencing the legal analysis, it must be acknowledged that no such thing exists as "the" law. Relevant law comes from many particular laws and, more importantly and with the exception of international law, what the law is varies from jurisdiction to jurisdiction. Differences exist between countries and also between states and territories within countries. The discussion of the law in this chapter will refer to international law and the law that has evolved from the English common law and will draw on the relevant laws of Australia, the United Kingdom, and the United States. Although the exact legal position will vary from jurisdiction to jurisdiction, the fundamental principles are likely to be similar.

15.1 INTERNATIONAL LAW

International law is suprajurisdictional. The subjects of international law are nation-states so that where states are bound by relevant international law, either by agreement or provisions of universal application, the law that applies will be the same for all.

Under international law the government of a disaster-affected state has primary responsibility for managing the response to the disaster and ensuring the relief of the affected communities. The United Nations

> *...recognizes the primary role of each State in caring for the victims of disasters occurring in its territory and stresses that all relief operations should be carried out and co-ordinated in a manner consistent with the priorities and needs of the countries concerned.*
>
> UNGA (1981); see also UNGA (1991).

Further, an "...analysis of General Assembly resolutions on disasters from 1981 until 2002 reveals an increasingly explicit emphasis on sovereignty" (Fidler, 2005, p. 472). Even though primary responsibility for disaster management lies with the government of the affected state, it does not follow that states are free to ignore the impact of disasters on their population or that the response to a disaster is not a matter of international concern. Apart from the humanitarian impulse, a legitimate concern exists with internationally recognized human rights. Key rights that can be affected by disasters include the right to "life," "economic, social and cultural rights," and the "right to a standard of living adequate for...health and well-being...including food, clothing, housing and medical care..." (UNGA, 1948). The International Covenant on Civil and Political Rights provides that everyone has the right to life (ICCPR, 1976). The International Covenant on Economic, Social and Cultural Rights affirms a right to work in safe conditions, the right to social security, an obligation to protect the family and to provide special protection for new mothers and children, and that it is "the right of everyone to (enjoy) an adequate standard of living for himself and his family, including adequate food, clothing and housing" (ICESCR, 1976).

A natural disaster can impact upon these rights by causing death, dislocation, destruction of homes and livelihoods, and the spread of disease. Where this happens and the affected population is not provided adequate assistance by their own government, the affected state is not abiding by its obligations. In terms of the Covenant on Economic, Social and Cultural rights that is an obligation to "...take steps, individually and through international assistance..." (ICESCR, 1976, art 2(1)) to allow the population to realize their rights. Under the Covenant on Civil and Political Rights the parties have agreed "...to ensure to all individuals within its territory and subject to its jurisdiction the rights recognized in the present Covenant..." (ICCPR, 1976, art 2(1)). A state fails in its international obligations if it fails to adequately provide for the health and well-being of its citizens following a natural disaster.

Too much faith can, however, be placed on international law:

There is a massive implementation gap in international law, even in relatively wealthy countries with the resources to meet fundamental rights. In addition, enforcement of international law is limited. Individuals cannot generally enforce their rights under international law unless it has been incorporated into national law. Enforcement action by the international community raises the question of whether, and to what extent, it is reasonable to dictate national priorities to countries and regions with their own distinctive cultural concerns and their own urgent priorities.

Handmer and Monson (2004), p. 48.

Even so, the international community has recognized that there is a shared "Responsibility to Protect" where an affected population is not, or cannot be, protected by their own government. The International Commission on Intervention and State Sovereignty argued that international intervention into the domestic affairs of a sovereign state may be justified where there are

...overwhelming natural or environmental catastrophes, where the State concerned is either unwilling or unable to cope, or call for assistance, and significant loss of life is occurring or threatened.

ICISS (2001), (4.20).

The United Nations General Assembly did not go so far. The General Assembly limited its endorsement of the Responsibility to Protect to cases where the affected population was subject to "genocide, war crimes, ethnic cleansing and crimes against humanity" (UNGA, 2005). Gareth Evans, Co-Chair of the International Commission on Intervention and State Sovereignty has since written that the response to natural disasters is "not normally, on the face of it, about protecting people from 'genocide, war crimes, ethnic cleansing, and crimes against humanity' but it is still the case that terrible natural catastrophes...cry out for humanitarian, protective response by the country's own government and by anyone else in a position to offer assistance" (Evans, 2008a, pp. 65–66).

If the response to a natural disaster does not reach the high legal threshold to trigger the "Responsibility to Protect" principles, the response to a disaster remains a matter of international concern (Evans, 2008a, p. 69), as shown by the response to the devastation of Burma by Cyclone Nargis in 2008. Following the cyclone, the ruling military junta was reluctant to recognize the extent of the devastation or to allow the delivery of foreign assistance. France and the United Kingdom argued that the circumstances may have been sufficient to trigger forced intervention under the "Responsibility to Protect" (Evans, 2008b); this, coupled with pressure from Burma's neighbors from the Association of Southeast Asian Nations, is credited with finally opening the door to allow international assistance to flow to the disaster-affected community (Evans, 2008a, p. 68). Even if the matter was not sufficient to trigger action under the Responsibility to Protect, "no one was to be heard saying that the Burma/Myanmar cyclone disaster, and the general's response to it, was none of the rest of the world's business" (Evans, 2008a, p. 68).

What follows from this brief review of international law is that the international community expects and requires governments to take primary responsibility for taking steps to reduce vulnerability to natural hazards in order to minimize the impact on recognized human rights (that is, to minimize the loss of life and infrastructure) and to respond to natural disasters. In the context of "shared responsibility" that is significant. International law makes it clear that the right-hand extreme in Figure 15.1, above, is not legally permissible; it is not open to governments to simply abandon their populations to their own devices and to suggest that their capacity to withstand the impact of natural hazards is entirely their own responsibility. Governments have responsibilities under international law and in particular international human rights law to take steps to protect their populations and to advance their rights to life, health, and shelter.

15.2 DOMESTIC LAW

When talking about domestic law in jurisdictions that have inherited their legal traditions from the English common law, two sources of law exist: legislation or statue law, and the judge-made common law.

15.2.1 LEGISLATION

The legislature (the Parliament or Congress) can pass laws that the executive arm of government, the relevant government departments and officials, must administer. Limits occur, however, on legislative freedom.

Constitutional limits are (reasonably) fixed points in domestic law (though Constitutions may be amended and are subject to interpretation by the domestic courts in particular by the High Court of Australia and the Supreme Courts of England and the United States). A Constitution, in particular written constitutions such as those in the United States and Australia, set limits on what governments and citizens can do and what they can expect from each other. The Federal Parliaments of both the United States and Australia are governments of limited jurisdiction; they may make laws only with respect to those subject matters assigned to them in their respective constitutions (Australian Constitution s 51; United States Constitution art 1). Negotiation on disaster risk reduction must take into account constitutional limitations, as stakeholders cannot expect governments to take measures that are beyond the government's constitutional competence, even where those measures would reduce risk or vulnerability.

Stakeholders can, however, expect governments to honor, and citizens may enforce, fundamental rights and guarantees. Handmer and Monson (2004) noted the difficulties in enforcing human rights in international law but recognized that

> *National (or domestic) law offers greater opportunities for vulnerability reduction. Firstly, there is far greater scope for the enforcement of rights in national law, because individuals can bring actions in the courts to enforce their rights under national law. Public law (which regulates relationships between governments and their citizens, and is distinct from private law, which governs relationships between private citizens) is particularly useful in terms of reducing vulnerability, because the rights enshrined in national public law can be enforced by individuals against national governments, who are often those with the clearest responsibilities and the greatest resources for reducing vulnerability.*
>
> Handmer and Monson (2004), pp. 48–49.

The United States has various rights enshrined into the Constitution by the Bill of Rights. The United Kingdom, although lacking a written constitution, has fundamental documents such as the Magna Carta 1215 and the Bill of Rights 1688 that remain, at least in part, part of the domestic law (Lyon, 2003). The United Kingdom is also committed to protecting human rights through its adoption of the European Convention on Human Rights and the passage of the Human Rights Act 1998 (UK). Australia has no national human rights guarantees but it has adopted a number of international conventions into domestic law (Australian Human Rights Commission Act 1986 (Cth)) and human rights legislation does exist in some jurisdictions (Human Rights Act 2004 (ACT);

Charter of Human Rights and Responsibilities Act 2006 (Vic)). These provisions help cast law's shadow by defining the commitments that governments have made to the well-being of their citizens.

Other law, that is local legislation, may provide remedies or require action to address risk. For example, the Civil Contingencies Act 2004 (UK) requires category 1 and 2 responders to prepare emergency plans and do other things in preparation for responding to an emergency. Those obligations could be enforced in local courts and help define responsibilities and obligations.

Domestic statutes are the output of the political compromise that is necessary in any law making. A particular Act that allows or requires action, such as buying back high-risk land, providing subsidized flood insurance, or mandating that people must evacuate in the face of an impending threat, all reflect a compromise on competing views and interests, views on the appropriate role of government, the appropriate role of personal autonomy, beliefs in the capacity of governments and individuals, the appropriate allocation of resources between various competing interests, etc. Within constitutional limits and obligations, governments may legislate to provide any number of risk reduction measures, or they may not. The choices they make reflect the political realities and cultural expectations of each society. These legislative actions are vitally important in developing resilience and reducing vulnerability but they are the *end point*, rather than the starting point, of the discussion on the limits of shared responsibility.

15.3 THE COMMON LAW

The other source of law, for common law countries, is the law developed by courts on a case-by-case basis. Historically, the Crown or government could not be sued but the Crown's immunity has been abrogated by legislation (Judiciary Act 1903 (Cth), s 64; Crown Proceedings Act 1947 (UK), s 1; Federal Torts Claims Act 28 USC §1346(b); 28 USC §2674 (1946)). Today, the common law applies to governments as it does to citizens so governments can be sued for negligence, nuisance, and other common law wrongs.

15.3.1 No Duty to Rescue

No observer would have any difficulty outlining the current state of the law throughout the common-law world regarding the duty to rescue. Except when the endangered person and the potential rescuer are linked in a special relationship, there is no such duty.

Weinrib (1980), p. 247.

People who are at risk from, or have been impacted by, a natural hazard are as much in need of rescue as a drowning person or a person trapped in a house fire, and, although we expect our government agencies to respond to those emergencies and use their very best endeavors to preserve life and minimize property losses, no common law duty exists to do so. The government and its services are not liable in negligence law for failing to respond or for failing to make good the rescue (Stuart v Kirkland-Veenstra (2009) 237 CLR 215; Warragamba Winery Pty Ltd v New South Wales (No 9) (2012) NSWSC 701; Capital and Counties PLC v Hampshire CC (1997) QB 1004; Camp v State (2010) 184 Cal.App 4111; Zelig v. County of Los Angeles (2002) 27 Cal.4th 1112; American Law Institute, 2012, § 37; but see cf Electro Optic Systems Pty Ltd & West v New South Wales (2012) ACTSC 184; Lowns v Woods (1996) Aust Torts Reports ¶81–276).

In the High Court of Australia it was said

> *The common law generally does not impose a duty upon a person to take affirmative action to protect another from harm... In principle a public authority...should not be regarded by the common law any differently from a citizen. It should not be considered to have an obligation to act. But the position of a public authority is not the same as that of a citizen... It has public functions and it has statutory powers which the citizen does not... The relevant concern of the common law is whether a public authority might nevertheless be considered to be under a duty of care which obliges it to exercise its powers in a particular way.*
>
> Stuart v Kirkland-Veenstra (2009) 237 CLR 215, para 127–129 (Crennan and Keiffel JJ).

As discussed above, national and state legislatures can and do pass laws creating government agencies and setting out their functions, powers, and responsibilities. Those statutes distinguish the statutory emergency service from the citizen and may give rise to a legal duty to aid others. In Australia, the United Kingdom, the United States, and across the world, state and national legislatures have established emergency response and emergency management agencies (Civil Contingencies Act 2004 (UK); Robert T. Stafford Disaster Relief and Emergency Assistance Act (Pub Law No 93–288) (US); Australia does not have national emergency management legislation, but see State Emergency and Rescue Management Act 1989 (NSW); Emergencies Act 2004 (ACT) and similar legislation in each state and territory). As the High Court judges noted, these agencies, equipped as they are with emergency powers and obligations, are different to citizens, but that does not prove that they owe a duty to come to the aid of individuals.

In Australia, the question of whether or not the common law will impose a duty of care requires a careful analysis of all the "salient features" (Graham Barclay Oysters v Ryan (2002) 211 CLR 540; Caltex Refineries (Qld) v Stavar (2009) 75 NSWLR 649) of the relationship between the parties to determine whether or not one party should be under a legal duty to aid the other. Tests vary across the United States but the California Supreme Court, for example, has said that deciding whether or not a duty of care is owed in particular circumstances

...involves the balancing of a number of considerations; the major ones are the foreseeability of harm to the plaintiff, the degree of certainty that the plaintiff suffered injury, the closeness of the connection between the defendant's conduct and the injury suffered, the moral blame attached to the defendant's conduct, the policy of preventing future harm, the extent of the burden to the defendant and consequences to the community of imposing a duty to exercise care with resulting liability for breach, and the availability, cost, and prevalence of insurance for the risk involved.

Rowland v. Christian 69 Cal. 2d 108 (1968).

In the United Kingdom, the question is whether it is "fair, just and reasonable" to impose a legal duty of care (Caparo Industries PLC v Dickman (1990) 2 AC 605).

Taking these considerations into account the courts have affirmed that the states' rescue and emergency management agencies *do not* generally owe a duty to those in need of rescue, other than a duty not to make the situation worse (Stuart v Kirkland-Veenstra (2009) 237 CLR 215; Capital and Counties PLC v Hampshire CC (1997) QB 1004; Zack v City of Alameda (Cal, RG12632015, 11 February 2013)). Several theoretical justifications exist for the refusal to find a duty of rescue:

The most common relies on the distinction between placing limits on conduct and requiring affirmative conduct. This distinction in turn relies on the liberal tradition of individual freedom and autonomy. Classical liberalism is wary of laws that regulate conduct that does not infringe on the freedom of others. Some commentators have argued that mandating a duty to rescue might cheapen acts of Good Samaritans. Some commentators also argue that it might be difficult to limit an affirmative duty to aid others in peril from becoming a general duty of self-sacrifice. And in those instances when there are many potential rescuers, there may be no basis for choosing one on whom to place a rescue duty... Finally, determining factual causation tends to be more difficult when the tortious act entails a failure to prevent harm from occurring.

American Law Institute (2012), § 37.

When it comes to the state and its emergency services, the public nature of those services suggests that they do not owe a duty to individuals. When a fire brigade turns out in response to a fire, their primary obligation is to stop the fire spreading to other properties for the community benefit; that is, to stop the street, village, or town being lost to fire, rather than to protect the interests of the person whose property is already on fire (Capital and Counties PLC v Hampshire CC (1997) QB 1004). Furthermore, the emergency services are funded by the community as they provide that community benefit. Historically fire services were funded by insurance companies to protect the property of their policy holders, but having the fire service extinguish the fire conveyed a benefit to others. Insurance companies realized that extinguishing the fire in the house next to the insured property was a more cost-effective way to reduce their exposure than waiting for the fire to spread and then combating it. Either way fire services brought a benefit to those that had not paid for it; in all the circumstances governments accepted that a coordinated community funded emergency service provided a whole of community benefit.

If the emergency services did owe a duty or care to individuals, it would affect the decisions they would have to make, including how to allocate resources and whether decisions could be made to forsake some properties for the greatest benefit. In dismissing an action against the New South Rural Fire Service (RFS) alleging negligence over their response to an uncontrolled wildfire on the outskirts of Sydney, Australia's largest city, the trial judge said,

> *The defendants could not have been expected to have concentrated their attentions on individual plaintiffs. They each had to consider the community as a whole. That included making choices, such as between fighting one fire or another, or between the assets of individuals, when deciding which house to save, and which house should be left to take the brunt of the fire so others could be saved.*
> Warragamba Winery Pty Ltd v New South Wales (No 9) (2012) NSWSC 701, para 708.

The need, in that case, to allocate resources across the state, to multiple fires, all pointed against a duty owed to particular individuals. That governments and agencies must make policy decisions on the allocation of resources both to the emergency services generally and in response to a particular emergency; they must set priorities for response and identify which assets to protect at the expense of others; and the fact that some decisions must be made in the face of a fast moving emergency, all move against a finding that a duty is owed to a particular individual (Graham Barclay Oysters Pty Ltd v Ryan (2002) 211 CLR 540; Leishman v Thomas (1958) 75 WN (NSW) 173).

15.3.2 Breach

Even if a duty exists to act for the benefit of an individual, the duty is not to guarantee safety but to take reasonable steps to respond to the risk (Gorman v Wills (1906) 4 CLR 764).

> *The perception of the reasonable man's response calls for a consideration of the magnitude of the risk and the degree of the probability of its occurrence, along with the expense, difficulty and inconvenience of taking alleviating action and any other conflicting responsibilities which the defendant may have. It is only when these matters are balanced out that the tribunal of fact can confidently assert what is the standard of response to be ascribed to the reasonable man placed in the defendant's position.*
> Wyong Shire Council v Shirt (1980) 146 CLR 40, pp. 47–48 (Mason J).

In some cases, "the response of prudent and reasonable people to many of life's hazards is to do nothing" (New South Wales v Fahy (2007) 232 CLR 486, para 7 (Gleeson CJ); para 123 (Kirby J)). Just because a risk, such as a risk of loss due to flood, fire, or other natural hazard, is foreseeable, it does not follow that those agencies with the statutory mandate to manage the response to the hazard have to ensure everyone's safety. They have to take reasonable steps and sometimes that means decisions must be made on issues such as the allocation of resources, the determination of priorities, and assessment of what can and cannot be saved with the consequences that foreseeable losses occur. In simple terms, a decision not to commit resources to fight a fire or remove a person from danger from flood waters may not be a breach of duty even if the result is property loss or death.

15.3.3 Policy

As noted at the start of this chapter, the move in emergency management is to adopt policies to develop resilient communities. Although a resilient community depends on the presence of resilient individuals, the two concepts are not synonymous. In the field of disaster risk reduction, resilience is defined as

The ability of a system, community or society exposed to hazards to resist, absorb, accommodate to and recover from the effects of a hazard in a timely and efficient manner, including through the preservation and restoration of its essential basic structures and functions.

UNISDR (2009), p. 24.

The commentary to the United Nations International Strategy for Disaster Reduction glossary adds that the "resilience of a community in respect to potential hazard events is determined by the degree to which the community has the necessary resources and is capable of organizing itself both prior to and during times of need" (UNISDR, 2009, p. 24). Critically what is being referred to are the resources and capacities of communities, not individuals. In a recent editorial on the response to wildfire it was said:

Among the priorities facing firefighters, defense and protection of human life and protection of private property ranks right at the top; protection of infrastructure, watersheds and cultural sites is next.

Bailey (2013), p. 6.

With respect, and with the exception of the primacy of human life, those priorities are in the wrong order. Protection of infrastructure, watersheds and cultural sites will enhance community resilience. If the infrastructure (e.g., utilities, roads, community assets such as schools, libraries, meeting places) and cultural sites are maintained, then the community will have better resources to recover. If those that are left can send their children to school, maintain their employment, resupply the shops, have building materials delivered, and retain their community identity through the culturally important assets that encourage tourists and locals to return, then the community may, in time, recover. If, on the other hand, the first priority is the protection of private property, then homes may be saved but without power, water, or those assets required to allow economic activity to return, the capacity to recover will be reduced. If that analysis is correct then the loss of some private property, and even the loss of some lives, is not inconsistent with the objective of developing resilient communities. Furthermore, if government and their agencies owe their primary duty to advance the community good (Graham Barclay Oysters Pty Ltd v Ryan (2002) 211 CLR 540; Capital and Counties PLC v Hampshire CC (1997) QB 1004) then their duty is to take those steps that will best serve the community, which may mean directing resources to protect "infrastructure, watersheds, and cultural sites" ahead of private property interests.

An example can demonstrate the difference between community and individual resilience. In the summer of 2010/2011, floods devastated the city of Brisbane, capital of the Australian state of Queensland. The floods were the worst since 1974. Approximately 22,000 homes and 7,600 businesses across 94 suburbs were flooded. In 2013, the estimated value of damage to council property alone exceeded A$355 million (Brisbane City Council, 2013). Notwithstanding this devastation it is clear that Brisbane has demonstrated resilience. Notwithstanding further flooding in January 2013, key economic data show that Brisbane has continued to thrive. As of January 2013, Brisbane contributes almost 50 percent of the Queensland economic activity, which in turn accounts for nearly 20 percent of the Australian Gross Domestic Product; employment in Brisbane has continued to grow at about 0.6 percent per annum; continued commercial growth of about 10 percent has occurred since 2011 (Quirk, 2013). Individuals may not be so resilient, some people will still be struggling to recover from the floods, some may have moved out of the Brisbane area, shops and businesses will have changed hands. Depending on how one defines "community" some communities may also remain affected, but at the macrolevel the evidence is that there was, and is, a resilient community; the city of Brisbane has been able to "absorb, accommodate to, and recover from the effects of a hazard in a timely and efficient manner" even if some individuals have not.

A government (whether local, state, or federal) committed to developing community resilience can show that community resilience does not require universal, individual resilience. It is not inconsistent with a policy committed to developing resilient communities to realize and even accept that some individuals will not be resilient to a particular hazard but a government, consistent with its policy choices, may choose to accept that trade-off.

15.4 DISCUSSION

We can now return to Figure 15.1. It has been argued, above, that the right-hand extreme of the shared responsibility spectrum is not tenable; international law and, where applicable, domestic guarantees of fundamental rights, means that governments cannot simply abandon their populations to their fate. The further analysis, above, shows that the left extreme is also unsupported;

governments and their agencies, even agencies that governments have chosen to create with the express purpose of preparing communities for, and then responding to emergencies, are not required to ensure the protection of everyone. As the flood waters rise, or the fire burns across the urban interface, the law only requires governments and their agencies to take reasonable steps to mitigate risks to the community; legal responsibility for the protection of private assets and for the consequences of decisions to respond to warnings or directions to evacuate and a number of other decisions belong to the individuals affected.

It is from these extreme points that negotiation on how responsibility for disaster risk management will be shared, can take place—negotiation that will occur in the shadow of the law. What this brief identification of the legal extremes has done is help identify where the shadow cast by the common law of Australia, the United Kingdom, and the United States falls. We can demonstrate this by modifying Figure15.1, to produce Figure 15.2.

It is open for stakeholders to negotiate on how responsibility is to be shared, that is, what aspects of risk reduction and response belong to the government, and what belongs to individuals or community. In those negotiations the parties know that governments cannot abandon their citizens to nature's fury, obligations exist that may be enforced (Handmer and Monson, 2004), these may require governments to provide minimum standards of disaster relief, to ensure housing standards take into account the hazard environment, that inappropriate development in hazardous areas is prohibited or restricted, and that citizens are given access to information to inform them of their risk and how to protect themselves. Equally the parties know that, if left to the common law, governments are not legally obligated to save everyone, nor are they legally obligated to make good all the damage that is done by natural hazards. Some level of responsibility for preparation, response, and recovery lies with each individual. Within those extreme

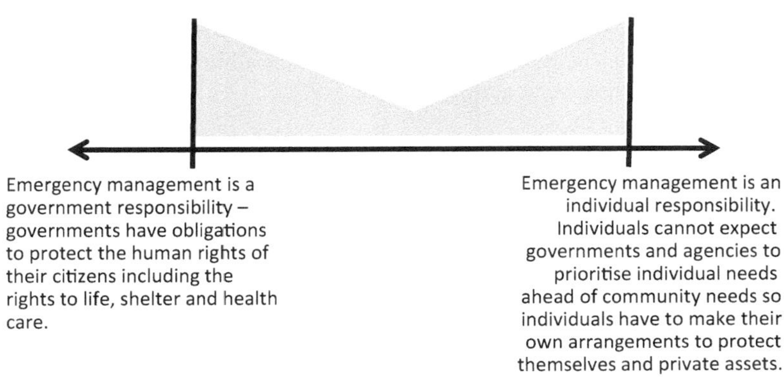

FIGURE 15.2 Shared responsibility in the shadow of the law.

points—within the shadow cast by the law—different countries, cultures, and communities can adopt a myriad of approaches to risk and risk management. The important lesson from this analysis is that the appropriate roles and responsibilities for governments and individuals are not, legally, self-evident.

Law is a tool that can be used by interest groups to advance what they perceive to be their best interests, whether that is advocating for their right to build homes in hazard-prone environments or advocating against such development because of the social costs of the impact of disasters. Balancing these competing values, interests, and demands on resources is a political process (Davis, 1993), which means that different stakeholders can use their political influence and ability to affect the outcomes of policy choices. Developers with the ear of government may be able to use their influence to weaken building codes and development rules to maximize development at some cost to community safety; advocates for ecosystem protection, or wealthy residents who enjoy the natural environment, may affect the policy choices that may be used to reduce hazards from wildfire; a desire to allow population growth may bring pressure to release land on the wildland–urban interface or the floodplain even though they bring their own risk. Whether those outcomes are "good" or "bad" depend on one's views, interests, and personal philosophy. If, as argued above, the appropriate roles and responsibilities for governments and individuals are not, legally, self-evident, then disaster risk reduction becomes *one* of the concerns that governments need to address (albeit an important one that comes with international and domestic legal obligations).

That reality could be disillusioning for those with a genuine interest in developing community resilience and protecting communities and individuals from foreseeable, inevitable natural hazards. Rather than be disillusioned, however, the message should be empowering; if those advocating for increased resilience and further risk reduction understand that they must engage in the political process and negotiate, in the shadow of the law, for desired outcomes, then they will be better informed and prepared to engage in the "art of the possible."[1] Emergency management professionals will, however, continue to be disappointed if they hope that claims that a certain policy choice will increase resilience or reduce disaster risk will trump all competing concerns or that their government *must* act on their advice.

15.5 CONCLUSION

This chapter has considered how law, in particular law based on the English common law tradition, addresses issues of risk and the responsibility of governments to manage risk reduction for the benefit of the community and individuals. It has been argued that the appropriate role of government is not self-evident. The Anglo-Australian-American common law does not require or

1. "Politics is the art of the possible" (Bismark, 1867).

oblige governments to guarantee the safety of its citizens or to bear the risk of natural hazards. On the other hand, fundamental norms of international law that have been incorporated into national constitutions and domestic law do not allow governments to simply abandon populations to their fate.

Apart from these broad parameters, the position is open to negotiation. Legislators need to be persuaded that actions are necessary to reduce risk particularly if the decisions come at the expense of other priorities and interest. This message could be disappointing; to describe decisions as political is usually to imply that the decision is improper or made for impure motives, but politics is an essential tool, particularly in Western democracies, to resolve issues of competing values and interests (Davis, 1993). Seen in that light, the message is empowering; if the law is allowing inappropriate development or fails to recognize the resilience of local populations, then, rather than seeing the law as a barrier or fixed rules that *must* be accommodated, getting political and changing the law may become the emergency managers role. Getting political, however, requires recognition that there are other competing viewpoints, and the argument that changing the law will reduce risk is unlikely, on its own, to carry the day. As with any negotiation engaging with the opposing views will be necessary, and sometimes compromise will be called for.

This chapter contributes to that negotiating process by making explicit the political nature of the compromise but at the same time identifying some of those (reasonably) fixed points of law that identify some limits of legal obligation. Those points cast the shadow within which negotiation for laws that appropriately shares responsibility for risk management can take place.

REFERENCES

American Law Institute, 2012. Restatement (Third) of Torts (Philadelphia).

Bailey, D., 2013. National dialogue needed about WUI fires. Wildfire 22 (5), 6.

Bismark, Otto von, 1867. http://en.wikiquote.org/wiki/Otto_von_Bismarck (viewed 17.12.13.).

Brisbane City Council, 2013. Fact Sheet – Brisbane City Council Three Year Flood Recovery Update – November 2013 (viewed 15.11.13.). http://www.brisbane.qld.gov.au/community/community-safety/disasters-and-emergencies/types-of-disasters/flooding/recovering-after-the-floods/3-year-flood-recovery-update/index.htm.

Cabinet Office, 2013. Improving the UK's Ability to Absorb, Respond to and Recover from Emergencies: Building a Resilient Society. Government of the United Kingdom, London.

Council of Australian Governments, 2011. National Strategy for Disaster Resilience. Commonwealth of Australia, Canberra.

Davis, G., 1993. Public Policy in Australia, second ed. Allen and Unwin, Sydney.

Depoorter, B., 2010. Law in the shadow of bargaining: the feedback effect of civil settlements. Cornell Law Rev. 95 (5), 957–987.

Evans, G., 2008a. The Responsibility to Protect: Ending Mass Atrocity Crimes Once and for All. Brookings Institution, Washington DC.

Evans, G., 2008b. Facing up to our responsibilities. The Guardian. May 12, 2008, (viewed 10.04.14.). http://www.theguardian.com/commentisfree/2008/may/12/facinguptoourresponsbilities.

Federal Emergency Management Agency, 2011. National Preparedness Goal. Department of Homeland Security, Washington DC.
Federal Emergency Management Agency, 2013. Whole Community: Shared Responsibility (viewed 10.04.14.). http://www.fema.gov/national-preparedness/whole-community.
Fidler, D.P., 2005. Disaster relief and governance after the Indian ocean tsunami: what role for international law? Melbourne J. Int. Law 6 (2), 458–473.
Handmer, J., Monson, R., 2004. Does a rights based approach make a difference? the role of public law in vulnerability reduction. IJMED 22 (3), 43–59.
ICCPR – See International Covenant on Civil and Political Rights.
International Commission on Intervention and State Sovereignty, 2001. The Responsibility to Protect. International Development Research Centre, Ottawa.
International Covenant on Civil and Political Rights 1976, opened for signature December 16, 1966, 999 UNTS 171, entered into force March 23, 1976.
International Covenant on Economic, Social and Cultural Rights 1976, opened for signature December 16, 1966, 993 UNTS 3, entered into force January 3, 1976.
Lyon, A., 2003. Constitutional History of the United Kingdom. Cavendish Publishing, London.
McAdams, R.H., Nadler, J., 2008. Coordinating in the shadow of the law: two contextualized tests of the focal point theory of legal compliance. Law Soc. Rev. 42 (4), 865–898.
Mnookin, R.H., Kornhauser, L., 1979. Bargaining in the shadow of the law: the case of divorce. Yale Law J. 88 (5), 950–997.
Quirk, G., 2013. Lord Mayor's Brisbane Economic Snapshot; January Quarter 2013. Brisbane City Council, Brisbane.
Stevenson, B., Wolfers, J., 2006. Bargaining in the shadow of the law: divorce laws and family distress. Q. J. Econ. 121 (1), 267–288.
UNISDR – See United Nations International Strategy for Disaster Risk Reduction.
United Nations General Assembly 1948, Universal Declaration of Human Rights, GA Res 217(III), 3rd sess, 183rd plen mtg, UN Doc A/RES/217(III) (1948).
United Nations General Assembly 1981, Strengthening the Capacity of the United Nations System to Respond to Natural Disasters and Other Disaster Situations, GA Res 36/225, UN GAOR 36th sess, 103rd plen mtg, UN Doc A/RES/36/225 (1981).
United Nations General Assembly 1991, Strengthening of the Coordination of Humanitarian Emergency Assistance of the United Nations, GA Res 46/182, UN GAOR, 46th sess, 78th plen mtg, Annex [2], UN Doc A/RES/46/182 (1991).
United Nations General Assembly 2005, 2005 World Summit Outcome, GA Res 1/60, UN GAOR, 60th sess, 8th plen mtg, UN Doc A/RES/60/1 (2005).
United Nations International Strategy for Disaster Risk Reduction, 2009. UNISDR Terminology on Disaster Risk Reduction (United Nations, Geneva).
Victoria, 2010. Final Report, 2009 Victorian Bushfires Royal Commission, Melbourne.
Weinrib, E.J., 1980. The case for a duty to rescue. Yale Law J. 90 (2), 247–293.

LEGISLATION

Australian Constitution.
Australian Human Rights Commission Act 1986 (Cth).
Bill of Rights 1688 (UK).
Charter of Human Rights and Responsibilities Act 2006 (Vic).
Civil Contingencies Act 2004 (UK).
Crown Proceedings Act 1947 (UK).

Emergencies Act 2004 (ACT).
Federal Torts Claims Act 28 USC §1346(b); 28 USC §2674 (1946).
Human Rights Act 1998 (UK).
Human Rights Act 2004 (ACT).
Judiciary Act 1903 (Cth).
Robert T. Stafford Disaster Relief and Emergency Assistance Act (Pub Law No 93−288) (US).
State Emergency and Rescue Management Act 1989 (NSW).
United States Constitution.

Case Law

Caltex Refineries (Qld) v Stavar (2009) 75 NSWLR 649.
Camp v State (2010) 184 Cal.App 4111.
Caparo Industries PLC v Dickman (1990) 2 AC 605.
Capital and Counties PLC v Hampshire CC (1997) QB 1004.
Electro Optic Systems Pty Ltd & West v New South Wales (2012) ACTSC 184.
Gorman v Wills (1906) 4 CLR 764.
Graham Barclay Oysters v Ryan (2002) 211 CLR 540.
Leishman v Thomas (1958) 75 WN (NSW) 173.
Lowns v Woods (1996) Aust Torts Reports ¶81−276.
New South Wales v Fahy (2007) 232 CLR 486.
Rowland v. Christian, 69 Cal. 2d 108 (1968).
Stuart v Kirkland-Veenstra (2009) 237 CLR 215.
Warragamba Winery Pty Ltd v New South Wales (No 9) (2012) NSWSC 701.
Wyong Shire Council v Shirt (1980) 146 CLR 40.
Zack v City of Alameda (Cal, RG12632015, February 11, 2013).
Zelig v. County of Los Angeles (2002) 27 Cal.4th 1112.

Chapter 16

Self-Care in Bangladesh: Local Level Resilience and Risk Reduction

Ross Edgeworth
Disaster and Development Network (DDN), Northumbria University, Newcastle, UK

ABSTRACT
An interest in human coping applicable to endemic disease environments, such as those occurring in Bangladesh, has identified the pivotal role played by local level disease resilience and risk reduction strategies. These approaches resonate strongly with the concept of self-care, a ubiquitous and multidimensional construct that encompasses the actions individuals take to stay healthy and manage minor and/or chronic conditions. The extent of self-care as a predominant response to ill health throughout the global north has been well documented, with a burgeoning body of evidence capturing self-care in the global south. This has drawn attention to the concept as a primary public health resource. Despite this, a lack of insight remains into how self-care is implemented as part of resilience and risk-reduction strategies among communities living within endemic disease environments. Therefore, this chapter discusses theses issues, and drawing on research from Bangladesh, advocates the role self-care can play in resilience to disease and its effectiveness as a disease risk-reduction strategy. As such self-care could provide a mechanism for achieving better disease mitigation and low-cost health responses without overriding the necessary structural changes that offer the potential to achieve improved disease management for the poor of Bangladesh.

16.1 INTRODUCTION

Evidence suggests ill health is both a cause and effect of poverty as disparities in the impact and incidence of endemic disease reveal a higher burden upon impoverished households (OECD, 2003; Bates et al., 2004; WHO, 2004). This is reinforced by evidence from poor households that identify health risks as the greatest threat to their lives and livelihoods (Sen, 1997; Narayan et al, 2000; Segall et al, 2002; Hulme, 2003; Falkingham, 2004; Krishna, 2004; Nopen and Kantor, 2004; Russell, 2004). Attempts to alleviate these risks have been compounded by the failure of health systems to deliver effective and equitable

Hazards, Risks, and Disasters in Society. http://dx.doi.org/10.1016/B978-0-12-396451-9.00016-0

service to the poor throughout many areas of the developing world (Cockcroft et al, 2004, WHO 2008). However, there has been acknowledgment that delivery system sustainability comes from the recognition of local level coping strategies and resilience (Adams et al., 1998), whereas the importance of people-centered risk reduction of infectious disease is implied in policy agendas such as the Hyogo Framework (WCDR, 2005). This is because people are not passive observers of events that unfold around them; they adjust to shocks such as those caused by illness through coping strategies and human agency orientated to reducing risk. Some evidence suggests collective resilience and coping to solve public health issues in the face of epidemics (Macfarlane et al., 2000), thus enabling minimal survival, sustenance (Davies, 1993), or primary subsistence (Collins, 1998). Resilience has also become a central feature of disaster discourse. The World Disaster Report (2004) placed resilience at the heart of calls for stronger approaches to risk reduction that required local knowledge and implied a paradigm shift toward people-centered approaches and understanding of how communities survive and cope with disasters (IFRC, 2004). This includes the onset of disease epidemics and contexts of endemic disease due to their damaging impact upon households' economic and social status (Zwi and Yach, 2002; Connelly et al., 2004).

As a result, the growing interest in human coping and resilience applicable to endemic disease environments such as Bangladesh is moving toward disease mitigation and management through self-care. This has been identified as a primary public health resource within the health care system (WHO, 2001). However, the immediate and wider implications remain limited in terms of how it may be facilitated to the benefit of sick individuals, the extent of choice behind self-care selection, its use as a cost-effective management strategy, and potential integration as a low-cost component of health delivery systems in some of the poorest countries. It is therefore important to understand and assess how self-care can be used to its full potential, both to benefit sick individuals and to provide wider impacts for local level resilience and risk reduction strategies in the face of endemic disease.

16.2 SELF-CARE: DEFINITIONS AND THEORETICAL PERSPECTIVES

Self-care is a multidimensional construct with numerous definitions and approaches influenced by different social, economic, political, and historical factors (Wilkinson and Whitehead, 2009). Although a ubiquitous practice (Dean, 1981), self-care is regularly documented as a widely adopted response to ill health (Phillips and Verhasselt, 2001; Edgeworth and Collins, 2006; Edgeworth, 2011), which is increasing in some countries (Ahmed et al., 2003). Changes in the pattern of disease from acute to chronic (Segall and Goldstein, 1989; Orem, 1991), increasing discontent with a depersonalized health care system (Illich, 1995), and shifts from a cure to a care philosophy in treatment provision (Health Canada, 2004) have all been suggested as possible causes for the rise in self-care utilization. The need to control escalating health care costs (Anderson, 1990), an increase in alternative health providers, and the growth in lay knowledge of health are also advocated as influential factors (WHO, 2000; Pagan et al., 2006). Current knowledge regarding self-care implies empowerment perspectives (Anderson, 1996; Anderson and Funnel, 2005) and coping strategies (Edgeworth and Collins, 2006) linked to resilience in order to reduce disease risks ensuring primary subsistence (Collins, 1998) or minimal survival and sustenance (Davies, 1993).

Self-care has been described as the first step to feeling unwell (Van der Geest, 1987) and can encompass actions related to disease prevention, illness treatment, health promotion, and chronic disease management and rehabilitation (Bhuyan, 2004). It often involves people functioning on their own behalf in the prevention and detection of disease. Taking increased rest and altering diet and exercise habits are common examples in developed countries (Punamaki and Aschan, 1994). The use of medicinal plants (Sowell et al., 1997), home remedies (Edgeworth and Collins, 2006; Edgeworth, 2011), oral rehydration solution (ORS) (Ahmed, 2005), and purchasing medications without professional consultation or prescription have been documented as self-care practices in several developing countries (Shankar et al., 2002; Edgeworth, 2011). The diversity of self-care is also reflected in the range of illnesses it can be used to treat, such as diabetes (Anderson et al., 1996), depression (Peden, 1994), minor illnesses in children (Cunningham-Burley and Irvine, 1987), arthritis (Shaul, 1995), schizophrenia (Baker, 1995), human immunodeficiency virus/acquired immunodeficiency syndrome (Sowell et al., 1997), and diarrheal diseases (Edgeworth, 2011).

Definitions of self-care are fraught with problems due to the nature in which self-care behavior can become intertwined with external assistance (both qualified and unqualified). Self-care has been defined as exclusively nonprofessional (Haug et al., 1989), interactive with the health care system (Hickey et al., 1986), or a continuum spanning from self-care managed alone at one end to self-care shared with professionals at the other end of the scale (Roger and Hay, 1998). Alternatively, Gantz (1990) identified four common characteristics specific to self-care: the "role of culture," the "capacity to act and make choices," the influence of "knowledge, skills, motivation, and values," and "a focus on aspects of health care under individual control" (as opposed to social policy or legislation). However, proponents of the individual exercising control and responsibility through self-care can place unrealistic expectations upon patients, particularly in resource-poor settings. This includes the notion that individuals will successfully make behavior changes for disease management and prevention, often in contexts such as Bangladesh where the health care system has been unable to manage the disease process (Redman, 2007). These changes are also likely to be expected without alterations to wider societal, economic, and political structures, which increase vulnerability to disease in the first place and restrict the ability to respond to cases of ill health (Wilson et al., 2007). Therefore, any shift toward the individual directs focus away from the structural constraints that may lie at the heart of social and economic inequities (Chapple and Rogers, 1999). Consequently, merely defining self-care on the basis of the individual's actions may be an inadequate conceptual and practical application in the context of Bangladesh. Berman and Iris (1998) acknowledge self-care as a "value-laden concept" grounded in value systems and judged by society. Therefore, self-care is defined within this chapter as the action individuals take for themselves and their families to stay healthy and manage minor and chronic conditions, based on their knowledge and the information available and working in collaboration with health care professionals where possible, desired, and necessary.

16.3 RESEARCH CONTEXT AND METHODS

In order to further understand the role of self-care within resilience and risk reduction it is imperative to take into account the local milieu where self-care takes place as "self-care actions do not take place in a political and societal vacuum" (Kickbush, 1989, p. 129). Bangladesh is one of the most disaster-prone countries, highly vulnerable to tropical cyclones and floods (DFID, 2005; ISDR, 2002). Severe flooding in Bangladesh can have substantial impacts on people's health as contaminated water accentuates the longer term risk of diarrheal disease, skin infections, and respiratory disease (Few et al., 2004). Stunted growth has also been reported due to interruptions in the access to food through crop damage (Buttenheim, 2006). Access to health facilities often becomes restricted due to infrastructural damage, restrictions on travel routes, and/or damaged roads (Beck, 2005), which presents potential scope for an increased likelihood of self-care use. It is estimated that illness-related expenditures and income loss due to disease caused by disasters are the most detrimental to rural households (GOB, 2003).

The country has also been severely affected by the burden of disease as preventable infectious and poverty-related diseases dominate the top 10 causes of morbidity within the country. The perpetuation of these problems within Bangladesh and the country's relatively poor health record in comparison to other developing countries in South Asia (World Bank, 2002; UNDP, 2003) makes health and population among the most important development issues for both public and private sector organizations. It is also a pertinent reason to address continuing problems within the health sector to research and identify systems, interventions, and policy to alleviate the burden of ill health for the impoverished and most disadvantaged populations of the country. Within this context, the reexamination of ongoing enteric disease causation, transmission, and response remains in high demand. The remainder of this chapter therefore attempts to illustrate the role of self-care with local level resilience and risk reduction strategies by drawing on research evidence obtained in Bangladesh between 2007 and 2009. Both quantitative and qualitative research was undertaken in Northwest, Central, and Southeast Bangladesh in the subdistricts of Domar, Matlab, and Chakaria, respectively. This process included structured surveys with 630 households, semistructured interviews with male and female participants (45), and focus group discussions (15), all of which were complemented by mapping (15) and photo diary exercises (10). Each methodological tool focused on understanding the nature and extent of self-care as well as identifying key factors that facilitated or restricted self-care adoption.

16.4 THE PREVALENCE OF SELF-CARE AT THE LOCAL LEVEL

A high propensity for self-care use existed in the targeted areas as this method was adopted by over 36 percent of the participants compared with the use of government health facilities (23 percent), traditional healers (22.9 percent), and unqualified practitioners (17.8 percent) (statistically significant $p < 0.05$). Self-care was primarily applied to minor illnesses, including uncomplicated fever (8.6 percent), jaundice (9.4 percent), skin infections (2.4 percent), and diarrheal disease (18.9 percent). More complex conditions likely to require qualified medical care to ensure correct diagnosis and appropriate treatment, such as malaria or cholera, were not readily treated through self-care methods (less than 1 percent of participants). These results reflected similar research findings from other developing countries (Leyva-Flores et al., 2001), including Bangladesh (Ahmed et al., 2003; Edgeworth and Collins, 2006). There was no statistically significant difference between the characteristics of those who utilized self-care, indicating that levels of resilience and risk reduction through self-care do not possess any significant gender, age, or socioeconomic dimensions. The equal use of self-care among men and women, young and elderly, and households dependent on menial labor and households that have a higher socioeconomic status highlights the widespread nature of local level resilience and coping among all households at risk of disease within these rural environments of Bangladesh. The wide and regular use of self-care across all members of the rural villages suggests a strong base for community participation. It also implies local level resilience and risk reduction to ill health through self-care can be attained by vulnerable groups such as the elderly and women, who are often restricted in their access to support from the health system (HelpAge International, 2000; Shaikh and Hatcher, 2005). Self-care therefore holds resonance with the ideals of primary health care and greater health equity drives currently advocated by the WHO (2009).

16.5 THE VALUE OF LOCAL KNOWLEDGE AND LOCAL PRACTICE

Very few studies have been directed toward the components of self-care. Interviews and focus groups provided a forum in which the types of self-care practices adopted could be detailed. Through this process a range of self-care behaviors for a variety of ailments and illnesses were identified from the three subdistricts. In particular, examination of what self-care involved highlighted a range of effective measures that offered an appropriate mechanism for responding to various illnesses in a disease endemic context. Natural, traditional, and biomedical practices were described by participants in descriptions of self-care methods. Many of these treatments were taken with differing frequency and at different points in symptom manifestation. This often depended on perceptions of illness severity, the influence of previous experience, and recommendations from family and neighbors. Table 16.1.1 summarizes many of the most frequently utilized self-care practices.

The use of these practices indicates that many households are able to manage both the risk and the impact of commonly encountered diseases. Respondents regularly recounted the successful use of the various methods detailed in Table 16.1, with no instances of self-care accentuating a particular illness. This positive response rate is reinforced by scientific evidence for many of the methods implemented at the local level. For example, the antidiarrheal agents found in banana (Faisant et al., 1995; Rabbani et al., 2001),

TABLE 16.1 Types of Self-Care for Different Illnesses in Rural Bangladesh

Illness	Treatment Description
Cough	Two local leaves (*Bashok pata* and *pouri pata*) crushed into a paste and taken with lime juice and/or ginger Rubbing mustard oil into the body
Cold	Wearing extra clothes and rubbing mustard oil into the body Local leaf (*Tulshi pata*) mixed with honey and mustard and taken orally Taking a mixture of oil and onion orally
Diarrhea	ORS, either purchased or made within the home using 1 l of water, salt, and sugar Coconut water taken with little sugar Mashed rice mixed with water, salt, and boiled (*chaler bori*)
Dysentery	Banana curry taken with rice (*kacha kola* or *bicchii kola*) *Thankuni pata* (edible bitter herb) ground and mixed with water
Fever	Pouring water on the head and put wet cloth on the head
Skin problem	A local leaf (*Nim pata*) rubbed directly onto the skin
Digestive problems	Eating sour foods, following previous prescriptions for a similar condition, using herbal remedies

Source: Edgeworth (2011).

and the effectiveness of rice-based diets in treating diarrhea (Roy et al., 1994) have been previously documented. Additionally, antimicrobial activity has been observed in the pharmacological properties of the *Hydrocotyle asiatica* plant, locally known as Thankuni, which are beneficial to the treatment of dysentery (Chowdhury et al., 1997). Other types of plant life in Bangladesh have been recognized as having medicinal qualities used to treat a range of illnesses including fever, jaundice, vomiting, and diarrheal diseases (Rahman et al., 2001).

Further scientific support is available for the use of ORS. This mixture of sodium, carbohydrate, and water was established in the 1980s and has become the cornerstone of many programs trying to control diarrheal disease (Victoria et al., 2000). Bangladesh has played a prominent role in developing and distributing ORS. The legacy of these efforts is thus apparent in households' management of diarrhea and their understanding of both when to apply ORS and in many cases how to make the rehydration solution using ingredients available within the home. As a result, the majority of participants across the research sites were well versed in the concept and application of ORS. This was explained by one participant: "they (Bangladesh NGO) told us when you have diarrhoea not to go to the doctor but at first to try the saline. They taught us how to make the saline because at that time the packet saline was not available…and asked at first to try this saline, if it doesn't work then try to give the patient chaler bori (mashed rice mixed with water and boiled, becomes a type of soup and eaten). But they also said that if you don't have the ingredients for saline then try to make chaler bori first rather than going to the doctor." Through the provision of adequate education messages and training at the local level, household resilience is strengthened to treat diarrheal disease through easily accessible low-cost means. Institutional support through local nongovernmental organizations' health programs also promotes alternative self-care solutions if initial attempts fail or ingredients are not available. By providing viable options, households' resilience is reinforced further through the range of self-care practices used in disease management.

The manner in which these measures are applied and the level of knowledge and understanding attributed to self-care practices by households demonstrate local level capabilities in managing disease and illness. Additionally, the rural context of Bangladesh offers an environment in which locally adopted disease management can thrive and be used to their maximum potential. Due to the predominance of self-care, the way in which many practices are ingrained in cultural and historical aspects of treatment and the regular success of many methods means strong levels of trust and faith are attributed to self-treatment. Combined with the understanding among many elements of the community that responsibility for treatment resides with the household (with some wider community support) a strong platform for self-care exists at community level. This receptivity toward self-care ownership, practice, and implementation resonates strongly with components of the

Hyogo Framework (WCDR, 2005) in terms of building on existing indigenous knowledge and natural resource management to further enhance people's empowerment in disaster and disease management. Further elements of the empowerment perspective can be identified through local level self-care use as described in the next section.

16.6 EMPOWERMENT AND DIGNITY

Analysis of the benefits self-care offered the rural household included the preservation of dignity and notions of empowerment. Dignity is viewed as central to theoretical constructs of empowerment and poverty alleviation as it represents an important goal for impoverished households marginalized by society (Ahmed, 2008). Maintaining dignity is therefore an important component in poor people's daily realities. Measures that are able to assist in this process are thus valued at the household level. The use of self-care methods that allow the patients to treat themselves within the security of their own home and/or without the need to seek external treatment from various health practitioners can help preserve feelings of dignity. Similarly, individuals who are able to use their own knowledge in the production and administration of herbal and home-based remedies are also likely to harbor notions of empowerment and dignity. The process of treating themselves rather than relying on external assistance was captured by one participant: "I feel good doing self-care it gives me good results.... I think I am taking responsibility for my health. I am taking care of myself and I feel good about that" and reinforced by the views of another participant: "I want myself to be as self-reliant and keeping a healthy life."

Ignorance has often been proposed by health practitioners, policy makers, and in some cases community members as an underlying reason for the extent of self-care adoption. There was awareness of this label among many participants and successful utilization of self-care in response to cases of ill health may also enhance feelings of dignity and empowerment in the light of accusations of ignorance. Overall, these influences were reflected in the survey findings as preservation of dignity was identified by almost 37 percent of participants as a benefit of self-care.

The notion of empowerment was also supported by cases of reduced gender stratification in which men and women worked together to administer self-care, as captured by one female participant: "It's a kind of sharing; I know something and he knows something. I might say which medicines we need and then he will go to the shop to buy them. If it's not available in the home then I ask him to buy it. Isn't this helping each other?" The decision to implement self-care, through self-medication in the example above, becomes a joint effort as opposed to the traditional division of responsibilities for health care normally held by the female. Although the manner in which the self-care is undertaken resorts to traditional gender roles, for example, mobility restrictions on women resulting in the male leaving the household to purchase required medicines. The overall self-care process within the research sites is a shared responsibility that takes place in a society where gender divisions of everyday tasks within the household are common place and far reaching. Self-care is perhaps one of the few tasks undertaken by the rural household that reduces this gender division.

Within these contexts there appears to be a case for self-care practices facilitating household empowerment and preserving and/or enhancing dignity among users. Dignity is widely acknowledged as an important goal for marginalized households, as well as a central component of wider poverty alleviation initiatives (Ahmed, 2008). Any measures that are able to assist in dignity preservation and enhancement should be highly valued. These measures hold particular salience in regards to the personal nature of many health problems and the "taboo status" associated with disease. Self-care can facilitate treatments within the household, thus providing a measure of sensitivity to local cultural issues such as Islamic laws, "conservative principles," and elements of patriarchy. The latter factor must be carefully addressed in the Bangladesh context, particularly with regard to any external intervention or development approach. The research findings demonstrated the differing circumstances in which men and women employed self-care strategies when responding to ill health. The context of these self-care decisions is reflective of broader social and cultural influences; however, other self-care practices challenged the status quo of gender dynamics in the rural Bangladesh context. By examining the less visible forces surrounding self-care uptake it is possible to identify examples of male and female partnerships in providing home-based care in response to ill health. Evidence also exists to suggest the use of self-care can minimize the impact of patriarchal relations upon some women and the manner in which men are able to maintain livelihood practices through self-care utilization.

Some of the greatest synergy between the use of self-care and local level resilience can be seen through the manner in which self-care draws upon local knowledge, creating notions of empowerment and dignity, rather than dependency on external assistance. Processes that add to other evidence indicating community health processes initiated by households themselves can achieve a greater impact than external top-down programs (Macfarlane et al., 2000). The wider lessons for community-based disaster management appear to be no more complicated than locally focused strategies that ensure high levels of local ownership. However, the health system does still have a role to play in achieving the maximum impact if such approaches are to be implemented.

16.7 SELF-CARE AS A LOW-COST, MANIFOLD STRATEGY

The economic burden of health care is widely documented as a key force in triggering impoverishment or pushing households into greater levels of destitution (Sen, 1999; McIntyre et al., 2006). With the majority of the Bangladesh population living below the poverty threshold (BBS, 2007), the economic cost of health care utilization can be devastating (Hulme, 2003). Financing treatment for common illnesses can vastly exceed average daily household income, with the cost of responding to fever, diarrhea, and skin infections up to five, six, and eight times, respectively, the average daily wage (Wahed and Mahmood, 2009). Low-cost methods that can tackle multiple disease risks therefore clearly hold several benefits for the household in terms of strengthening cost-effectiveness, reduced opportunity costs, and efficacy of responding to illness, all of which strengthen local level resilience to disease and increase the likelihood of adoption among the poorer and more marginalized communities and households.

Self-care practices offered substantial savings in comparison to different forms of external treatment (both formal and informal) for a number of different illnesses. Traditional remedies using natural ingredients collected from the locality of participants' households did not require any cash payment. Other widely adopted forms of self-care, such as ORS or self-medication with antibiotics, are relatively cheap in comparison to the expenses incurred through consultation and travel to hospitals and qualified doctors. These differences were summarized by one participant: "If we visit the doctor we have to pay but if we do it ourselves we can do it for maybe only five or ten taka. The doctor will cost maybe 50 taka or maybe 200 taka and then there is also the rickshaw, more money, more money we don't have, so it is difficult. But if we do it ourselves the cost is almost nothing, maybe only a little money for medicines...if we use the herbal leaves it is free for us."

The use of self-care in these contexts therefore enables households to manage the impact of ill health without incurring significant financial outputs, reduces the potential loss of earnings, and facilitates coping with ongoing disease risks. The manner in which self-care methods are produced and administered also reduces many opportunity costs associated with other forms of treatment response. These two factors represent a clear advantage for many households in their decision to adopt self-care. The wider implication from this finding is for development agencies and other external institutions to provide similar means to reduce cost and time factors for impoverished groups in their attempts to minimize disease risks and illness. Support strategies for disease risk management and community-based disaster management would also do well to recognize and replicate the manner in which self-care can respond to multiple disease risks. This has application not only as a prevention and response strategy, but also in the way certain self-care strategies are able to tackle a range of diarrheal diseases in comparison to some other disease reduction initiatives that are only able to tackle one disease at a time.

The potential of self-care use could also have implications for policy makers in terms of weighing the potential benefits of greater self-care utilization at the community level as an effective disease management strategy against alternative interventions that mitigate the impact of other diseases endemic in the Bangladesh environment. Considerable investment has been channeled into the increased provision of clean water and adequate sanitation facilities, often as part of wider development strategies supported by multilateral and bilateral development agencies (Hutton and Haller, 2004; Fewtrell et al., 2005; Clasen et al., 2006). A number of other initiatives have targeted individual and household behaviors in order to sustainably modify hygiene practices. Education and hygiene promotion campaigns have been at the forefront of these programs designed to increase levels of health education, influence hand washing practices such as the uptake of soap, alter water and sanitation behaviors, and improve food handling (Curtis et al., 2000; Curtis and Cairncross, 2003; EHP et al., 2004; Frewtrell et al., 2005). These programs have not always significantly affected diarrheal disease morbidity and mortality rates but they have highlighted a range of coping strategies employed to mitigate the impacts of disease on household health and livelihoods. These health initiatives should be considered along with any moves toward self-care support and integration into health care provision in setting policy priorities with respect to supporting local level resilience and disease risk management in resource poor contexts.

16.8 COPING WITH ENVIRONMENTAL HAZARDS THROUGH SELF-CARE

In the current policy environment of community participation and empowerment in mitigating health disasters, the balance between self-treatment and governmental and other institutional responsibility to provide support is yet to be calculated. This requires knowing how self-care is interpreted and utilized in terms of disaster vulnerability and response. Findings indicated the potential value of self-care practices for some households during times of disasters such as flooding. The use of herbal remedies or treatments produced within the home offered an immediate method of response that circumnavigated potential difficulties in accessing external health care providers and facilities. A key component of this is natural resource availability, which has been previously documented as an influential component in people's use of self-care in rural Bangladesh (Edgeworth and Collins, 2006). This was supported in this research through qualitative work and survey findings in which 73.4 percent of the respondents stated that access to natural materials was a requirement for them to be able to adopt self-care strategies. Therefore, adopting certain self-care measures during the onset of natural hazards can offer an effective response strategy.

However, when access to natural resources is circumscribed, it can have a detrimental effect on the adoption of self-care. Growth in population density, increased use of available land, and modernization of farming practices, such as fertilizer use, were frequently expressed as reasons for the decline in natural resource availability. Flooding also posed a constraint by obstructing access or destroying vegetation utilized in traditional remedies. A number of people believed that any attempt to try and establish their own natural resource base in the household for medicinal plants was futile due to the extent of flooding. Although historically flooding has presented risks throughout Bangladesh, the increasing regularity of flash floods and the prolonged duration of seasonal flood waters increased the likelihood of vegetation destruction in these locations. Nevertheless, many participants demonstrated resilience in response to these contexts. Several people described investing extra time and effort in their traditional remedy preparations by seeking ingredients from local markets or alternative locations less affected by population and agricultural changes or environmental hazards. As a result, plant life remains relatively undisturbed and thus provides a wealth of differing natural products that people use in various herbal remedies. Although these alternative sites presented some drawbacks in terms of distance, opportunity costs, and the additional time burdens placed on self-care preparations, participants were prepared to adapt in order to ensure that self-care could take place. "It can take some time to collect the ingredients that you need, if the patient is here then we will go and collect and come back and prepare it. Earlier we could find what we needed, but now it is not so available, so it can take 10 or 20 minutes depending on what we need."

This context clearly represents the notion stated earlier in the chapter that people are not passive observers of events that unfold around them; they adjust to differing environmental changes and occurrence of environmental hazards in a manner that enables self-care continuity. It suggests that people are committed to local level resilience through the use of local strategies which can be utilized in the home. It also indicates the resilience of households to changing circumstances in order to ensure self-care adoption.

16.9 CONCLUSION

This chapter has discussed the role self-care plays in local level resilience and risk reduction within a disease endemic environment. In doing so, the chapter contributes to the debate of self-treatment and how it is applied (Chapple and Rogers, 1999; Wilkinson and Whitehead, 2009) and the wider notion of people-centered risk reduction of infectious disease (WCDR, 2005; IFRC, 2009; WHO, 2009). This has highlighted the crucial role self-care can play in coping with the insurmountable endemic disease risk environment in rural Bangladesh. Resilience and risk reduction is salient to examples of households accessing finite natural resources, seeking cost-effective health care solutions, or utilizing support from some local health programs. The findings also suggest that local levels of resilience and risk reduction through self-care do not possess any significant gender, age, or socioeconomic dimensions, indicating the potential for facilitating treatment equity for certain illnesses now and in the future. Given the current levels of health care provision and socioeconomic characteristics in rural Bangladesh, it is likely that self-care will continue to remain one of the predominant responses to common acute and chronic illnesses. Although self-care offers the means to achieve better resilience to disease it is not a panacea for the disproportionate impact of ill health upon the poor. As such it should be viewed in conjunction with other approaches that can tackle the burden of ill health. A greater focus on supporting self-care to ensure ongoing local resilience to disease should not override necessary structural changes that have greater potential to achieve health equity and justice for the poor of Bangladesh.

REFERENCES

Adams, A.M., Cekan, J., Sauerborn, R., 1998. Towards a conceptual framework of household coping: reflections from rural West Africa. Africa 68 (2), 263–283.

Ahmed, S.M., 2005. Exploring Health-seeking Behaviour of Disadvantaged Populations in Rural Bangladesh (Ph.D.). Karolinska Institute.

Ahmed, S.M., et al., 2003. Changing health-seeking behaviour in Matlab, Bangladesh: do development interventions matter? Health Policy Plan. 18 (3), 306–315.

Ahmed, F.E., 2008. Hidden opportunities: Islam, masculinity and poverty alleviation. Int. Fem. J. Politics 10 (4), 542–562.

Anderson, J.M., 1990. Home care management in chronic illness and the self-care movement: an analysis of ideologies and economic processes influencing policy decisions. Adv. Nurs. Sci. 12 (2), 71–83.

Anderson, J.M., 1996. Empowering patients: issues and strategies. Soc. Sci. Med. 43 (5), 697–705.

Anderson, R.M., Funnel, M.M., 2005. Patient empowerment: reflections on the challenge of fostering the adoption of a new paradigm. Patient. Educ. Couns. 57 (2), 153–157.

Baker, C., 1995. The development of self-care ability to detect early signs of relapse among individuals who have schizophrenia. Arch. Psychiatr. Nurs. 9 (5), 261–268.

Bates, I., et al., 2004. Vulnerability to malaria, tuberculosis and HIV/AIDS infection and disease. Part 1: determinants operating at individual and household level. Lancet Infect. Dis. 4 (5), 267–277.

BBS (Bangladesh Bureau of Statistics), 2007. Statistical Pocketbook of Bangladesh 2006. BBS, Planning Division, Ministry of Planning, Government of Bangladesh, Dhaka.

Beck, E., 2005. The UCSD student run free clinic project: transdiciplinary health professional education. J. Health Care Poor Underserved 16 (2), 207–219.

Berman, R., Iris, M., 1998. Approaches to self-care in late life. Qual. Health Res. 8 (2), 224–236.

Bhuyan, K.K., 2004. Health promotion through self-care and community participation: elements of a proposed programme in the developing countries. BMC Public Health 4 (11), 1–12.

Buttenheim, A., 2006. Flood Exposure and Child Health in Bangladesh. Californian Centre for Population Research and UCLA School of Public Health.

Chapple, A., Rogers, A., 1999. Self-care and its relevance to developing demand management strategies: a review of qualitative research. Health Soc. Care Community 7 (6), 445–454.

Chowdhury, A.K.A., et al., 1997. Antibacterial activity of *Pergularia daemia*. Banglad. J. Life Sci. 8 (2), 71.

Clasen, T., Roberts, I., Rabie, T., Schmidt, W., Cairncross, S., 2006. Interventions to improve water quality for preventing diarrhoea. Cochrane Database Syst. Rev. Issue 3. Oxford: Update Software.

Cockcroft, A., Milne, D., Anderson, N., 2004. The third service delivery survey 2003: Final Report. CIET Canada and the Ministry of Health and Family Welfare, Government of Bangladesh, Dhaka.
Collins, A., 1998. Environment, Health and Population Displacement: Development and Change in Mozambique's Diarrhoeal Disease Ecology. Ashgate, Aldershot.
Connelly, M.A., et al., 2004. Communicable diseases in complex emergencies: impact and challenges. Lancet 364 (9449), 1974–1983.
Cunningham-Burley, S., Irvine, I., 1987. And have you done anything so far? An examination of lay treatment of children's symptoms. Br. Med. J. 295 (6600), 700–702.
Curtis, V., Cairncross, S., Yonli, R., 2000. Review: domestic hygiene and diarrhoea—pinpointing the problem. Trop. Med. Int. Health 5 (1), 22–32.
Curtis, V., Cairncross, S., 2003. Effect of washing hands with soap on diarrhoea risk in the community: a systematic review. Lancet Infect. Dis. 3 (5), 275–281.
Davies, S., 1993. Are coping strategies a cop out? IDS Bull. 24 (4), 60–72.
Dean, K., 1981. Self-care responses to illness – a selected review. Soc. Sci. Med. A 15 (5), 673–687.
DFID (Department for International Development), 2005. Disaster Risk Reduction: A Development Concern. Policy Briefing Paper. DFID, London.
Edgeworth, R., Collins, A.E., 2006. Self-care as a response to diarrhoea in rural Bangladesh: empowered choice or enforced adoption? Soc. Sci. Med. 63 (10), 2686–2697.
Edgeworth, R., 2011. Disease resilience in rural Bangladesh: the adoption and practice of curative self-care. Local Environment 16 (5), 461–472.
EHP, UNICEF & WSP, 2004. The Hygiene Improvement Framework: A Comprehensive Approach for Preventing Childhood Diarrhea [Online]. Available at: http://www.ehproject.org/PDF/Joint_Publications/JP008-HIF.pdf.
Faisant, N., et al., 1995. Banana starch breakdown in the human small intestine studied by electron microscopy. European Journal of Clinical Nutrition 49 (2), 98–104.
Falkingham, J., 2004. Poverty, out-of-pocket payments and access to health care: evidence from Tajikistan. Soc. Sci. Med. 58 (2), 247–258.
Few, R., Ahern, M., Matthies, F., Kovats, S., 2004. Floods, Health and Climate Change: A Strategic Review, Tyndall Centre for Climate Change Research Working Paper, 63.
Fewtrell, L., Kaufmann, R.B., Kay, D., Enanoria, W., Haller, L., Colford Jr, J.M., 2005. Water, sanitation, and hygiene interventions to reduce diarrhoea in less developed countries: a systematic review and meta-analysis. Lancet Infect. Dis. 5 (1), 42–52.
Gantz, S.B., 1990. Self-care: perspectives from six disciplines. Holist Nurs. Pract. 4 (2), 1–12.
Government of Bangladesh (GOB), 2003. Conceptual Framework of Health, Nutrition and Population Sector Programme (HNPSP) July 2003–June 2006. GOB. Planning Wing. Ministry of Health and Family Welfare, Dhaka.
Haug, M.R., Wykle, M.L., Namazi, K.H., 1989. Self-care among older adults. Soc. Sci. Med. 29 (2), 171–183.
Health Canada, 2004. Supporting Self-care: The Contribution of Nurses and Physicians [Online]. Available at: http://www.hc-sc.gc.ca/hcs-sss/pubs/hhrhs/1997-self-auto-contribut/literature_e.html.
HelpAge International, 2000. Uncertainty Rules over Lives: The Situation of Older People in Bangladesh. HelpAge International, London.
Hickey, T., Dean, K., Holstein, B., 1986. Emerging trends in gerontology and geriatrics: implications for the self-care of the elderly. Soc. Sci. Med. 23 (12), 1363–1369.
Hulme, D., 2003. Thinking "small" and the understanding of poverty: Maymana and Mofizul's story. In: Institute for Development Policy and Management University of Manchester United Kingdom, Working Paper, vol. 22, pp. 1–21.

Hutton, G., Haller, L., 2004. Evaluation of the Costs and Benefits of Water and Sanitation Improvements at the Global Level. World Health Organisation, Geneva.
IFRC, 2004. The World Disaster Report: Focus on Community Resilience. IFRC, Geneva.
IFRC, 2009. World Disasters Report – Focus on Early Warning, Early Action. IFRC, Geneva.
Illich, I., 1995. Limits to Medicine. Medical Nemesis: The Exploration of Health. Marion Boyars, London.
ISDR, 2002. Living with Risk: A Global Review of Disaster Risk Reduction Initiatives. ISDR, Geneva.
Kickbush, I., 1989. Self-care in health promotion. Soc. Sci. Med. 29 (2), 125–130.
Leyva-Flores, R., Kageyama, M.L., Erviti-Erice, J., 2001. How people respond to illness in Mexico: self-care or medical care? Health Policy 57 (1), 15–26.
Macfarlane, S., Racelis, M., Muli-Musiime, F., 2000. Public health in developing countries. Lancet 356 (2), 841–846.
McIntyre, D., et al., 2006. What are the economic consequences for households of illness and of paying for health care in low- and middle-income country contexts? Soc. Sci. Med. 62 (4), 858–865.
Narayan, D.C.R., Shah, M.K., Petesch, P., 2000. Voices of the poor: Crying out for change. World Bank, New York.
OECD, 2003. Poverty and Health. DAC Guidelines and Reference Series.
Orem, D.E., 1991. Nursing: Concepts of Practice. Mosby, St Louis.
Pagan, J.A., et al., 2006. Self-medication and health insurance coverage in Mexico. Health Policy 75 (2), 170–177.
Peden, A., 1994. Up from depression: strategies used by women recovering from depression. J. Psychiatr. Ment. Health Nurs. 1 (2), 77–83.
Phillips, D.R., Verhasselt, Y., 2001. Health and Development. Routledge, London.
Punamaki, R.L., Aschan, H., 1994. Self-care and mastery among primary health care patients. Soc. Sci. Med. 39 (5), 733–741.
Rabbani, G.H., et al., 2001. Clinical studies in persistent diarrhoea: dietary management with green banana or pectin in Bangladeshi children. Gastroenterology 121, 554–560.
Rahman, S., et al., 2001. Pharmacological evaluation of Bangladeshi medicinal plants – a review. Pharm. Biol. 39 (1), 1–6.
Redman, B.K., 2007. Responsibility for control: ethics of patient preparation for self-management of chronic disease. Bioethics 21 (5), 243–250.
Rogers, A., Hay, S., 1998. Self-care and Access to Health Care in Opportunities for Improving Care. Open University Press, Buckingham.
Roy, S.K., et al., 1994. Persistent diarrhoea: efficacy of a rice-based diet and role of nutritional status in recovery and nutrient absorption. Br. J. Nutr. 71 (1), 123–134.
Segall, A., Goldstein, J., 1989. Exploring the correlates of self-provided health care. Soc. Sci. Med. 29 (2), 153–161.
Segall, M., Tipping, G., Lucas, H., Dung, T.V., Tam, N.T., Vinh, V.X., Huong, D.L., 2002. Economic transition should come with a health warning: the case of Vietnam. J. Epidemiol. Commun. Health 56 (7), 497–506.
Sen, B., 1997. Health and poverty in the context of country development strategy: a case study on Bangladesh. Macroeconomic and Health Development Series No. 26. World Health Organisation, Geneva.
Sen, A.K., 1999. Health in development. Bull. World Health Org. 77 (8), 619–623.
Shaikh, B.T., Hatcher, J., 2005. Complementary and alternative medicine in Pakistan: prospects and limitations. Evid. Based Complement Alternat. Med. 2 (2), 139–142.

Shankar, P.R., Partha, P., Shenoy, N., 2002. Self-medication and non-doctor prescription practices in Pokhara valley, Western Nepal: a questionnaire-based study. BMC Fam. Pract. 3 (17), 1–7.

Shaul, M., 1995. From early twinges to mastery: the process of adjustment in living with rheumatoid arthritis. Arthritis Care Res. 8 (4), 290–297.

Sowell, R.L., et al., 1997. Self-care activities of women infected with human immunodeficiency virus. Holist Nurs. Pract. 11 (2), 18–26.

UNDP, 2003. Human Development Report: Bangladesh. www.undp.org/hdr2003/indicator/cty_f_BGD.html.

Victoria, C.G., et al., 2000. Reducing deaths from diarrhoea through oral rehydration therapy. Bull. World Health Org. 78 (10), 1246–1255.

Wahed, T., Mahmood, S.S., 2009. Cost associated with utilisation of health care services. In: Bhuiya, A. (Ed.), Health for the Rural Masses: Insights from Chakaria. ICDDR, B, Dhaka, pp. 95–113.

WCDR, 2005. The United Nations Hyogo Accord. Kobe, Japan.

WHO, 2000. World Health Report – Health Systems: Improving Performance. WHO, Geneva.

WHO, 2001. Improving Health Outcomes of the Poor: The Report of Working Group 5 of the Commission on Macroeconomics and Health. World Health Organisation, Geneva.

WHO, 2004. World Health Report, "Changing History". World Health Organisation, Geneva.

WHO, January 7–9, 2009. Regional Consultation on Self Care in the Context of Primary Health Care. Thailand, Bangkok.

Wilkinson, A., Whitehead, L., 2009. Evolution of the concept of self-care and implications for nurses: a literature review. Int. J. Nurs. Stud. 46 (8), 1143–1147.

Wilson, P.M., Kendall, S., Brook, S., 2007. The expert patients programme: a paradox of patient empowerment and medical dominance. Health Soc. Care Community 15 (5), 1–13.

World Bank, 2002. Bangladesh: Improving Governance for Reducing Poverty. World Bank, Washington DC.

Zwi, A.B., Yach, D., 2002. International health in the 21st century: trends and challenges. Soc. Sci. Med. 54 (11), 1615–1620.

Chapter 17

Culture: The Crucial Factor in Hazard, Risk, and Disaster Recovery: The Anthropological Perspective

Susanna M. Hoffman
Hoffman Consulting, Telluride, Colorado, USA

ABSTRACT

Throughout the decades of research into disasters, aspects of how risk is constructed, how calamities unfold, and why recovery only sometimes succeeds have often proven enigmatic. More and more agencies have come to realize the importance of incorporating the key element of a people's culture into their work and understanding. Culture is the core explanatory tool of the field of anthropology. From the anthropological perspective, the chapter covers the definition and profoundness of culture; the importance of a people's local traditions and knowledge; variations in risk perception; differing arrangement of time and space; different styles of thinking; the influence of a culture's expressive images, including mythology, art, and legend; various social structural and kinship systems, including other methods of segmenting the human continuum, such as by gender; alternative economic modes; the importance of cultural flexibility or fixedness; land tenure and attitudes toward property; place attachment; history; legal systems; religious mandates; and more. Although local in many deterministic elements, almost all cultures are also participants in one of the world's Great Traditions and the emergence of a pan-global culture with its endeavor to spread universal humanitarian principles over and above the relativity of differing societies.

17.1 INTRODUCTION

Throughout the decades of research into disasters, questions of how they are constructed, how they unfold, how a people sculpt their perception of risk and create vulnerability, why recovery sometimes succeeds and sometimes fails have often proven enigmatic. By way of reply, in recent years, more and more agencies and organizations worldwide have come to realize the importance of

Hazards, Risks, and Disasters in Society. http://dx.doi.org/10.1016/B978-0-12-396451-9.00017-2

incorporating into their programs the key element of culture, something anthropologists in the disaster field have been promoting for a long time.

Since the inception of the field of anthropology, the concept of culture has been its main innovation and core explanatory tool. It emerged as academics, and others began to recognize that many people around the globe lived, thought, and behaved in diverse manners, and they began to investigate why such differences occurred. The concept came to refer to these distinctions. In its deep and holistic framework, it embraces kinship, subsistence, economics, politics, dress, food, exploitation of place, scheduling of time, arrangement of space, classification of humans, social roles and relationships, child raising, laws, rules, and more. It entails a people's cosmology, how they see their gods, ancestors, the universe, Earth, and stars. It speaks to beliefs, spirituality, explanations, and how various groups envision their origins and purpose. Very significantly, culture also functions as a guide to a people's perception, how they see colors, smell odor, hear sounds, and feel touch, and thus how they filter and format information. Every human from childhood up absorb the culture of the society in which they are raised and by and large live their entire lives by its grid (Tyler, 1871; Kroeber, 1952, 1963; Boas, 1911, 1940).

Over time, the concept of culture has provoked various definitions and redefinitions and considerable discourse as to its nature and inclusiveness, but perhaps, the most fundamental is the first simple definition given by Edward Tylor, the father of anthropology in 1871. Culture is "that complex whole which includes knowledge, beliefs, arts, morals, laws, custom, and any other capabilities and habits acquired by a (human) as a member of society." (Tylor, 1871).

This does not mean that culture does not change. It does, as many in the disaster realm have seen. It is, however, sluggish and conservative. Nor does it mean that all the people of a society live exactly according to their culture's dictates. They do not. Still, the degree of their variance is rather like the stretch of a rubber band, limited in range and manner. It also does not mean that at times people do not contest their culture. They do. Even so, their contestation emerges from the reflective comprehension of their own life ways. Ironically though, it is that very contestation that often leads to cultural change (Steward, 1955; Frake, 1962; Harris, 1968; Firth, 1959).

With their core concept in mind, anthropologists began their foray into the study of disaster early on. Anthony Wallace in 1956 described *a tornado in Worcester* (Wallace, 1956) and Schneider *a typhoon in Yap* in 1957 (Schneider, 1957). Since then, anthropological inquiry into disasters has expanded into an ever-growing number of researchers and an extensive body of literature. As anthropologists, along with other social scientists, have developed their insight into the fabric from which natural and technological calamities are woven, the theoretical and methodological pertinence of anthropology's perspective and the necessity of considering a people's culture on both local and overarching levels have become yet more manifest. As well as contributing the concept of culture, only anthropology takes into its reckoning the three planes that interface in catastrophe: environmental, biological, and sociocultural. The developmental and comparative perspectives of anthropological research also uniquely merge with disaster studies as disasters are the result of long and complex processes. The cultural, historical, social, and physical phylogenies of calamities are manifold, a particular purview of anthropology. Anthropology further examines the small against the large, contrasts one society against another, and addresses social continuity as well as change—all pivotal topics in disaster. Added to all these approaches, anthropology has contributed its fundamental methodology: on the ground ethnographic fieldwork. Indeed, ethnographic fieldwork greatly, sometimes singularly, elucidates how disasters are constructed, what recovery entails, and what the often enshrouded elements are that lead to a people's vulnerability. In addition, anthropological study encompasses communities worldwide (Hoffman and Oliver-Smith, 2002; Hoffman, 2010; Oliver-Smith, 1996).

Unfortunately over time, although anthropologists have continued to explore the depth and nuance of culture from the small district to the greater comprehensive level, the concept, being such a facile way of accounting for diversity, has been coopted by institutions such as corporations to distinguish not societies, but their idiosyncratic organizational credos. Culture as an explanatory tool has also been co-opted by disciplines beyond anthropology; other social sciences; and also humanities, journalism, governmental studies, and more. They have tended to grab onto the concept in a way that could be called "culture lite." They assume that a quick grasp of language, costume, religion, or laws gives them the complete understanding. So borrowed, over time, the premise has lost its deep-reaching implications. It became "defanged." Nonetheless, it is culture in its most discerning sense that has proven the key to understanding why and how various people worldwide deal with risk and undergo disaster (Hoffman and Oliver-Smith, 2002).

It is a culture that determines how a particular people calculate peril, experience catastrophes, and recover from them, or do not recover, or do not protect themselves. Its deterministic influence on a group's life ways has been identified as the essential underpinning of vulnerability and how that is manufactured and manifested. Yet, instilling the knowledge of culture in its profound connotation into much disaster prevention and aid, into the policies and practices of international, national, and nongovernmental agencies dealing with disaster has proven stubborn, often leading to the severe detriment of survivors.

17.2 SOME ESSENTIAL WAYS IN WHICH CULTURE MATTERS

One of the key operative factors to risk reduction worldwide is the fact that members of disparate societies simply do not perceive reality and environment in the same way.

All sorts of components are involved. For example, few people realize that members of different cultures do not see color the same way. That is, all people perceive the entire color spectrum, but from one culture to another they merge the colors into different groupings they consider essentially the same. Some people see the whole spectrum as only two distinct colors, black and white, or maybe three colors, black white, and red. People hear sound and see sights differently too, and the doors determining the walls restricting perception close about the age of 14. After that crucial point, they are difficult to alter (Berlin and Kay, 1969).

People do not perceive divisions of time or segments of space in the same way either, and the treatment of time and space is cardinal to disaster. Nor do people think in the same way. Some people think and calculate time and other matters in a linear fashion. For instance, they see life unfolding as a progression, straight and direct though occasionally punctuated with episodes that they try to predict. Others, such as the Hindu, Tibetan, and ancient Mayan, think and assess time and events in a cyclical fashion, meaning all manner of things, including disasters, transpire on a reoccurring calendar no matter what intervenes. The implication for practitioners is that linear thought and time imply hope of change. Mitigation can alter a straightforward progression. With cyclical thinking, often mitigation is dismissed as useless as all catastrophes are preordained (Friedel et al., 1993; Nicholson, 1967; Hoffman, 2002). Still, styles of thinking are not always as simple as they appear on a surface level. Occasionally, people such as those in Western societies, who think or see time in a lineal fashion, still indulge a covert view that disasters are in fact cyclical. They, some of them scientists, claim a calamity to be "the 100-year flood" or "the 50-year earthquake." Doing so, unfortunately, can instill a notion that such a similar calamity is alleviated for a specified time. However, in truth, the same sort of disaster can happen unpredictably over and over again (Hoffman, 2002).

Some people further think in a more comprehensive manner, something like a "surround sound" of happening. All things occur at once, with a piece here and a piece there. As illustration, many lineal-thinking folk expected the Greek Olympic and the Brazilian World Cup facilities would never be completed on time, the stadiums were so incomprehensibly being built a little bit here and a little bit there. The events would end up taking place in unfinished fiascos. Yet proceeding in an encompassing manner, all the structures were finished on time.

Additionally, not all people share the Westerners' distinction of what is domestic space versus wild. Westerners tend to think of nature, from which many risks stem, as something removed and untamed (Levi-Strauss, 1966). But in truth, such a separation does not exist, and no such thing exists as nature. Nature is merely a term and idea used to distance and separate ourselves from the controlled and contained environment in which we erect and station our lives. In reality, wherever a person stands at any one moment is "nature" and what comes from nature can visit any time, so in anthropology to avoid confusion, we tend to say "physical plane" rather than nature. Still, not by any means do all people share the distinction between the natural and what is the cultural milieu. To some, the physical plane is a day-to-day, moment-to- moment, part of their lives and what comes from it is simply part of its essence.

The two versions of nature, however, bear a consequence in terms of risk reduction. Those who separate nature as distinct, often attempt to control it, generally with physical means. They erect levees, fire walls, or redirected river. But since nature does not actually exist separately from ourselves, harnessing it to achieve safety is, as many in the disaster field recognize, not the answer. Risk reduction is a social—cultural and perceptual problem, not a physical one, albeit it is a solution hard to instill in those who think nature can be managed.

A further factor concerning space that plays a rule in determining vulnerability is culturally how a people arrange the space they inhabit (Feld and Basso, 1996; Low and Lawrence-Zuniga, 2003). Some people consider any and all the space around them available for occupation, whether it could be deemed safe or not. Other groups strictly define what can be lived on and what must be left alone according to sacred definition or perhaps assessed legitimized codes. Cultures also narrate the density of human habitat. Some cultures evince a preference for clustered villages and cities, thus potentially making more persons vulnerable to mudslides or firestorms. Others mandate people to live spread out. They value large yards and distance, which can make people less physically vulnerable, depending on surrounding hazards.

Far more crucially, it is a culture's particular stance that determines a people's perception of what is a peril. Danger and risk, after all, are not the same thing. Danger is an extant, factual hazard that exists manifestly in the environment: a floodplain, an earthquake fault, a leaky factory. Risk sits between people and danger much like a water faucet (Paine, 2002; Ingold, 1992). The faucet can be completely open, so a people comprehend all the dangers around them, flood, fire, and avalanche. Or the faucet can be only partly open so that people only acknowledge some of the surrounding hazards, but are blind to others. Many Californians are good examples. Californians tend to perceive that earthquakes are a threat and prepare for them. Yet, they are in almost total denial over firestorms, floods, and avalanches, which are in fact more common, kill more people, and destroy more property every year than earthquakes do.

Alternatively, a culture's perceptual faucet can be turned off completely so that the people recognize no danger at all. My favorite example comes from the Greeks who live on the island of Santorini. Santorini is an active volcano seven times larger than Krakatoa. Despite the fact that the volcano erupted twice in the twentieth century, caused a tremendous earthquake in 1956, plus three little ones within recent years, the inhabitants voice no alarm and refute any exposure. A young woman there asked me what I did. After I told her, she declared, "Thank goodness we don't have problems like that here." Another condensed catastrophic occurrences set into mythological history, complete with assurances of divine safety. He told me a 50,000-year old lava pour that had created a red-hued mountain top had been blown out by the Virgin Mary and everyone saved, though of course, no people lived on the island at the time (Hoffman, 1976). Changing a people's recognition of peril is very demanding. In short, the faucet takes a lot of greasing. Still, occasionally people see the risk of exploiting an environment as too high and move, as have some on the Mississippi River flood bank. Others see that an industry as too dangerous and reject it, as the people of Bhopal, India, actually did to no avail (Rajan, 2002, 1999).

Part of a people's risk perception also lies in their expressive culture, something few disaster researchers consider. In every culture, disasters have been the subject of mythology, legend, art, and literature for millennia. Hanging up on a wall, told around a campfire, or read in books, the portrayals carry profound impact. They set forth an illustration of potential risks and tell what a disaster will be like. They forge shared understandings, function as oracle, and speak of consequence. As most of the portrayals are greatly repeated in a culture's lore, their continued depiction compounds their fixedness and seep into the culture's capital like an echo of Jung's collective conscience. Such stories range from Sodom and Gomorrah, Noah, the Eddas, Emperor Yao, Manu, Atlantis, and Pompei. Depending on the culture, they at times invoke catastrophic-carrying personas like Damballah, Kali, and even Typhoid Mary. Likewise, art depictions and literature, including futuristic portrayals and illustrations, inculcate images of disaster. In modern times, movie and television depictions have added to the cultural impressions. In addition, a person's prior experiences of disaster and stories of events told by parents, grandparents, and others create a depiction of what any forthcoming disaster will unfold. This was certainly true of people's expectations and response to Hurricane Sandy on the eastern coast of the United States in 2012 and also the highly destructive Hurricanes Katrina and Rita of 2005 (Hoffman, 2002, 2005). Yarns told of former disaster kept people in their houses and contributed to denial of weather reports.

Both the anthropologist George Lakoff, and Daniel Kahneman, the Nobel Prize winning psychologist, tell us preformed concepts tend to become the formats by which people filter happenings and act upon them, whether those formats take into account actuality or not. The images presented in a culture's stories and art and legends told do just that. They tender a template by which people recognize, disregard, or readjust factualness. Furthermore, when a compelling impression clashes with fact, it is the impression, not the fact, that usually prevails. Kahneman calls it the "illusion of validity." Memory, he says, automatically provides the tale about what is going on, true or not, and suppresses alternatives. A peoples' disaster art and stories thus affirm two old adages: "a picture is worth a thousand words," and the power of a tale "told a thousand times" (Lakoff and Johnson, 1980; Kahneman, 2011).

To put it another way, a people that experiences or anticipates disaster stand on a pivot differentiating what their culture provides as a model "of" disaster and what thus becomes a model "for" what risk and disaster are. A model "of" is the symbolic presentation of the existing relationship between them and their physical, social, or spiritual environment. A model "for" presents an image of a postulated reality complete with meanings that could be realized with any disaster to come. In short, whatever the culture has told them about their risk or what a disaster is like, they will perpetuate.

17.3 OTHER UNDERLYING CULTURAL FACTORS AND THEIR IMPACT

Many other cultural factors, probably too many to list out, contribute to a society's construction of risk, their vulnerability, and their response. In fact, each and every local region, commonly quite tiny, embodies a singular set of cultural nuances affecting their disaster mind-set that only deep ethnography reveals. Within any larger society backdrop, the various components that impact disaster may vary in relevance, or even in existence, according to different locations and threatening hazards. Yet everywhere, they effect what could be called "different shares of safety" of the people imperiled and what constitutes recovery, or possibly successful resettlement, should a people undergo catastrophe.

Among the most crucial elements is a society's social structure, that is, its social arrangements. Probably the most pivotal is people's calculation of blood relationship, that is, their kinship system. Indeed, in most societies, it is blood relationships that define a person's major alliances and specifies care takers and also identity. It is kinship that most everywhere operates as a society's underlying "glue."

By and large, Westerners tend to think that every people's version of kinship reiterates their own, generally a system called "bilateral." In this, blood relationships are calculated equally through both father and mother. This assumption illustrates another salient issue. Disaster researchers and workers are themselves inculcated in their own culture and have a tendency to apply their culture to others. But the assumption that all people reckon their blood ties the same way is not so. Some people, such as the Muong of Southeast Asia, the Hausa of Africa, to a large extent those following Arabic tradition, and many groups in India see themselves related by blood only through their father's line, an arrangement called "patrilineal." One's mother and her relatives are in essence in-laws. In other places, quite the opposite, only one's mother and her kin may be considered essential kinship group, such as among the Navajo, some peoples in Tibet and Indonesia. A father and his kindred are merely affines. They are "matrilineal." Even when kinship is calculated through both parents, often different values, inheritances, and obligations run according to the different lines. Whatever the kinship system, it can greatly impact disaster scenarios.

When kinships are calculated only through one parent, for example, it creates defined factions in a society, that is "clans," in which the uncles, aunts, or cousins who belong are clearly delineated by who is related by the male or female line. Such segments determine unity and opposition, loyalty, authority, leadership, and distribution of goods. In bilateral kinship with relatives on both sides recognized, everyone's set of uncles, aunts, and cousins is diverse and overlapping. Clear units and loyalties cannot be drawn and extend from both sides of any person's kinship. Nonetheless, no matter what the kinship system, everywhere a person's blood ties carry with them culturally delineated rules of behavior that include protocols of obedience, duty, avoidance, joking, shares, and sharing.

Every group of people further incorporates other ways of dividing up the human continuum. Along with segmenting space and time, humans everywhere seem unable to deal with the entire spectrum of people without creating some sort of division. The types of segmentation include by class, caste, ethnicity, race, shade of skin, age, neighborhood, religious sect, gang, and more. Indeed, the devises used to divvy up the human community are almost endless and all contribute to an "us" versus "them" deportment. Associations based on special interest, such as labor union, clubs, church also occur, as do fellowships based on town, province, and country. All entail conventions of correct conduct and responsibility and all mandated by the culture.

All of a society's various arrangements may bear varying weights. Sometimes, these are manifest, and sometimes latent. In the United States, for example, most people nowadays consider friendships to be their main social alliance, much preferred over family. But when disaster strikes, to the dismay of many, friendships often prove evanescent, whereas ties of old fashioned kinship reemerge to reveal their underlying durability. In some societies, various sorts of association can also be combined. Best friends in Mexico are culled from among one's relatives, particularly from first cousins, or "primos." Of import is that all convey loyalty in times of stress.

Age group is another common factor employed in social stratification. Again, its pertinence is dictated by a people's particular customs. In some places, the young are more esteemed and the older discounted, but in quite a number, the elders of the society are not only honored, but must be obeyed. In other societies, the association looked to for cohesion and aid stems from occupation, religion, neighborhood, color, language, or class. No matter which, all entail elements of identity and unity. All bear import in predisaster and postdisaster actions, and carry consequences in terms of accord and contest, recompense, and political agenda.

The social divisions of certain cultures are rigid and inflexible whether there is disaster or not. Class remains class, religion, color, and family remain fixed with little or no motility possible. People cannot shift. Quite possibly, the divisions are made concrete by bigotry and discrimination. Among other cultures, the partitions are more porous, particularly postcalamity. In all, from kinship to club, class to creed, these relationships provide fuel for favoritism and come with notions of who should cooperate in acts of prevention, who deserves aid, wealth, and shelter when reconstruction or resettlement looms.

To put it another way, according to a person's culture, each association carries content, and people's actions and behavior toward one another depend upon how they are proscribed by their affiliations, and the associations and substance. All too often, network studies, especially in terms of resilience, elude the all-important factor of what each connection entails, for instance, which bears gravity and is likely to endure and which does not, especially in postdisaster recovery. Yet everyone within a society knows the rules, whether the aid giver does or does not. Each person knows who they must respect, who they can dismiss, to whom they owe help, with whom they must vie, and when they must make a show of presence or give presents. The various associations also usually imply subtleties of power. A certain association may transmit dominance or include directives, that is, how to act toward labor boss, mayor, patron, or who should follow bidding. Sometimes, the potency of a relationship within a particular cultural format is quite subtle. My favorite example is that in some cultures, in fact none too few, for example, mothers-in-law gain power by only giving food and goods, never receiving them. In others, mothers-in-law are by and large ignored and shunted as soon as possible to age group and infirmity homes.

Yet another key way exists of segmenting the human community that is almost universal, and that is by gender. What the gender guidelines are varies from culture to culture, but everywhere, though, gender rules are particularly resistant to chance. Along with racial segregation, differential treatment by gender seems immediately verifiable by visual perception. Used as justification, biological fact is culturally abstracted to create strictures that are not at all biological. The constraints range from dress code to behavioral dictates. They dictate a woman's—or ethnic group's—place in space, use of time, occupations, and standing in the spiritual world. As is well documented, a culture's gender rules can severely affect vulnerability and ability to recovery. Far more often than for men, women are forced to live or work in more dangerous circumstances, as in Bangladesh. They are restrained by dress and behavioral decree in their movement, such as in the Aceh, Sumatra, where 80 percent of the deaths were reported for women. They are frequently deprived of economic viability and property, as for a long time occurred in India. Depending on cultural edits, sometimes women are not allowed money or jobs, or they endure pay rates lower than men's. Universally, women are more among the poorest populations. Their wealth may be in blankets or bracelets rather than in currency, and is thus not recognized or replaced by agencies. What work they are allowed is frequently only informal, such as being maids or care takers, or they experience a glass ceiling. At times, they are not allowed the skills needed for survival. In many nations seeming freed of gender proscriptions, after an event, old customs defining roles resurface once again constraining women, or there emerges a return crisscross of gender within a cultural dichotomy of public and private domain. In consequence, after a disaster, one finds men assuming all the public roles while women are relegated to homeware stores. And everywhere, domestic violence rises severely (Enarson and Morrow, 1998; Hoffman, 1994).

Cultures also hide certain members of their societies, usually the ones deemed shameful and disparaged anyway; rag pickers, sex workers, trans-dressers, and transgendered. All too often, ready to see only what the culture outwardly presents, NGOs and other agencies are content to ignore these people, too (Hoffman, 2004). Yet with cultural knowledge the people who do not just fall through the cracks, but live in the cracks, can be served as well.

The matter of a people's economics tends to gain the most agency attention predisaster and especially postdisaster, but the economic systems of various human groups may be quite different from those of agencies. Most disaster policies regard the matter of economics in terms of cash, continuity of commerce, persistence or replacement of stores and businesses, and employment. However, the economics of a people does not necessarily operate merely by the exchange of goods and services via money. It can instead function, or partially function, according to an intricate and shared system of reciprocity. Or, subsistence can depend on culturally arranged modes of redistribution. Among many groups, the economy also runs on the exchange of people through systems of arranged marriage that function to bind families, goods, and holdings together. These culturally defined mechanisms often become disrupted by a calamity. They can be further disrupted by the policies, practices, and shortsightedness of outside organizations. An aid worker cannot expect that money and goods given to one person will not be redistributed to either extended family members or others, nor that a woman will not give loans specifically meant for her to her husband or father. If corruption has infiltrated an economy, along with perhaps a political system, that is cultural too, as are the leaders, how they are chosen, and governmental systems (Mauss, 1954). In general, people will act according to their culture's economic and social mandates.

Along with other economic features, a people's labor might well be culturally bound. Work is not by any means always a matter of choice, opportunity, or training. In many communities, a person's labor is determined by legacy passed down through families for generations. Other times, the choice of labor is bound by societal constrictions defined by social strata and gender. Land tenure, education, and status play a part in economic viability. Positions may be distributed by long embedded traditions of nepotism or mechanisms of patronage, which individuals, dependent for livelihood and bound by tradition, are hesitant to shed. Also, certain cultures engender economic exploitation or breed financial cunning as part of admired character. Such economic contrivances can be quite obscure, yet still deterministic, and in most places, one way or another, labor overlaps with power and stratification. Again, all of these are deep cultural shading that takes profound familiarity to unravel.

Land use and ownership intermesh with a culture's economics, descent system, place attachment, and even spirituality. In Japan, the idea of *furusato*, or "home" or "ancestral" town, has great symbolic identity and often economic value even for those who were not born in the town or may never have seen it. Possessing fields to farm, owned or even if leased, holding grazing rights, possessing houses and lots are rights often so tenaciously held that people prefer the danger of clinging to them rather than to safety. Again, such holdings and prerogatives are arranged in all sorts of cultural ways. Ownership may or may not be held by individuals but rather families or clan, or whole communities. Shares may be distributed by various arrangements. In Botswana, the land is officially owned by the nation state, yet each villager knows whose farm is what or can place a house where. As well, boundary lines may be formally written or simply known by social consensus, as in the Ninth Ward of New Orleans (Altman and Low, 1992; Gill, 2013; Gill et al., 2013).

The spatial arrangement of homes, whether cluster or scattered, and whether fields or work arenas are near or far from residences are culturally regulated. As well as land tenure, how the land can be modified falls within cultural definition and most certainly contributes to vulnerability. Can trees be felled, jungle cleared, water diverted, canals dug, roads plowed through? Can land be purchased, squatted on, or only passed down families? Along with those agreements, sitting on top of the land are people's structures. The materials for homes, styles, and spatial arrangements are also a matter of customs. Sometimes, the styles are rigidly determined, so that altering them to new or safer methods may be rejected. Other times, the people may be amenable to new materials and techniques. Yet everywhere, building materials and styles contribute to a people's risk. As Anthony Oliver-Smith notes in *The Martyred City* (1992), the original building material among the indigenous population of Yungay, Peru, was light thatch, and homes were dispersed in recognition of the many earthquakes. The Spanish conquerors, however, forced the people to duplicate their culture's version of housing material and spatial arrangement, so the natives were compelled to build thick adobe walled homes with heavy tiled roofs and live collected into towns. Thirty thousand of them died in the 1970 earthquake, partly due to the change of buildings and their living in a constellated community.

Along with the concepts of land and fixed property come notions of nonfixed or material property. Different cultures define whether objects should be permanent or can be fleeting. Culture all sets out whether objects are linked to a people's identity, such as types of cars, or whether all members should have the same type of objects or unique ones. The English collect both heirlooms and antiques and suffered greatly over their loss in the floods of 2007 and 2014. The Greeks despite having both bequeathed the word and concepts of nostalgia have less binding to material belongings, and do not seem to especially regret their loss after their numerous earthquakes and fires. They, however, are so attached to their land they are loathed to leave their plots after a calamity. They will not go to a disaster center. Rather, they camp on their land. Once more, each culture is different.

The most mysterious of cultural factors in risk construction and vulnerability emerges over property and locus, and that is "place attachment." The why of it remains obscure. Is it a leftover from evolution, though now translated to hunting for a beer at the local tavern or gathering groceries in a known store? Whatever it comes from, it is strong, and seemingly globally. Everywhere, people want to go back to where they once were. The places they occupy may be perilous, sometimes chronically, but they will not leave or return again and again. To describe place attachment, I have come to borrow a term from epicurean vocabulary, *terroir. Terroir* describes the loam of the earth, the texture of climate, and waft of air that engenders a special nature to what grows there, such as wines. The marrow of a peoples' life is also a product of the ground they occupy, the atmosphere they breathe, and the way they live. Denizens of a place feel it, smell it, taste it, as much as maneuver it and manipulate it, and in some sense they themselves–their accents, dress, gestures—like vines are a product of *terroir.* Despite the risks, they desire to remain where they are or, and after a disaster, they attempt to return. Along the gulf coast of Mississippi after Hurricanes Katrina, Rita, Ike, and Gustav, where people cannot build on their land again, they have turned house foundations into picnic platforms to which they return every weekend. Place attachment seemingly still calls to people when a whole community has voluntarily moved for safety. Slowly but surely, residents "drift back" to their former locations. The mystery of place attachment calls to question if culture is embedded in or linked to location, and thus, those retaining a culture seek to reroot it in its fermenting soil.

In whatever spot, old or new, some communities tend to share a pervasive cultural unity where all the members have in common much the same background and values. Other communities are composed of a number of factions and evince considerable disunity, some displaying no conflict; other with a history of contestation. Many communities, especially the great megacities that are expanding, are marked by criminality and violence. Some amalgamate settlements house shabby ghettos and sections of diverse wealth and ethnicity. One of the problems of such programs as Habitat for Humanity is that their resettlement villages often mix people from a variety of communities with differing ethos. All these ingredients influence strength and fragility. Postdisaster, the suburb of Ano Liosia outside of Athens, Greece, versus its neighbor serves as an example. Ano Liosia lay near the epicenter of Athen's earthquake of 1999 and suffered considerable damage. The community, however, was made up of almost entirely Greeks from the Pontos region near the Caspian Sea where their ancestors had dwelled for millennia.

Understanding their culture well, and being a member himself, the mayor initiated various recovery programs based on his local knowledge. He hired only disaster victims. He insisted that houses be reconstructed on their prior locations, not in a new disaster resettlement. He established a temporary trailer camp that reflected previous neighborhood and family arrangement rather than a random distribution and had the camp run by survivor women. The community recovered speedily. The suburb of Menidi adjacent to Ano Liosia, on the other hand, was comprised an amalgam of persons, some Greeks, some Romani; some well off, others quite poor, and it struggled to recover. Years later, it remains under duress. At the same time, occasionally even wealthy and homogenous communities can be endangered. Malibu, California, is situated on cliffs that are that crumbling and where fires threaten.

It matters whether a people have at some point in their history been colonized, like Haiti, and in what condition the colonization left them (Schuller and Morales, 2012; Schuller and Farmer, 2012). It is also significant as to how many times a place has experienced calamity, and whether the people view disasters as a rather normal happening or if the catastrophe is viewed to be singular and dire, for it is when disasters are embedded in an environment, but occur only intermittently that leads to most denial. Indigenous studies show that when a people view disruptions as a normal happening and have strategies to deal with them, the concept of disaster eludes them. In short, the history of a place in all its shading affects tenacity, and it behooves an aid giver or agency to learn the whole chronology. It is the back story that reveals the evolution of cultural innuendo. It matters also if a community emerged from a single settlement and remains consistent, or whether it has absorbed changing or immigration populations. Some enclaves seriously devastated by Hurricane Sandy were established originally by particular groups, some of them occupational and of shared ethnic and religious background, with mostly Irish and Italian Catholic firemen. They remained intact over a number of generations. When Sandy struck, those enclaves rallied to revive themselves more rapidly and better than those with more mixed jumbled histories. History, and length of it, also contribute to place attachment and willingness to move.

Different religious edicts and moral standards also prevail from one culture to another and profoundly effects behavior and recovery. Doctrinal decree may assert that certain castes or high-born families have the rights to special benefits or hold incontestable prerogatives to status and office. Nor do all societies share the same ideas of ethic actions. What may be stealing, lying, or cruelty to one may not constitute reprehensible behavior to another. Outsiders converging on a strange culture cannot expect their moral standards to be replicated by their clients. One group might esteem honesty, another might admire slyness and "getting away with something." Many cultures simply do not embody the concept of charity and are perplexed by the idea. They find organizations that practice charity neither understandable nor trustworthy. Numerous societies completely lack the existence and institution of civil society organizations. Throughout their history, they only relied on family or government.

Diverse societies also have quite different legal systems, Common law, Napoleonic, Sharia, and others idiosyncratic ones. As well, the legal system operating within a particular society may or may not reflect their actual codes of behavior, so both need to be considered. Law often reinforces formal cultural mandates, but not the informal yet shared ones, such as when to say please or return borrowed objects. Moreover, as a society changes, inevitably the formal legal system lags behind, sometimes drastically, so that perhaps new principles of risk reduction lack reinforcement. Meanwhile, local officials and judges may or may not recognize state systems over their local ones, or any formal or informal laws at all, and rather act according to their tenets, class, or ethnicity.

It matters how information is distributed within a culture and to whom. Perhaps only officials and not the general population are allowed cautionary understandings. Language use, literacy, and education effect a culture or cultural segment's vulnerabilities. Among some people, only men speak and understand modern parlance. Women are restricted to old linguistic forms. The language used, whether it depicts objects in a pictorial manner or more abstractly, also impacts perception. A particular faction may or may not be allowed education. As well, differing cultures hold differing ideas of what constitutes health and illness, wholeness and injury, and how to act when ill or injured, along with traditional and adhered to methods of medical treatment. All can impact life, death, and safety.

Without a doubt, many cultures demonstrate along with all else, what could be called a shared character. It perhaps colors the attitude and paints the actions of almost all the members. Sometimes, the trait is telltale enough to be known to outsiders. Perhaps, the culture instills a bold posture or a demure one. All the members tend to be rebellious, augmentative, submissive, cooperative, democratic, selfish, noisy, withdrawn, scrappy, or dramatic. Such a temper might function merely as a superficial gloss or it might pervasively permeate every aspect of the people's disposition and conduct toward one another, and toward the outer world.

Deeply embedded within a culture also is whether it allows for flexibility and is therefore amenable to change, or if it is inflexible. Those that are pliant more readily adapt when faced with new circumstance, although their malleability may entail limitations or occur only in certain arenas. Other cultures rigidly resist alteration. They fix ever more firmly their way of life when threatened. Under threat or in resettlement, they entrench their time-honored customs rather than adjust (Hoffman, 1999a).

Finally, every culture includes varying kinds and degrees of pressures obliging its members to conform in their comportment and to their rules. These include gossip, criticism, disapproval, reputation, and shunning. The operation of social conformity mechanisms and their power over persons without a doubt effects risk reduction and disaster recovery. Any practitioner oblivious to such influences is bound to encounter frustration.

17.4 THE TWO LEVELS OF CULTURE AND SOME CULTURAL UNIVERSALS

Although much risk reduction and disaster recuperation work has increasingly come to recognize the import of local traditions and wisdom in every region, especially concerning the environment, it is also highly significant to recognize that most people of the world are also descendants of and participants in what have been called "great traditions." These traditions also profoundly affect ideologies, cosmologies, structures, rules, and behaviors. Thus, culture almost everywhere actually has two aspects, one regional, one overarching.

Among the great traditions affecting great masses of people worldwide are Indo-European, Arabic, Chinese, Ural Altaic/Turkic, Bantu, and Southeast Asian. Long-standing cultural histories and traditions also exist of a somewhat smaller scale, and among them are Maya-Aztecan, Australian First People, Pacific Island Polynesian, Melanesian, and Micronesian, various North American tribal and linguistic groups, Pueblo, Iroquoian, Athabascan, Asian Mongolian/Manchu Tungus, Tamil, and this is very much a short list. Each of these carries millennia of language, family, customary social structures, mythologies, and cosmologies. Corresponding religious traditions with all their pillars alone include Christianity, Islam, Buddhism, Judaism, Hinduism, and many sorts of native belief systems. Most of the traditions also have a literature detailing origin, heroes, enemies, and afterlife.

The impact of cultural great tradition must be considered along with the regional in all truly nuanced understanding of any people's culture. Indians faced with ominous threat will go to the god Ganesha to remove obstacles. Christians will assume an afterlife is forthcoming no matter their risk behavior. God is far away in Bantu concept but spirits abide everywhere. How space and time are calculated, people are sorted, unity outlined, and other perceptual arrangements are impacted as much by long legacies as small ecozone, economy and region. They are interwoven.

Furthermore, in terms of risk reduction, it cannot be said that people everywhere grasp or enact risk reduction, but after suffering a disaster all victims and survivors everywhere, no matter where the place, or what the disaster, earthquake, fire, flood, tsunami, cyclone, drought, contamination, experience certain universal experiences, albeit in every place they manifest in that particular culture's ways.

Following a calamity, all people undergo a state of psychological liminality. They all experience the emergence of mutual aid and all important self-help. All survivors everywhere unite and develop an agreed upon enemy, usually the government or aid giver. New leaders arise out of seemingly nowhere. New political agendas and economics emerge (Hoffman, 1999a,b,c).

All start out thinking that they will return to the way they were before. All undergo a cascade of emotions ranging from horror and dread, love, grief, anger, and resignation. These stages are pertinent to any intervention and help and should be recognized.

In ending, a nod must be given to the fact that in recent decades the concept of culture itself has begun an evolution from the local to global. The global rendition speaks of an emerging and, hopefully, spreading set of principles, or universal morals, to be applied to all people everywhere. Some call them "humanitarian" principles. These include the valuing and upholding of every version of human life, and the banning of such practices as genocide, violence, gender inequality, racism, and classism. The term "cultural relativity" means that each and every culture must be honorable for its own rules and life ways. However, quite clearly, at the same time, the emerging global principles sometimes conflict with what particular cultures espouse.

Each individual must determine for himself or herself their position on the ecumenical all versus the particulars of diverse societies, but in our ever-expanding global times, no denying, it is a line we all must tread.

REFERENCES

Altman, I., Low, S. (Eds.), 1992. Place Attachment: Human Behavior and the Environment, vol. 12. Sage Publications, Thousand Oaks, CA.

Berlin, B.O., Kay, P.D., 1969. Basic Color Terms. University of California Press, Berkeley, CA.

Boas, F., 1911. The Mind of Primitive Man. Macmillan, New York.

Boas, F., 1940. Race, Language and Culture. Macmillan, New York.

Enarson, E., Morrow, B. (Eds.), 1998. The Gendered Terrain of Disasters: Through Women's Eyes. Greenwood Publishing Group, Westport, CT.

Feld, S., Basso, K. (Eds.), 1996. Senses of Place. School of Advanced Research Press, Santa Fe.

Firth, R., 1959. Social Change in Tikopia: Restudy of a Polynesian Community after a Generation. Allen and Unwin, London.

Frake, C.O., 1962. The ethnographic study of cognitive systems. In: Anthropology and Human Behavior. Society of Washington, Washington, DC.

Freidel, D., Schele, L., Parker, J., 1993. Maya Cosmos. Morrow, New York.

Gill, T., 2013. This spoiled place: people, place and community in an irradiated village in Fukushima Prefecture. In: Japan Copes with Calamity. Peter Lang Publishers, Bern, pp. 201–234.

Gill, T., Stegar, B., Slater, D. (Eds.), 2013. Japan Copes with Calamity. Peter Lang Publishers, Bern.

Harris, M., 1968. The Rise of Anthropological Theory. Routledge and Kegan Paul, New York.

Hoffman, S., 1976. The ethnography of the islands: Thera. In: Dimen, M., Friedle, E. (Eds.), Regional Variation in Modern Greece and Cyprus: Toward a Perspective on the Ethnography of Greece. New York Academy of Sciences, New York, pp. 328–340.

Hoffman, S., 1994. Up from the embers: a disaster survivor's story. In: Clinical Quarterly, National Center for Post-Traumatic Stress, vol. 4, 2, 15–16.

Hoffman, S., 1999a. After atlas shrugs: cultural change or persistence after a disaster. In: Oliver-Smith, A., Hoffman, S. (Eds.), The Angry Earth. Routledge, New York, pp. 302–326.

Hoffman, S., 1999b. Anthropology and the angry earth: an overview. In: Oliver-Smith, A., Hoffman, S. (Eds.), The Angry Earth. Routledge, New York, pp. 1–16.

Hoffman, Susanna, M., 1999c. "The Worst of Times, The Best of Times: Toward a Model of Cultual Response to Disaster". In: Oliver-Smith, A., Hoffman, S. (Eds.), The Angry Earth. Routledge, New York, pp. 134–155.
Hoffman, S., 2002. The mother and the monster: the symbolism of disaster. In: Hoffman, S., Oliver-Smith, A. (Eds.), Catastrophe and Culture. School of Advanced Research Press, Santa Fe, pp. 113–142.
Hoffman, S., 2004. Hidden victims of disaster. In: Cutter, Susan (Ed.), Environmental Hazards, vol. 5. Elsevier, pp. 67–70.
Hoffman, S., 2005. Katrina and Rita: a disaster anthropologist's thoughts. Anthropol. News 46 (8), 19.
Hoffman, S., 2010. Of increasing concern: disaster and the field of anthropology. Anthropol. News 51 (7), 4–5.
Hoffman, S., Oliver-Smith, A. (Eds.), 2002. Catastrophe and Culture. School of Advanced Research Press, Santa Fe.
Ingold, T., 1992. Culture and the perception of the environment. In: Cross, E., Parkin, D. (Eds.), Bush, Base: Forest Farm. Routledge, London, pp. 39–55.
Kahneman, D., 2011. Thinking Fast and Slow. Farrar, Strauss and Giroux.
Kroeber, A., 1952. The Nature of Culture. University of Chicago Press, Chicago.
Kroeber, A., 1963. Anthropology: Cultural Patterns and Processes. Harcourt, Brace and World, New York.
Lakoff, G., Johnson, M., 1980. Metaphors We Live by. University of Chicago Press, Chicago.
Levi-Strauss, C., 1966. The Savage Mind. Wiedenfeld and Nicholson, Paris.
Low, S., Lawrence-Zuniga, D., 2003. The Anthropology of Space and Place: Locating Culture. Blackwell, Oxford.
Mauss, M., 1954. The Gift. Free Press, London, Glencoe, Il.
Nicholson, I., 1967. Mexican and Central American Mythology. Paul Hamlyn, London.
Oliver-Smith, A., 1992. The Martyred City: Death And Rebirth in the Andes. Waveland Press, Prospect Heights, IL.
Oliver-Smith, A., 1996. Anthropological research on hazards and disasters. Annu. Rev. Anthropol. 25, 303–328.
Paine, R., 2002. Danger and the no-risk thesis. In: Hoffman, S., Oliver-Smith, A. (Eds.), Catastrophe and Culture. School of Advanced Research Press, Santa Fe, pp. 67–90.
Ravi Rajan, S., 1999. Bhopal: vulnerability, routinization, and chronic disaster. In: Oliver-Smith, A., Hoffman, S. (Eds.), The Angry Earth. Routledge, New York, pp. 257–277.
Rajan, S. Ravi, 2002. Missing expertise, categorical politics, and chronic disasters: the case of Bhopal. In: Hoffman, S., Oliver-Smith, A. (Eds.), Catastrophe and Culture. School of Advanced Research Press, Santa Fe, pp. 237–260.
Schneider, D., 1957. Typhoons on yap. Hum. Organ. 16 (2), 10–15.
Schuller, M., Farmer, P., 2012. Killing with Kindness. Rutgers University Press, New Brunswich, New Jersey.
Schuller, M., Morales, P. (Eds.), 2012. Tectonic Shifts: Haiti since the Earthquake. Kumarian Press, Sterling, Virginia.
Steward, J., 1955. Theory of Culture Change. University of Illinois Press, Champaign, IL.
Tylor, E., 1871. Primitive Culture Volumes I and II. John Murray, London.
Wallace, A., 1956. Tornado in Worcester (Disaster Study #3). Committee on Disaster Studies, National Academy of Sciences-National Research Council, Washington, DC.

Chapter 18

Risk, Resilience, and Readiness: Developing an All-Hazards Perspective

Douglas Paton
School of Psychology, University of Tasmania, Launceston, TAS, Australia

ABSTRACT
This chapter discusses how readiness contributes to developing community resilience. It does so from the perspective of understanding how people interpret risk, make disaster risk reduction (DRR) choices, and implement actions to manage their risk under conditions of uncertainty. It proposes that the development of DRR strategies that facilitate sustained readiness must accommodate two issues. The first is to identify the atypical demands and challenges people need to be resilient to and/or adapt to and that change as people negotiate the impact, response, and recovery phases of disaster experience. The second issue derives from identifying how personal, community, and societal factors interact to predict whether and to what extent people and communities proactively develop sustained resilient and adaptive capacity. The chapter draws on empirical studies undertaken in Thailand (2004 Indian Ocean tsunami) and New Zealand (2011 Christchurch earthquake) to illustrate the nature of resilience. Studies of earthquake, tsunami, and wildfire readiness serve to illustrate how theory supports an all-hazards approach to increasing readiness in DRR strategies. The chapter concludes with a discussion of issues that need to be incorporated in a comprehensive theory of readiness and the need to accommodate people at different levels of readiness when developing DRR strategies.

18.1 INTRODUCTION

The loss and disruption that natural hazard activity (e.g., from volcanic, wildfire, storm, flooding, tsunami, and seismic processes) can inflict on societies and their members have been well documented. However, evident differences in the ability of people and communities to draw on and use their knowledge, skills, and relationships to mitigate the impact of hazard events and hasten their recovery have identified a need to determine what makes some

Hazards, Risks, and Disasters in Society. http://dx.doi.org/10.1016/B978-0-12-396451-9.00018-4

people and communities more resilient than others. The importance of identifying the sources of this resilience and how it can be developed and sustained was acknowledged by the Hyogo Framework for Action 2005–2015 (International Strategy for Disaster Reduction (ISDR), 2005) pinpointing the development of resilience as a pivotal aspect of any disaster risk reduction (DRR) strategy.

Variability in the resilient or adaptive capacities of disaster-affected communities and the fact that this variability can be traced, at least in part, to the community characteristics and competencies that develop from people's experiences and expectations (e.g., Paton et al., 2014) identifies where information about developing resilience can be found. This chapter discusses how readiness and readiness development strategies can be incorporated into comprehensive DRR planning in ways that can contribute to realizing the kind of benefits that led to the ISDR calls for the need to develop resilience within DRR strategies designed to manage risk.

In its most basic guise, risk can be represented as the product of the likelihood of hazard exposure and the consequences of that exposure. Although the likelihood of a natural hazard event *occurring* cannot be influenced by DRR strategy, it is possible to influence the nature, severity and duration of hazard consequences. The ISDR call for developing resilience to play a prominent role in DRR derives from its reflecting people's ability to "bounce back" or adapt to the consequences of hazard impacts (e.g., Berkes et al., 2003; Carpenter et al., 2001; Klein et al., 2003; Norris et al., 2008; Paton and Johnston, 2006; Pelling and High, 2005). The inclusion of an adaptive component in this conceptualization is important. The reason for this is outlined next.

When impacted by hazard activity, societies and their members suddenly find themselves having to deal with demands that differ considerably from anything they would encounter under normal conditions and in circumstances in which normal societal functions and resources are marked by their absence (at least during the early period of hazard impact). People, communities, and societies have to bounce back from or adapt to *what has been experienced*. The fact that the specific mix of hazard characteristics and how they interact with the social, built, and ecological context is impossible to predict (for a given event) identifies resilience, in practice, as context dependent; it can only be defined in relation to people's experience of significant levels of disruption from hazard events (Carpenter et al., 2001; Gaillard, 2007; Klein et al., 2003; Paton and Johnston, 2006). Further, because the demands and challenges community members experience fall well outside normal experience, how well people respond and recover (to sudden and significant changes in their circumstances) will be a function of whether they have developed the knowledge, skills, and relationships required to respond or how well they can adapt by developing appropriate competencies and relationships in situ.

People can experience significant changes to their circumstances with no or little warning. Consequently, their response and recovery will be expedited by their developing appropriate capabilities prior to the occurrence of hazard events. It is the need to develop appropriate (and novel) knowledge, skills, and relationships (with neighbors, community members, government agencies, etc.) to deal with atypical and challenging circumstances that identify the interface between resilience and readiness.

Understanding how resilience/adaptive capacity and readiness are related requires a knowledge of two broad areas (Norris et al., 2008; Paton and Johnston, 2006; Paton et al., 2014). The first concerns identifying what people need to know and be able to do (individually and collectively) to cope with and adapt to the primary, secondary, and tertiary hazard consequences that emerge and evolve over the course of the response, recovery, and rebuilding phases of disaster. The second is to identify why some people develop these capabilities and others develop these less so or not at all.

Readiness strategies are intended to increase the likelihood of people and communities being in a position to be able to "respond" in planned and functional (resilient and adaptive) ways to complex, challenging, emergent hazard experiences, and demands rather than having to "react" to them in ad hoc ways. Although people can and do cope using ad hoc reactions, being prepared expedites response and recovery and reduces the psychological, social, and physical costs associated with response and recovery. The importance of being able to adapt derives from the inclusion of "emergent" in this conceptualization and the fact that some response and recovery demands are more readily anticipated than are others.

For example, although the effectiveness of any response strategy depends on things like the specific intensity of a hazard event (e.g., a given earthquake), ensuring the physical integrity of the house and storing water and food prepares people for the anticipated effects of ground shaking on buildings and on the loss of utilities such as water and power. The former not only reduces the risk of injury and death to its inhabitants but it also increases the likelihood of people having shelter and reduces demands for temporary accommodation, and makes it more likely that people will remain in an area and be able to participate in social (mutual aid, social support), economic, and environmental recovery activities. Preparing for loss of utilities increases people's self-reliance during periods of disruption to normal life. These actions, which can be developed (and ideally need to be developed) before hazard events occur, will safeguard against predictable consequences (e.g., the impact of ground shaking on the built environment).

However, other consequences are less easy to anticipate. This uncertainty derives from diversity in both the permutations of intensity, magnitude, duration, and location of hazard effects that can occur from unpredictability in how hazard characteristics will interact with the characteristics of the built, social, and ecological contexts people inhabit (Paton and McClure, 2013). Although all of these parameters can be estimated, it is not possible to predict the specific mix (e.g., how much loss will occur, how it is distributed, how long it will last, whether people need to be temporarily or permanently relocated) that will characterize a specific hazard event. Thus, the specific mix of consequences people will experience cannot be specified in advance and so creates the unpredictable, emergent, evolving characteristics people need to respond to. Hence, it is beneficial for community members to develop their knowledge, skills, and relationships with neighbors, community members, and civic agencies necessary to expedite local response and recovery initiatives in the context of unpredictable, emergent hazard–environment consequences. In this context, readiness strategies function with the intent of developing these capabilities and relationships in advance of hazard events occurring. Finding considerable diversity in the existing levels of readiness introduces the second issue discussed in this chapter, the need to account for this diversity.

This chapter postulates that the source of this variety arises from differences in the (preevent) resources and competencies available to people and communities to make DRR choices under conditions of uncertainty (Norris et al., 2008; Paton and Johnston, 2006). Pulling these areas together, the relationship between resilience and readiness can be expressed in terms of mobilizing community knowledge, expertise, competencies, and social capital in ways that facilitate people's ability to anticipate; cope with; adapt to; recover from; and learn from the demands, challenges, changes encountered before, during, and after hazard events (Norris et al., 2008; Paton and Johnston, 2006; Pelling and High, 2005). Realizing the benefits of readiness in DRR first requires conceptualizing readiness.

18.2 CONCEPTUALIZING READINESS

Russell et al. (1995) developed a readiness typology that comprised structural, survival, and planning categories. Structural actions comprise activities that secure the house (e.g., secure house to foundations, creating a defensible space) and its contents (e.g., securing water heaters and tall furniture) to prevent contents from injuring inhabitants (e.g., from ground shaking accompanying earthquakes). Survival actions facilitate people's capacity for self-reliance during (short–immediate impact) periods of disruption (e.g., ensuring a supply of water/dehydrated or canned food for several days, having a radio with spare batteries). Finally, planning includes, for example, developing household hazard plans and attending meetings to learn about hazards and how to deal with their consequences. The latter introduces a social dimension into how readiness is conceptualized. The subdivision of readiness activities into structural, planning, and survival categories was reiterated in a study of wildfire preparedness (Prior and Eriksen, 2012). However, a subsequent factor analytic study (Lindell et al., 2009) proposed a two-factor solution comprising Direct Action (e.g., learned how to shut off utilities, have four-day supply of canned food, strapped heavy objects) and Capacity Building (e.g., joined an earthquake-related organization, attended meetings about earthquake hazards) factors. These studies identify a need for readiness strategies to facilitate the development of structural, survival, planning, and capacity building (social) activities. The importance of doing so has been reinforced by studies of what members of disaster-affected communities identified as being required to increase their readiness (for future events).

For example, research in Christchurch (New Zealand) with residents affected by the 2011 earthquake reinforced the importance of developing (preevent) structural and survival readiness and household emergency planning (Paton et al., 2014). Christchurch respondents reported how their lack of survival and planning readiness made coping with the loss of essential societal utilities (e.g., loss of water, power, and sewerage services) more challenging than was necessary. To the latter were added calls for readiness programs to address dealing with loss of or disruption to employment (both directly from damage to place of work or indirectly from being relocated or injured), developing psychological readiness, adapting to changes in living conditions and loss of and disruption to family and social relationships and to accommodate the fact that the period of loss and disruption people have to confront can extend over periods of months or years. Survivors also called for more effort to be directed to developing social relationships and competencies to deal with emergent issues within this recovery context (see above).

Respondents discussed the benefits that would accrue if readiness programs developed the capacity for neighborhood and community groups to be able to confront local physical and social demands (e.g., removing rubble, providing mutual support, setting up community meeting places, taking care of those with special needs, organizing local efforts to repair homes, and identifying and meeting local needs) under challenging conditions when access to normal societal resources and functions is absent. They identified a need for readiness programs to develop, for example, leadership, social inclusion, prioritizing problems, collective problem-solving, and decision-making skills. Similar capacities (e.g., the ability to identify community needs and the development of community-based mechanisms capable of securing the resources people required to meet their emergent needs) were identified as important by members of Thai communities dealing with the impact of the 2004 Indian Ocean tsunami (Paton et al., 2008b).

Respondents in the Paton et al. (2014) study of people's views of what contributed to their resilience in the aftermath of the 2011 Christchurch earthquake also suggested that social readiness should include developing community capacity to represent their information and resource needs to diverse agencies (e.g., to nongovernmental organizations (NGOs) such as the Red Cross, the Police) and government departments (e.g., the Christchurch Earthquake Recovery Authority). Being able to interact effectively with agencies helped community members secure the resources they needed if they were to be able to take responsibility for their own recovery. The fact that community groups experienced their relationships with agencies and government departments as either facilitating or marginalizing community action identified a need for agency and organizational development to ensure agencies complemented and empowered community action (Figure 18.1). The perceived benefits of proactively collaborating with the government and aid agencies to meet local needs were also evident in Thai communities confronting the consequences of the 2004 Indian Ocean tsunami (Paton et al., 2008).

In Christchurch, it was evident that what people had to contend with changed over time (e.g., from loss of utilities to making neighborhoods safe to dealing with government agencies and insurance companies) as they negotiated physical, social, and institutional demands that changed over the course of a recovery period that spanned years. Recognition of the fact that recovery not only takes years but also presents people and communities with diverse and emergent needs that will change over time identifies a need to develop and implement readiness strategies that can be delivered at several points during recovery to facilitate people's capacity to cope with and adapt to demands and challenges that change over the course of what can be a long recovery and rebuilding period.

Taken together, this discussion highlights a need for readiness strategies to encompass structural, survival, planning, and community capacity capabilities. However, the Christchurch work introduced a need to complement these functional categories with those addressing livelihood, psychological and community–societal (government, NGO, business, etc.), readiness (summarized in right-hand column, Figure 18.2). While the Christchurch research discussed what people realized they needed during a disaster, the focus of readiness research is on developing these capacities and relationships prior to the occurrence of a hazard event. A need to research the latter reflects the finding that comprehensive readiness is the exception rather than the norm.

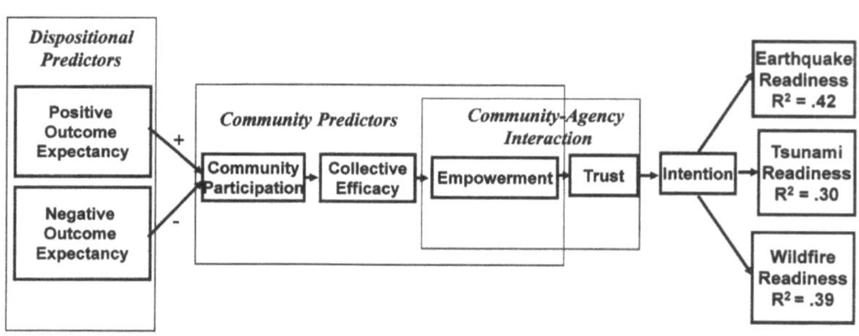

FIGURE 18.1 Summary of relationship between dispositional factors, structural factors, trust, and intention to adopt readiness measures. *(Adapted from Paton (2013).)*

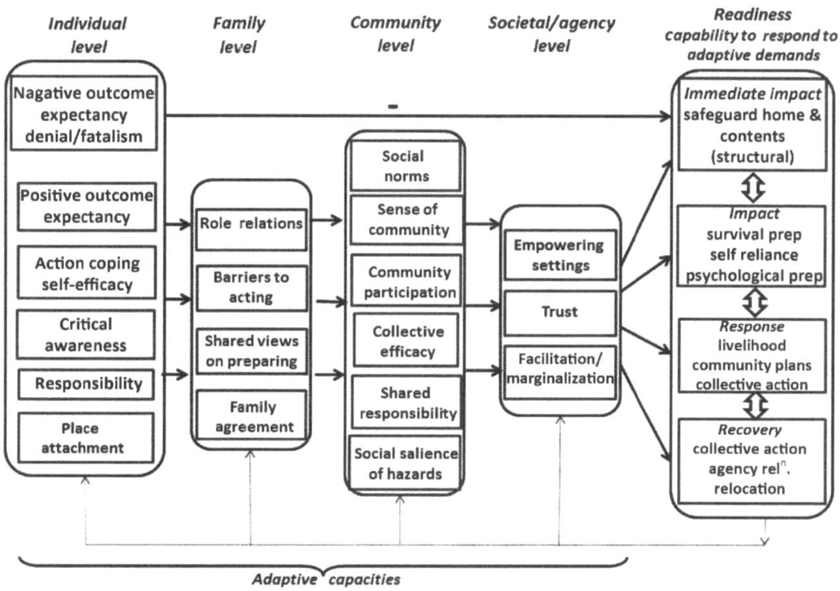

FIGURE 18.2 Summary of the predictor (adaptive capacities) variables and readiness measures (capability to respond to adaptive demands presented by actual hazard events during impact, response, and recovery).

18.3 ACCOUNTING FOR DIFFERENCES IN READINESS

Readiness research has consistently identified that, even when people acknowledge their risk, there is considerable diversity in levels of readiness. This prompted the development of theories to account for this (e.g., Duval and Mulilis, 1999; Lindell et al., 2009; Paton et al., 2005; Paton, 2008). A theory needs to demonstrate its validity across a range of hazards that differ with respect to their characteristics and behaviors (e.g., earthquake readiness encompasses preparing for ground shaking, volcanic preparedness includes readiness to deal with ashfall hazards) and their consequent implications for what being prepared means. That is, any theory must have "all-hazards" applicability (Eiser et al., 2012; Paton, 2013). Much readiness research has focused on hazard-specific work (e.g., earthquake preparedness). Testing all-hazards applicability requires a different strategy. One approach derives from focusing less on the hazard and more on a common denominator that all hazards present to those at risk, uncertainty.

People face uncertainty with regard to, for example, when they might experience a hazard event, what magnitude or intensity of hazard activity they could experience, what actual consequences will be experienced and for how long, and how long it will take for recovery to occur. The infrequent nature of hazard activity introduces additional uncertainty regarding knowledge of what can be done to manage their risk and how effective measures are likely to be. Under these circumstances, citizens cannot rely on experience. Rather, they have to rely on interpretive and decision-making competencies and relationships developed and used to deal with the more regularly occurring trials and tribulations of life (Norris et al., 2008; Paton and Johnston, 2006).

When faced with uncertainty, people turn to others to help them reduce their uncertainty and decide what to do to manage their risk (Eng and Parker, 1994; Earle and Cvetkovich, 1995; Marris, Langford and O'Riordan, 1998; Rippl, 2002; Paton, 2008). These others can be civic agencies (e.g., emergency management), and they can also be family members, neighbors, and members of the communities (e.g., workplaces, social and sporting clubs, and churches) with whom people interact regularly. When making choices under conditions of uncertainty, when people are heavily reliant on others for information and advice, an important predictor of whether information and advice is used to make decisions is peoples' assessment of how much they trust their source of information rather than information about hazard risk per se (Earle and Cvetkovich, 1995; Siegrist and Cvetkovich, 2000). Because it affects people's interpretation of other's motives and their competence and the credibility of the information they make available, trust acts to reduce uncertainty and facilitate action in circumstances in which people are dependent on others for information (Earle, 2004; Siegrist and Cvetkovich, 2000). Trust will thus play a pivotal role in a theory of readiness.

Paton (2008) developed a theory that identified how trust mediated the relationship between the personal and social characteristics that influenced the social construction of risk and hazard readiness. The theory describes how interaction between outcome expectancy beliefs (i.e., whether people believe that personal actions can mitigate risk), community characteristics (community participation and collective efficacy) that articulate risk beliefs and facilitate the formulation of plans for action, and people's beliefs about whether they have been empowered by agency sources predicted trust, with intentions mediating the relationship between trust and hazard preparedness. The intention measure comprised items that assessed people's intention to acquire hazard knowledge, increase actual preparedness, and to work with other people/civic agencies to develop knowledge and capability (Paton et al., 2005). This theory is summarized in Figure 18.1.

Empirical confirmation of the all-hazards utility of this theory (Paton, 2013) indicates a need for readiness strategies to facilitate the development of outcome expectancy beliefs and social interpretive (community participation, collective efficacy, empowerment, and trust) processes (Figure 18.1). Positive outcome expectancy beliefs can be developed by presenting people with a small number of readiness items initially (e.g., storing canned food and water), starting with relatively easily adopted items, and introducing progressively more complex preparedness tasks (e.g., developing neighborhood emergency plans) over time (Paton and Wright, 2008). By presenting specific explanations about how readiness actions reduce risk and mitigate hazard consequences progressively over time, sustained adoption is more likely.

Confirmation of roles for community participation, collective efficacy, and empowerment (Paton, 2013) highlight the importance of facilitating social interaction and relationships with those who share similar beliefs and values to construct risk beliefs and to facilitate the development and use of the planning and problem-solving competence required to enact risk beliefs in readiness strategies that will meet community needs. This illustrates the benefits that can occur when DRR strategies are based on community engagement principles.

The all-hazards utility of this work is supported by its ability to account for differences in levels of readiness for hazards (e.g., tsunami, earthquakes, and wildfire) that differ in the consequences they create, and thus the kinds of readiness activities required (Paton, 2013). However, given that the theory accounted for 30%, 42%, and 39% of the variance (R^2) in levels of preparing for tsunami, earthquake, and wildfire hazards, respectively (Paton, 2013), a need to expand the constructs canvassed in order to develop a more comprehensive understanding of readiness processes is clear. This issue is considered in the next section and commences with potential candidates that relate to personal characteristics.

18.4 INDIVIDUAL PREDICTORS

Several individual-level predictors have been identified. Some, such as fatalism, reduce the likelihood of preparing. Others influence how people make judgments about who is responsible for their readiness. Yet other factors increase the likelihood of people preparing (Figure 18.2).

People who are *fatalistic* about hazard activity are unlikely to prepare (McClure and Williams, 1996; Turner et al., 1986). However, fatalistic attitudes can be reduced by asking people what can be done to help specific vulnerable groups, such as people living in unsafe buildings or young children in schools (McClure et al., 2001; Turner et al., 1986). When people deliberated about readiness in vulnerable groups, they became less fatalistic and were more likely to perceive readiness actions as more manageable.

Another dispositional characteristic that acts to reduce readiness is *Denial* (Crozier et al., 2006; DeMan and Simpson-Housley, 1988). If people believe that they have no control over a hazard and/or its activity, they can attempt to cope with this by denying the seriousness of the risk. Denial can be countered by providing information that focuses people's attention specifically on how readiness actions can mitigate the consequences of hazard activity and increase a degree of control over them (Lehman and Taylor, 1987). The degree to which people prepare is also affected by how their interpretation of threat influences their beliefs about the distribution of responsibility for community readiness.

Risk compensation (e.g., Etkin, 1999) describes how people's motivation to take "Responsibility" for their readiness is influenced by their perceptions of how personally threatening the environment is. One factor that influences perceived environmental threat is people's interpretation of the actions of agencies that have a responsibility for managing risk have undertaken to mitigate risk (Paton et al., 2000). Paton and colleagues found that receipt of information about what scientists and civic agencies were doing to mitigate volcanic risk (which they did not have or know beforehand) resulted in people perceiving the environment as being less threatening to them and reduced their sense of being personally responsible for their own safety or readiness. They transferred this responsibility to risk management agencies. This cognitive bias can be reduced by explicitly informing people that societal mitigation and personal readiness complement one another (rather than being substitutable) and by explaining that personal action is required to cover the fact that societal resources cannot cater for all eventualities (Paton and Wright, 2008). That is, risk communication and community outreach must facilitate the personalization of risk to increase the likelihood that people understand how their specific circumstances (over which societal agencies have no control) influence their risk (e.g., the degree to which their house is structurally capable of withstanding certain levels of ground shaking, or how topographical factors introduce household-level differences in wildfire risk). Encouraging the personalization of risk and outreach information in this way increases the likelihood that people will understand how risk has localized characteristics that they need to take personal responsibility for. Personal characteristics can also increase readiness (Figure 18.2).

Research on community response to adversity identified *Critical Awareness* (the extent to which people think and talk about hazards with others) as an important precursor of actions to deal with a community threat or opportunity (Dalton et al., 2001). Critical awareness was identified as an important predictor (Paton et al., 2005). The latter work introduced a need to understand how people's discussions of hazard-related issues could both increase or decrease the likelihood of people preparing. This underlines the value of community engagement-based approaches to risk communication and community outreach. Engagement-based approaches are more likely to be able to identify social processes that may undermine preparedness and adapt strategies to reverse this.

Paton et al.'s (2005) work also identified how people's belief in their ability to act influenced preparedness. This was assessed using the *Self-efficacy* construct. An important contribution made by self-efficacy to understanding readiness derives from its influence on the number and quality of action plans people develop, and the amount of effort and perseverance people invest in risk reduction behaviors (Paton, 2003). The importance of encouraging perseverance derives from its role in supporting preparedness that needs to be sustained during potentially prolonged periods of hazard quiescence.

A role for *Action Coping* as a precursor of preparedness has been recorded in several studies (Duval and Mulilis, 1999; Lindell and Whitney, 2000; Paton et al., 2001), as has the degree to which people feel that they are embedded within their social–ecological environment. The degree to which people feel they are psychologically embedded in their environment increases their emotional investment in their community (Hummon, 1992). Recognition of this has led to *Place Attachment* acting as a predictor of readiness (Frandsen et al., 2012; Paton et al., 2008). Readiness is also affected by social characteristics.

18.5 FAMILY AND COMMUNITY PREDICTORS

Family and community factors influence readiness (Figure 18.2). For example, Goodman and Cottrell (2012) discuss how gender *Role relationships* within families influence perceptions of responsibility for readiness and response decision making (e.g., acting as *barriers to planning*, inhibiting developing *shared views* on readiness needs). This echoes Paton and Buergelt's (2012) finding that the presence of conflicting views about the need for preparing within the family (particularly when husbands attributed a lower priority to readiness than other family members) was a significant impediment to preparing. They also identified how *Conflict* with or a lack of willingness to collaborate with neighbors to manage risk acted to reduce the likelihood of preparing, with this being explained using Fault Line theory. Consistent with the tenets of Fault Line theory, Paton and Buergelt found that the process of facilitating community readiness could activate divisions in communities that would remain dormant under any other circumstances. This highlights the need for an adaptive approach to risk communication and one that accommodates changes in community dynamics over time. They also identified how preexisting community processes and beliefs could positively influence preparedness.

Paton and Buergelt (2012) found that active *Community Participation* and the *Social Salience* of risk (e.g., sharing stories about hazards within their community) enhanced hazard knowledge, risk acceptance, and the importance of preparing. It also provided community members with information about what to do to prepare and increased trust in the effectiveness of preparedness measures. This social construction of risk beliefs and readiness facilitated the development of a sense of *Shared responsibility* for preparing and a *Sense of community*.

Sense of community has been implicated as a predictor of readiness (Paton et al., 2001). When the sense of community is high, others within that community are more likely to be regarded as credible and trusted sources of information (Lasker, 2004; McGee and Russell, 2003). When these others have local hazard experience and knowledge, they become an important resource for developing shared representations of risk and its management and for facilitating the development and maintenance of locally relevant risk management activities (Frandsen et al., 2012). That is, it contributes to the development of normative expectations.

If people believe that their significant others (parents, spouses, friends, peer group, etc.) hold favorable attitudes toward a particular behavior or that performance of a specific behavior is likely to be interpreted favorably by significant others, they are more likely to perform these actions (Smith and Terry. 2003). Prevailing *Social norms* regarding preparedness expectations of significant others, and personal motivation to act in ways consistent with these expectations, have a positive effect on readiness (McIvor and Paton, 2007).

18.6 CONCLUSION

This chapter discussed how readiness planning informs the development of resilience and adaptive capacity within DRR strategies. The discussion covered two areas. These are summarized in Figure 18.2. The first addressed the need for comprehensive measures of readiness to encompass structural, survival, planning, community, livelihood, psychological, and community–agency functional categories. Measures of structural, survival, planning, and community categories are well developed (if not always used to inform practical intervention). However, complementary work on the livelihood, psychological, and community–agency functional categories should be included in future research agendas. Future research and debate are required to identify how a mix of hazard-specific (e.g., to accommodate differences in the physical consequences posed by different hazards) and generic all-hazards measures can be developed to facilitate theory development and testing. Additional debate should also be directed to considering the need to include readiness strategies in recovery planning and intervention to accommodate and facilitate people's ability to respond to emergent demands over the prolonged period of disaster recovery and not just in the immediate impact period.

The second area addressed focused on how developing theory capable of accounting for differences in levels of readiness plays a pivotal role in understanding how readiness contributes to resilience. This discussion proposed that theories identify person-, family-, community-, and societal-level predictors and how they interact to explain differences in levels of readiness across a range of hazards (Figure 18.2). The issues canvassed in relation to this second area are important from the perspective of developing practical DRR strategies intended to facilitate sustained readiness in communities that coexist with hazardous circumstances. Bridging the theory–practice gap introduces some additional issues. One relates to the content of intervention strategies, and the other to the nature of intervention strategies used.

A need for additional research into the content of intervention strategies derives from the possibility that different predictors may play different roles in different facets of the preparedness process. Support for this contention comes from studies that identify readiness as a process in which people and communities progressively develop more comprehensive readiness over time. Work in this area has identified how the relative salience of predictors changes as people negotiate this developmental process (see Paton and McClure, 2013 for a review). Paton and McClure discuss how strategies to get people started (e.g., facilitating people's ability to personalize risk, using information from those in similar circumstances) differ from those designed to encourage adoption (e.g., providing information about risk, providing cause–effect information about hazard consequences and how specific actions are effective) and from those intended to sustain readiness (e.g., providing detailed hazard characteristics information, community-based planning). The transition between these levels describes a developmental process that should be mirrored in readiness strategies if they are to facilitate sustained readiness (Paton and McClure, 2013).

A need for additional attention to be directed to understanding the relative salience of preparedness predictors derives from the possibility that each functional category (e.g., structural, survival, relationship, and psychological) differs with regard to its knowledge, skill, and relationship antecedents. For example, decisions about survival preparedness can be made by a person/family independently. However, developing community preparedness makes different resource (e.g., giving time to work with others) and skill (e.g., social skills, planning skills, sense of community) demands on people. This contention has been supported by some preliminary work on this relationship, with evidence suggesting that survival and community categories are influenced by different predictors (Paton et al., 2014).

Additional work on the design of intervention strategies is also called for. The rational for this suggestion comes from finding that, in the studies cited above, the predictors of readiness reflected mainstream (as opposed to being specifically hazard-related) community processes and competencies. That is, the social context (community participation, collective efficacy, empowerment, and trust) variables identified as predictors tapped into preexisting community characteristics and competencies and patterns of community–agency relationships that reflect people's day-to-day experiences. This highlights the benefits that could accrue from integrating risk communication and community development processes rather than conceptualizing risk communication as a standalone process delivered independently from other community initiatives. Consequently, DRR programs that integrate risk management and community development activities in ways that specifically encourage discussion of hazard issues, develop community members' problem-solving competencies, and encourage agencies to engage with communities in ways that empower them will increase the likelihood of readiness strategies contributing to realizing the ISDR goal of developing sustained community resilience.

REFERENCES

Berkes, F., Colding, J., Folke, C., 2003. Navigating Social–Ecological Systems: Building Resilience for Complexity and Change. Cambridge University Press, Cambridge.

Carpenter, S., Walker, B., Andries, J.M., Abel, N., 2001. From metaphor to measurement: resilience of what to what? Ecosystems 8, 941–944.

Crozier, M., McClure, J., Vercoe, J., Wilson, M., 2006. The effects of land zoning information on judgments about earthquake damage. Area 38 (2), 143–152.

Dalton, J.H., Elias, M.J., Wandersman, A., 2001. Community Psychology. Wadsworth, Belmont, CA.

DeMan, A., Simpson-Housley, P., 1988. Correlates of response to two potential hazards. J. Soc. Psychol. 128, 385–391.

Duval, T.S., Mulilis, J.P., 1999. A person-relative-to-event (PrE) approach to negative threat appeals and earthquake preparedness: a field study. J. Appl. Soc. Psychol. 29, 495–516.

Earle, T.C., 2004. Thinking aloud about trust: a protocol analysis of trust in risk management. Risk Anal. 24, 169–183.

Earle, T.C., Cvetkovich, G.T., 1995. Social Trust: Towards a Cosmopolitan Society. Praeger, Westport, CT.

Eiser, J.R., Bostrom, A., Burton, I., Johnston, D.M., McClure, J., Paton, D., van der Pligt, J., White, M.P., 2012. Risk interpretation and action: a conceptual framework for responses to natural hazards. Int. J. Disast. Risk Reduction 1, 5–16.

Eng, E., Parker, E., 1994. Measuring community competence in the Mississippi Delta: the interface between program evaluation and empowerment. Health Educ. Q. 21 (2), 199–220.

Etkin, D., 1999. Risk transference and related trends: driving forces towards more mega-disasters. Environ. Hazards 1, 69–75.

Frandsen, M., Paton, D., Sakariassen, K., Killalea, D., 2012. Nurturing community wildfire preparedness from the ground up: evaluating a community engagement initiative. In: Paton, D., Tedim, F. (Eds.), Wildfire and Community: Facilitating Preparedness and Resilience. Charles C. Thomas, Springfield, Ill, pp. 260–280.

Gaillard, J.-C., 2007. Resilience of traditional societies in facing natural hazards. Disast. Prev. Manag. 16, 522–544.

Goodman, H., Cottrell, A., 2012. Responding to a fire threat: gender roles: dependency and responsibility. In: Paton, D., Tedim, F. (Eds.), Wildfire and Community: Facilitating Preparedness and Resilience. Charles C. Thomas, Springfield, Ill, pp. 281–299.

Hummon, D.M., 1992. Community attachment: local sentiment and sense of place. Hum. Behav. Environ. Adv. Theor. Res. 12, 253–278.

International Strategy for Disaster Risk Reduction (ISDR), 2005. Hyogo Framework for Action 2005–2015: Building the Resilience of Nations and Communities to Disasters. International Strategy for Disaster Reduction, Kobe, Hyogo, Japan.

Klein, R., Nicholls, R., Thomalla, F., 2003. Resilience to natural hazards: how useful is this concept? Environ. Hazards 5, 35–45.

Lasker, R.D., 2004. Redefining Readiness: Terrorism Planning through the Eyes of the Public. The New York Academy of Medicine, New York, NY.

Lehman, D., Taylor, S.E., 1987. Date with an earthquake: coping with a probable, unpredictable disaster. Pers. Soc. Psychol. Bull. 13, 546–555.

Lindell, M.K., Arlikatti, S., Prater, C.S., 2009. Why do people do what they do to protect against earthquake risk: perception of hazard adjustment attributes. Risk Anal. 29, 1072–1088.

Lindell, M.K., Whitney, D.J., 2000. Correlates of household seismic hazard adjustment adoption. Risk Anal. 20 (1), 13–25.

Marris, C., Langford, I.H., O'Riordan, T., 1998. A quantitative test of the cultural theory of risk perceptions: comparisons with the psychometric paradigm. Risk Anal. 18, 635–647.

McClure, J., Allen, M.W., Walkey, F., 2001. Countering fatalism: causal information in news reports affects judgements about earthquake damage. Basic Appl. Soc. Psychol. 23 (2), 109–121.

McClure, J.L., Williams, S., 1996. Community preparedness: countering helplessness and optimism. In: Paton, D., Long, N. (Eds.), Psychological Aspects of Disaster: Impact, Coping, and Prevention. Dunmore Press, Palmerston North, pp. 237–254.

McGee, T.K., Russell, S., 2003. "It's just a natural way of life..." an investigation of wildfire preparedness in rural Australia. Environ. Hazards 5 (1–2), 1–12.

McIvor, D., Paton, D., 2007. Preparing for natural hazards: normative and attitudinal influences. Disast. Prev. Manag. 16 (1), 79–88.

Norris, F.H., Stevens, S.P., Pfefferbaum, B., Wyche, K.F., Pfefferbaum, R.L., 2008. Community resilience as a metaphor, theory, set of capacities, and strategies for disaster readiness. Am. J. Community. Psychol. 41, 127–150.

Paton, D., 2003. Disaster Preparedness: A social-cognitive perspective. Disaster Prevention and Management 12, 210–216.

Paton, D., 2008. Risk communication and natural hazard mitigation: how trust influences its effectiveness. Int. J. Global Environ. Issues 8, 2–16.

Paton, D., 2013. Disaster resilient communities: developing and testing an all-hazards theory. J. Integrated Disast. Risk Manag. 3, 1–17.

Paton, D., Buergelt, P.T., 2012. Community engagement and wildfire preparedness: the influence of community diversity. In: Paton, D., Tedim, F. (Eds.), Wildfire and Community: Facilitating Preparedness and Resilience. Charles C. Thomas, Springfield, Ill, pp. 241–259.

Paton, D., Buergelt, P.T., Prior, T., 2008a. Living with bushfire risk: social and environmental influences on preparedness. Aust. J. Emerg. Manag. 23, 41–48.

Paton, D., Johnston, D., 2006. Disaster Resilience: An Integrated Approach. Charles C. Thomas, Springfield, Ill.

Paton, D., Johnston, D., Mamula-Seadon, L., Kenney, C.M., 2014. Recovery and development: perspectives from New Zealand and Australia. In: Kapucu, N., Liou, K.T. (Eds.), Disaster & Development: Examining Global Issues and Cases. Springer, New York, NY.

Paton, D., Gregg, C.E., Houghton, B.F., Lachman, R., Lachman, J., Johnston, D.M., Wongbusarakum, S., 2008b. The impact of the December 26th 2004 tsunami on coastal Thai communities: assessing adaptive capacity. Disasters 32 (1), 106–119.

Paton, D., McClure, J., 2013. Preparing for Disaster: Building Household and Community Capacity. Charles C. Thomas, Springfield, Ill.

Paton, D., Millar, M., Johnston, D., 2001. Community resilience to volcanic hazard consequences. Nat. Hazards 24, 157–169.

Paton, D., Smith, L., Johnston, D., 2000. Volcanic hazards: risk perception and preparedness. New Zeal. J. Psychol. 29 (2), 86–91.

Paton, D., Smith, L.M., Johnston, D., 2005. When good intentions turn bad: promoting natural hazard preparedness. Aust. J. Emerg. Manag. 20, 25–30.

Paton, D., Wright, L., 2008. Preparing for bushfires: the public education challenges facing fire agencies. In: Handmer, J., Haynes, K. (Eds.), Community Bushfire Safety. CSIRO Publishing, Canberra, pp. 117–128.

Pelling, M., High, C., 2005. Understanding adaptation: what can social capital offer assessment of adaptive capacity? Global Environ. Change 15, 308–319.

Prior, T., Eriksen, C., 2012. What does being 'well prepared' for wildfire mean? In: Paton, D., Tedim, F. (Eds.), Wildfire and Community: Facilitating Preparedness and Resilience. Charles C. Thomas, Springfield, Ill, pp. 190–206.

Rippl, S., 2002. Cultural theory and risk perception: a proposal for a better measurement. J. Risk Res. 5, 147–165.

Russell, L.A., Goltz, J.D., Bourque, L.B., 1995. Preparedness and hazard mitigation actions before and after two earthquakes. Environ. Behav. 27 (6), 744–770.

Siegrist, M., Cvetkovich, G., 2000. Perception of hazards: the role of social trust and knowledge. Risk Anal. 20, 713–719.

Smith, J.R., Terry, D.J., 2003. Attitude-behaviour consistency: the role of group norms, attitude accessibility, and mode of behavioural decision-making. European J. Soc. Psychol. 33 (5), 591–608.

Turner, R.H., Nigg, J.M., Paz, D.H., 1986. Waiting for Disaster: Earthquake Watch in California. University of California Press, Los Angeles.

Chapter 19

Interpretative Frameworks of Disaster in Society Close-up

Ryo Morimoto

Brandeis University, Waltham, MA, USA; International Research Institute of Disaster Science at Tohoku University, Sendai, Japan; Institute of Comparative Culture, Sophia University, Tokyo, Japan

ABSTRACT

This chapter aims to examine the question of how disaster researchers can use maps to locate "culture" and "local knowledge" in supposedly objective representations of disaster. As an anthropologist studying disaster, I attempt to present a close-up view of the cultural layering processes of interpreting maps of a disaster to argue that using maps achieves both the naturalization of the disaster and essentialization of hazards by habituating an interpretative framework in metaphorically containing and literally delimiting an emergent unknown. By juxtaposing five types of maps materialized in interpreting the Fukushima nuclear disaster, I argue that maps are both interpretative habits and material objects that presuppose and entail the lived experience of the disaster containment. This observation suggests the importance of historicizing the location of our study—predisaster of a disaster—in gaining a more holistic view of the disaster. For disaster researchers, map(s), as a historically meaningful series of signs of society and actions in society, are a useful analytical tool to use in identifying the work of *semiotic reterritorialization*—a materialization technology of a belief in which a propositional representation of an abstract layering of signs is presented and acted as if it is the only tangible, thus, plausible evidence of the belief.

Today abstraction is no longer that of the map, the double, the mirror, or the concept. Simulation is no longer that of a territory, a referential being, or a substance. It is the generation by models of a real without origin or reality: a hyperreal. The territory no longer precedes the map, nor does it survive it. It is nevertheless the map that precedes the territory - precession of simulacra – that engenders the territory...

Jean Baudrillard, Simulacra and Simulation

The map is open and connectable in all of its dimensions; it is detachable, reversible, susceptible to constant modification. It can be torn, reversed, adapted to any kind of

Hazards, Risks, and Disasters in Society. http://dx.doi.org/10.1016/B978-0-12-396451-9.00019-6

mounting, reworked by an individual, group, or social formation. It can be drawn on a wall, conceived of as a work of art, constructed as a political action or as a mediation.
Gilles Deleuze and Félix Guattari, A Thousand Plateau: Capitalism and Schizophrenia

19.1 INTRODUCTION: (RE)PRESENTING A DISASTER

In today's globalized world, representations of disaster such as diagrams, maps, images, and footage have become the predominant means by which the public and experts experience and learn about the hazards and disasters transpiring around the world. Generated by various spatiotemporal scales and social–historical and political perspectives, those representations are not only windows into the ever-increasing hazards and disasters in the world but they also indicate paralleling cognitive maps of the particular places and societies mediated through different disciplinary and interest-based lens. For example, many of us have learned the name "Fukushima" and the geography of Japan for the first time through the aerial maps of the Fukushima Daiichi Nuclear Plant. Regardless of how familiar one is with the physical location and social–historical context of a specific hazard or disaster under discussion, technologically mediated, abstract representations have brought disasters ever closer to our awareness, while keeping them physically distant. Effectively, disaster can be witnessed and understood theoretically without physically being at its site. Such a doubling of consciousness leads us to ask, alongside Baudrillard (1994[1981]), whether or not representations precede concrete phenomena in the context of a disaster, that is, are plentiful representations of disaster predominantly "hyperreal" and only partially a real?

Of those mediated visions of society, maps are of particular interest to disaster researchers, whose work concerns both hard facts and idiosyncratic narratives of eyewitnesses. With an interdisciplinary effort to integrate a plethora of accounts regarding a disaster, our theorizing necessarily and simultaneously engages our participation in the pragmatics of a map, that is, in the use and the consequence of map making and interpretation. According to Kurgan (2013, p. 14 and 17), maps today not only construct "physical, propositional, discursive, political, archival, and memorial spaces" but they also "omit, according to their conventions, those invisible lines of people, places, and networks that create the most common space we live in today."

This politics of the construction and omission of projective dots and lines follows the development in the early twentieth century of the literal "top-down" views enabled by aerial photography that supplied social scientists a distant perspective into social relationships. This led to metaphorically opposed "bottom-up" perspectives focused on mapping social space. Thus, the concept of social space gained significance by its *use* against what was viewed as the politically estheticized pixels depicted in these distant views from above (Haffner, 2013). In instances of both the twentieth and twenty-first centuries, the so-called reality is augmented by what is revealed to be the relativity of visions, or "disciplinary eyes" (Daston and Galison, 2007), so that what is "real" multiplies and vanishes in the eye of a beholder as the perceiving subject itself dissolves its objectifying statistical operations (see the idea of "view from nowhere" in Nagel, 1989) or is made asymmetrically perceivable as in the Marxian false consciousness. Starling (1998, p. 107, emphasis added) contends that "mapping as a mode of representation has an intimate relationship with what we *habitually* regard as the real," suggesting not only maps' mere articulation of and intimacy with authoritative power of the state (Wood et al., 2010) but also a dialogic coordination—that is, performative containment—of terra incognita with lived presuppositions where the center is always disoriented because the principal reference point of habit itself constantly shifts. Whichever perspective is favorably taken in a given context, maps have been used as tools to provide a holistic view or a "*vue d'ensemble*" (Haffner, 2013) to a set viewpoint of a given sociohistorical and technopolitical context. What seems productive therefore is to posit the multiplicities of interpretative frameworks or interpretative habits (cf. Parmentier, 1997) vis-a-vis disaster as a premise rather than a conclusion, in order to better focus an interdisciplinary way to read maps in understanding disaster.

This chapter aims to examine the question of how disaster researchers can use maps to locate "culture" (cf. Silverstein, 2013) and "local knowledge" (Geertz, 2000[1983]) in supposedly objective representations of disaster. As an anthropologist studying disaster, I attempt to present a close-up view of the cultural layering processes of interpreting maps of a disaster—what Galison (2012, p. 257) calls "territorial palimpsest"—to argue that using maps achieves, contra Baudrillard, both the naturalization of the disaster and essentialization of hazards by habituating an interpretative framework in containing and delimiting an emergent unknown. Exanimating maps comparatively points to a technohistory of signs that reveals multiple vessels of sociocultural containment (cf. Gusterson, 2011).

The specific cases are drawn from the 2011 Great East Japan Earthquake (3.11); the focus rests on five types of maps, including maps of evacuation, of monitoring nuclear radiation spread, of harmful rumor, and of tsunami lines as well as a cognitive map inhibited by the locals. In these maps of the nuclear disaster in Fukushima prefecture, abstract lines are drawn to "reterritorialize" the prefecture with invisible, though concretely lived, walls. I will mobilize the maps available mostly online to practice mapping as any interested mind could. Then, I will merge them with my own on-going ethnographic research from Minamisoma city, Fukushima, as the focalized mesh parameterization of the chapter. Although the example is not an inherently natural disaster, the nuclear disaster can illuminate the "semiotic" work of reterritorialization in which maps diagram the parallel process of abstracting invisible threats and risks on a map(s) of a society, on the one hand, and concretizing them so as to produce visible, tangible, social, and political boundary lines in the society on the other hand. By juxtaposing five types of maps of the nuclear disaster, I argue that maps are both interpretative habits and material objects that presuppose and entail the lived experience of the disaster. Maps are a potent technology that articulates signs in and of society.

19.2 BASE LAYER: THE BACKGROUND OF THE NUCLEAR DISASTER

On March 11, 2011, at 14:46 p.m., a magnitude 9.0 earthquake struck just off the coast of Miyagi prefecture, 231 miles northeast of Tokyo. Within an hour, a devastating tsunami swept along the coast, leaving >18,490 people dead and missing as of September 2014. Due to the fatal sequence of these two natural disasters, the Fukushima Daiichi Nuclear Plant (1F) operated by Tokyo Electric Company (TEPCO) was severely damaged, releasing undetermined amounts of radionuclides from its containment structure. What follows is a brief list of the spatiotemporal shifts of evacuation/exclusion zones in order to provide an abridged summary of the disaster vis a vis the development of various abstract circles (Figure 19.1) and nonlinear, data-based, lines on the physical geography (Figure 19.2).

1. On March 11 at 19:03 pm, the Japanese government released a *Declaration of a Nuclear Emergency Situation* of the 1F in Ōkuma town, while denying any leakage of radioactive materials and the need to evacuate immediately. As a safety measure, however, the government announced the first evacuation order of people in the 2-km radius area at 20:50 p.m.
2. On the same day at 21:23 p.m., the evacuation order was extended to the 3-km radius area of the 1F, which included parts of Futaba town and Ōkuma town. There was also an order urging residents within the 10-km radius zone to stay inside their houses (parts of Namie town, Tomioka town ~48,272 people). This zoning followed the conventional protocol of the Urgent Protective Action Planning Zone (Sato, 2012).
3. On March 12, at 5:44 a.m., the government extended its evacuation order to the 10-km zone. On the same day at 15:36 p.m., Reactor One exploded.

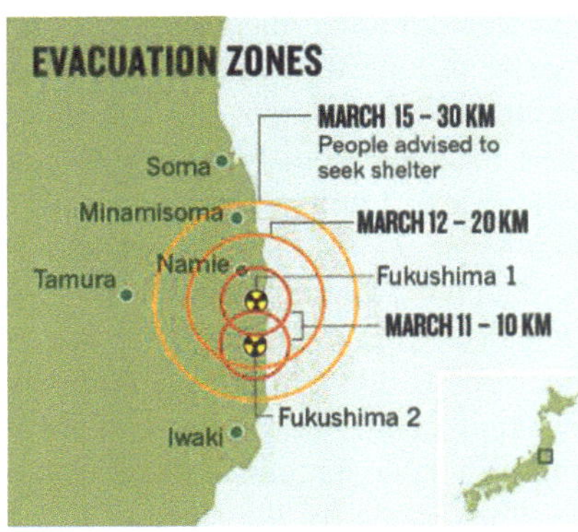

FIGURE 19.1 Shifting Evacuation Zones taken from Nature News Blog March 17, 2011. *These are presented here as images to make a point rather than high resolution mapping.*

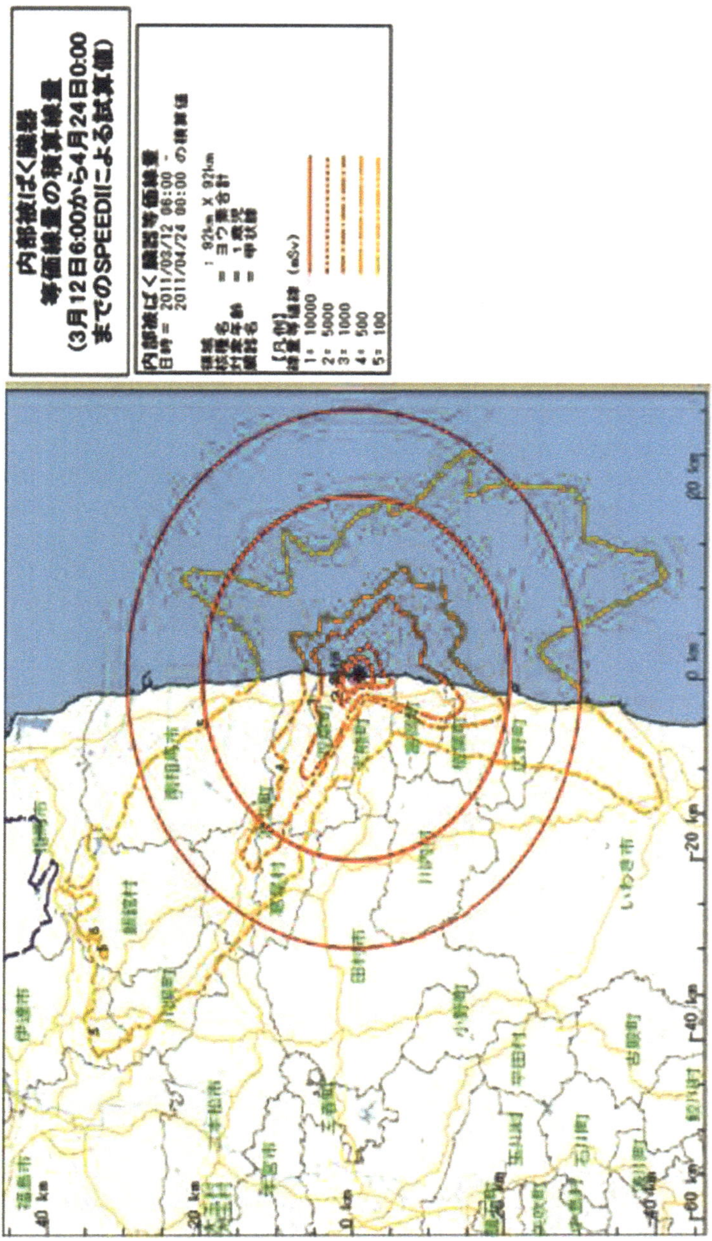

FIGURE 19.2 The estimation of radiation exposure between March 12, 6 a.m. to April 24, 12:00 a.m. by the SPEEDI Ministry of Education, Culture, Sports, Science, and Technology in Japan (MEXT). *These are presented here as images to make a point rather than high resolution mapping.*

Responding to this event, at 18:25 p.m., the evacuation order was extended to include the 10- to 20-km radius zone (a large part of Naraha town, Tomioka town, a part of Minamisoma city (Odaka ward) plus the above-mentioned towns, estimated around 78,200 people total) and the 0- to 10-km radius zone around the Fukushima Daini Plant (2F) in Tomioka town. At this point, the evacuation of the residents within the 3-km radius was completed, and only about 150 people remained within the 10-km radius.

4. On March 14, around 10:00 am, the then-Prime Minister Kan alleged that the evacuation order, "beyond 20 km radius is for the worst case scenario, and the specialists are in agreement that 20 km is enough distance" (Prime Minister, 2012). At 11:01 a.m., Reactor Three exploded, though there was no additional order from the government following this explosion.
5. On March 15, between 06:00 and 06:10 a.m., both Reactors Two and Four exploded. At 11:00 a.m., an evacuation preparation order (sheltering indoors) was given to the 20- to 30-km radius zone (including a small part of Iitate village and Iwaki city, and a large part of Minamisoma city). This set the total evacuation target at around 140,400 people.
6. On March 19, the US. Department of Energy sent aerial radiation monitoring data, taken between March 17 and 19, to the Ministry of Foreign Affairs of Japan. The data were relayed to other ministries, but the Japanese government did not use this information in determining its evacuation order. Similarly, the government had its own simulations of the radiation spreads from March 12 but did not release them until much later. Many would come to refer to the withholding of these facts as an act of government secrecy.
7. On April 21, the zoning relating to the 2F was adjusted to an 8-km radius, which converged the evacuation zone of 2F with the 10- to 20-km zone of the 1F.
8. On April 22, the government engaged in rezoning. As a result, the 10- to 20-km zone was designated as "alienation zone" (Area 3) and 30-km zone as "evacuation preparation zone" (Area 1), with a voluntary evacuation advisory for children and pregnant women. Additionally, Iitate village, which is mostly outside the 30-km zone, was designated as a deliberate evacuation zone (Area 2) in which residents would not be permitted to live. The 0- to 10-km zone remained closed (Ministry of Economy, Trade, and Industry (METI) Reference; Figure 19.3).[1] This reterritorialization order

---

1. Rearranging restricted areas, areas to which evacuation orders have been issued, and other areas (Ministry of Economy, Trade and Industry (METI http://www.meti.go.jp/english/earthquake/nuclear/roadmap/pdf/20120330_01b.pdf)). The Areas are determined by the 1 mSv/year standard posited by the Japanese government following the public dose limit suggestion by the International Commission on Radiological Protection. However, after the suggestion of the International Atomic Energy Agency in October 2013, the Nuclear Regulation Authority under the Ministry of the Environment is considering to lift the standard up to 20 mSv/year (Yomiuri, 2013, pp. 11-08), which is the equivalent amount of exposure allowed for a nuclear plant worker.

<Reference> Practical operations for designating areas to which evacuation orders have been issued as newly designated areas

	Basic Definition of the Areas	Details of Practical Operations
Area 1 Areas to which evacuation orders are ready to be lifted	Areas where it is confirmed that the annual integral dose of radiation will definitely be 20mSv or less	1. People concerned can pass through the areas along main roads, return home temporarily (staying overnight is prohibited), and enter the areas for the purpose of public benefit. 2. People concerned can : [a] resume businesses such as manufacturing, but regarding such businesses for the residents in the areas such as hospitals, welfare facilities, or shops, work is limited to that for preparation for resuming their businesses; [b] resuming farming,* and [c] start other work involving [a] and [b], such as conducting maintenance, repair, or transport-related activities. 3. People are not required in principle to take or carry out protection measures, such as screening or measures to control the radiation dose when they enter the areas temporarily. * This depends on the degree of limitation on rice planting and the extent to which radiation has been removed from the ground.
Area 2 Areas in which residents are not permitted to live	Areas where the annual integral dose of radiation is expected to be 20mSv or more and where residents are ordered to remain evacuated in order to reduce the risk of radiation exposure.	1. The operations applied to the deliberate evacuation areas are also basically applied to these areas. 2. People can temporarily return home in the areas (but staying overnight is prohibited), pass through the areas along main roads, and enter the areas for the purpose of public benefit, such as for repairing the infrastructure or conducting disaster prevention-related work.
Area 3 Areas where it is expected that residents will face difficulties in returning for a long time	Areas where the annual integral dose of radiation is expected to be 20mSv or more within five years and the current integral dose of radiation per year is 50mSv or more.	1. People are legally required to evacuate from the areas, for which physical barriers to entry such as barricades are placed at the boundaries of the area. 2. People may temporarily return home to meet domestic needs and requirements as far as possible, while those who are in charge thoroughly screen people for radiation, control individual doses of radiation, and require the people entering the zone to wear protective gear.

FIGURE 19.3 METI reference.

subjected around 146,716 people (Area 1 = ~58,506; Area 2 = ~10,010; and Area 3 = ~78,000)[2] to evacuation.

9. On September 30, the government lifted its evacuation order in regions designated previously as Area 1.

Immediately after the government's declaration of a nuclear emergency, what became critical—aside from the politics of blame focused on identifying the cause of the nuclear disaster[3]—was to visualize the urgent threat of radiation in the air. With dramatic images of explosions streaming through media channels worldwide, people questioned: how much and where?

In order to answer those life-preserving questions, various types of maps became crucial eyes for the government and the people to see the invisible spread of radioactive substances. It is important to point out that the first maps issued concerning the nuclear disaster emerged in the context of an unprecedented emergency. One of the victim in Minamisoma city, Fukushima, illuminates the vagueness of the situation in a poetic verse: "no parent, no child, no house, what's there is the invisible radioactivity" (Watanabe, 2012, p. 178). The very first desperate attempt at visualizing the imperceptible took a form in which *distance* became the central point of reference. People's pressing queries about how much and where were relayed back with an added question: "how far is my place from the nuclear plant?" As a result, the nuclear disaster made the physical distance between the nuclear plant and one's place of residence more salient than ever before.

Given this background information, the next three sections trace the metamorphosis and layering of the lines drawn on maps around the nuclear plant. In doing so, we will observe how, initially, radioactivity and radiation were visually "contained" on the maps within narrow parts of Fukushima and then sealed within Fukushima as a whole. The ramifications for containing the disaster within Fukushima manifest today as an essentializing harmful rumor or prejudice, which is coded as Fukushima, unsafe and not Fukushima, safer. I will discuss this prefectural territorial mapping as a projective containment of the disaster in the fourth section. Simulations of the laying processes of maps shed light on the practice of concretizing the disaster via the (re)presentations of the relative distance (by "convention") and of the empirical breadth (by "science") from the source of the disaster.

---

2. The detailed account of those shifts is available in Yoshida and Harada (2012). Fukushima Prefecture, Public Works Section "The Revitalization of and Return to Furusato." http://www.pref.fukushima.jp/aizu/kensetsu/pdf/fukkyuu_fukkou/daisinnsaikirokusi_zanntei_00zentai.pdf. For Minamisoma city's info at Minamisoma city http://www.mext.go.jp/b_menu/shingi/chousa/kaihatu/016/shiryo/__icsFiles/afieldfile/2013/06/24/1336650_1_1.pdf.
3. Even today, the direct cause of the nuclear disaster—whether it was the earthquake or the tsunami—is debated (Japan, 2013).

19.3 SPATIOTEMPORAL LAYER: DRAWING THE DISTANCES OF THE NUCLEAR DISASTER

The first conceptualization of the radiation spread was based on varying distances from the nuclear plants. From the development of this particular type of map (Figure 19.1), we can see how the government changed their initial estimation of the range of the disaster by manipulating the relative distance from the plants. This shifting of the evacuation/exclusion zones to correlate with the acknowledged severity of the "accident" made distance into the concrete measuring unit by which to assess risk: the greater the spatial extent of the mandated evacuation/exclusion, the more serious people perceived the disaster to be. For example, the US perceived the disaster more seriously than Japan did, ordering their citizens residing in Japan to evacuate to beyond the 80-km radius zone. The US decision bred skepticism among the Japanese public regarding their government's estimation of the disaster's severity, and many who were living around the edge of the government-regimented line of 30 km decided to voluntarily evacuate after learning of the US interpretation. An individual living in Sendai, >80 km away from the nuclear plant, reported that many people in her neighborhood, herself included, did not want to dry their laundry outside for a long time. This behavior, she noticed, coincided with 3.11, led to the growing fear of airborne radiation traveling from the plant. Obviously, then, the government was unsuccessful at containing the radioactive materials and their derivatives within the designated zones, as it claimed.

Equally significant is the fact that the relative distance from the plant was connected to the degree of dissemination of information regarding the evacuation/exclusion order, which was asymmetrically available to the residents within different zones. For instance, the people of Minamisoma city, which extends over three zones inside the 10–20 km, 20–30 km, and outside the 30 km, did not hear about the seriousness of the disaster until much later. A Minamisoma resident who lived about 18 km from the nuclear plant recalled that he learned of the need to evacuate by witnessing a group of people from Namie, a town a few kilometers south of his residence, escape a few days after the explosion of the first reactor. Thus, distance made the explosion as well as the order undetectable. In contrast, residents within the 10-km zone, especially the residents of Ōkuma town, Futaba town, and Tomioka town in Futaba district, evacuated immediately. Some residents outside 10-km zone sometimes attribute this difference to the closer tie to and/or benefits (*onkei*) from TEPCO. Others, in an attempt to attribute some hidden convention to the distance, reported that only those who went to schools within a 20-km zone participated in a school field trip to the nuclear plant. The distance to and from the plant, therefore, became both a convention and a stipulated measurement of risk.

Under the swirl of uncertainties, the distance map reframed the original abstract question of how far from the plant to "how close is one's place of residence from the plant?" The map with three circles of distance (Figure 19.1) made salient that distance mattered not only as a means of gauging the necessity to evacuate but also as a measure of the "symbolic capital" required to obtain necessary information to know what kind of action to take.[4]

19.4 SCIENTIFIC LAYER: VISUALIZING THE ARBITRARY BREADTHS OF THE NUCLEAR DISASTER

Those doubts and the related discontent among the people were not unfounded, as testified by another view offered by different maps made available to the general public about a month and a half after the disaster. Unlike the distance map, these radiations spread simulation and monitoring maps traced the noncircular spread of radioactive particles (Figures 19.2 and 19.4).[5] These maps that are in congruence with each other indicated the *actual*, empirical mapping of radiation in Fukushima prefecture and beyond. The verisimilitude of the supercomputer simulation of the System for Prediction of Environmental Emergency Dose Information (SPEEDI) was later confirmed via aerial monitoring completed by the Japan–US survey team (Sato, 2012). However, the government released only partial data on May 23 to the locals and did not officially release this information fully until April 26, that is, after the first reterritorialization enacted on the 22. Thus, the matching of the hypothetical model and actual observation, a typical scientific procedure, ironically problematized the initial secrecy of the existence of the simulation by the government.

With these more scientific projections now superimposed over the earlier and somewhat absolute circular lines, dire questions regarding the evacuation procedure that had been dictated by the distance map began to surface. For example, Iitate village, most areas of which are currently categorized as a deliberate evacuation area as of November 2013, is marginally included within the 30-km zone in the distance map; thus, the local municipality did not make the decision to evacuate until May 15, 2011. The consequence of the delay in sharing the SPEEDI data had huge reverberations to many evacuees. Famously, the residents of Namie town moved with its municipality in

4. The closeness of the distance to the plant was not necessarily what determined the delay in disseminating the emergency information. According to the record, the SPEEDI data were sent to the Fukushima prefectural office on March 12 though the office had no electricity due to the earthquake. On March 13, the prefecture received 32 facsimiles of the SPEEDI data but did not release the information (Fukushima Minpo, 2011 http://www.minpo.jp/pub/topics/jishin2011/2011/12/post_2755.html).
5. The SPEEDI data were made available on April 26 over a month and a half after March. The government later described later delay as a method of avoiding panic of the people. However, it is clear that the government itself was in panic, as the number of people who needed to be evacuated according to the SPEEDI data was much more than the ones enforced by the government. The concept of elite panic (Clark and Chess, 2008; Solnit, 2009; Tierney, 2008, p. 131) is applicable here to illuminate the process.

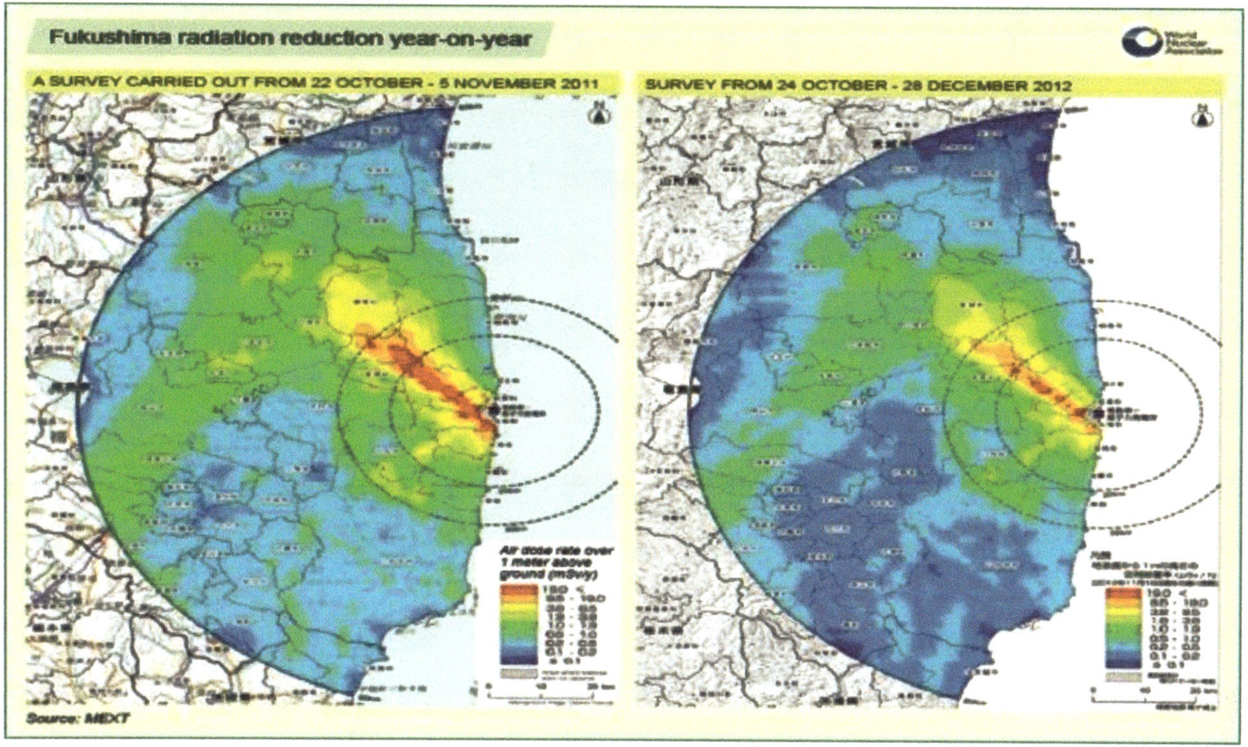

FIGURE 19.4 Aerial Measuring Results by MEXT and National Nuclear Security Administration at the US Department of Energy taken from World Nuclear Association: http://www.world-nuclear.org/info/Safety-and-Security/Safety-of-Plants/Fukushima-Accident/.

Tsushima ward, Namie, between March 12 and 15 where the amount of radiation was very high (Kahoku, 2011).

Therefore, the emergence of a more scientific perspective via the breadth map not only revealed the inaccuracy of the distance map but it also revealed the inability of the 30-km circle to fully contain the coverage of radiation. Since their publication, a temporal series of breadth maps have been used to further reterritorialize cities and towns (Figure 19.5), and to large extent the degree of expected radiation exposure has been used as the underpinning logic as to how to move from the Areas 3 to 1 in the METI reference (Figure 19.3), which determines the amount of radiation exposure allowed per year. Unlike the distance map, this new series of monitoring maps indicates that the breadth

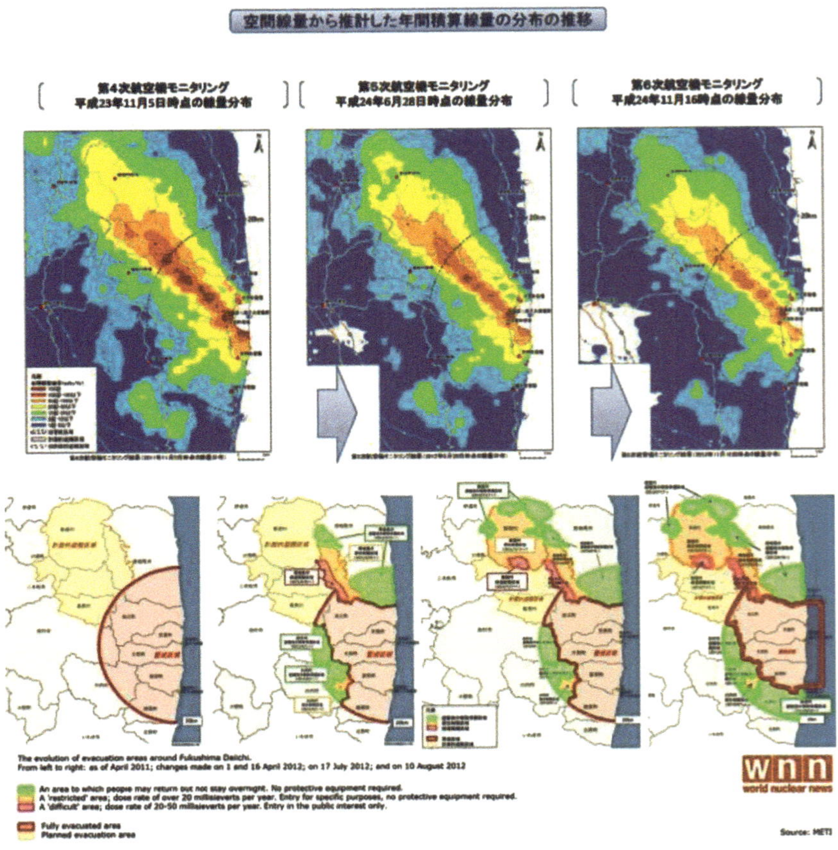

FIGURE 19.5 The relationship between the estimation from the aerial radiation monitoring results and the shifts in reterritorialization of evacuation/exclusion zones between November 2011 (above right) to November 2011 (above left) taken from METI at http://www.meti.go.jp/earthquake/nuclear/pdf/131009/131009_02a.pdf and April 2011 (below right) to July 2012 (below left) taken from World Nuclear News at http://www.nirs.org/fukushima/crisis.htm.

and amount of radiation are changeable over time. This vision enabled the idea that decontamination could help reopen Areas 2 and 3. Most notably, the principal evacuation/exclusion criterion is now based on the degree of contamination in the area, which is susceptible to change over time; this stands in opposition to the earlier, distance-based criteria, a model whose guidelines will remain mostly absolute. With this new type of real-time, fact-based maps, the new question became "how contaminated is my place of residence?" This question facilitated the local use of personal Geiger Miller counters to scrutinize the government's monitoring of the radiation level. Such a conceptual shift in maps from convention to facts made the radiation less containable within smaller regions and by the top-down order and generated another type of map I will discuss in the next section.

The last two sections have covered disaster responses represented in the first two types of maps. Importantly, the sequence in which the maps were generated is not reversible. The distance map would not have made any sense if the breadth map were available in the first place. The conceptual merging of the two maps in sequence operated here illuminates the presupposed "social containment vessels" (Gusterson, 2011) or the safety myth surrounding nuclear technology, in which the government's effort and struggle to contain the invisible radiation threat within maximally limited areas on the one hand, and, on the other hand, varying ideologies attempted at justifying evacuation/exclusion, its potential scale and associated body counts. Examining the two maps chronologically illustrates the materialization of "scientific" accuracy in visualizing the disaster, but the accuracy did not necessarily translate to the feeling of safety since the information was asymmetrically disseminated and areas of containment appeared to be irregular and spotty (cf. Phillips, 2013). In both cases, however, different maps functioned simultaneously as interpretative frameworks (conventional or scientific) and tangibly distinct objects (different pictures) for containing the invisible threat of radiation; cartographic representations were, at times, the eyes to see as well as the tool to contain the imperceptible. These efforts to contain by visualizing both the estimated and detected radioactivities and radiation spread not only concerned the residents in Fukushima but also those who lived outside of Fukushima.

19.5 SOCIOPOLITICAL LAYER: CIRCULATING RUMORS AND ENCIRCLING THE INVISIBLE THREAT

This section discusses how the disaster was imagined and mapped by outsiders and how this view reverberated back to many Fukushima residents. The third type of map is what I call the "harmful rumor (*fūhyō higai*)" map that materializes how Fukushima became encircled by radiation (Figure 19.6). this type of map is a hybrid of the circles from the distance map and the radiation spread data from the breadth map discussed above. It is used neither to willfully express nor to segregate Fukushima prefecture; rather, this segregation is

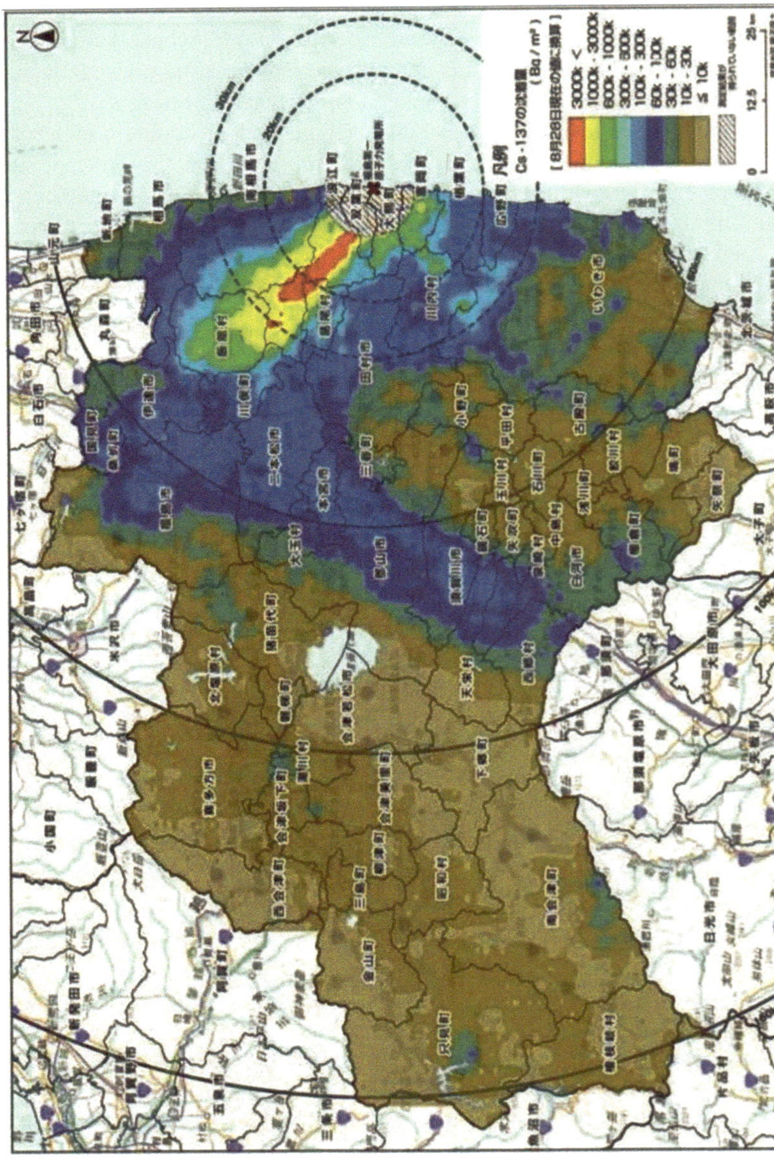

FIGURE 19.6 Radiation Rumor Map Taken from Sugenoya (2013). This particular map was originally taken from the MEXT aerial monitoring of Fukushima prefecture. Although there are other maps that include larger parts of Japan, this was selected to make the specific argument of Fukushima as bad as or worse than Chernobyl.

a by-product of its representational work. The particular map is taken from a book that compares the Fukushima disaster with the Chernobyl disaster to show how the Fukushima disaster has caused a wider spread of radioactive substances than was the case with Chernobyl (Sugenoya, 2012). However, the map used to compare the two nuclear disasters reflects how Fukushima is conceptualized as the only place in Japan where contamination is prominent. The question, asked by the people inside the distances and breadths, of how far and how much of the radiation shown in the previous maps, is coded here as the binary "either Fukushima or not Fukushima." In this particular politics of contamination geography, *anything* Fukushima, name and location, is subjected to contamination. Just like the distance map, this rumor map is not an accurate representation of the radiation spread depicted by the breadth map (since, in the rumor map, places neighboring Fukushima are represented with no radiation). At the same time, this map materializes, just like the other maps, an image of containment of the invisible threat (whose coverage was made haphazard by the breadth map) by rendering the prefectural boundary visible, salient, and more "real" than ever before. This newly materialized image in the map has been concretized through the production of real consequences for the people in Fukushima.

In Shinchi town, Soma district about 50 km away from the Fukushima Plant, some evacuees from the exclusion zones began to cultivate rice. Although 50 km is a safe enough distance according to the zones set by the government, farming anywhere in Fukushima has been very difficult. The struggle not only comes from the fact that evacuated farmers had to find new locations to farm but also stems from the negative reputation the name "Fukushima" gained since the disaster. Since 2012, for marketing purposes and for personal consumption, all rice in Fukushima has to be tested for the amount of radioactivity that might be contained in a bag. Given the disaster and the results of the aerial monitoring, this testing regime seems reasonable from the standpoint of food security. In reality, farmers themselves would like to know if it is safe to consume foodstuffs they have produced. An owner of an agricultural cooperative in Shinchi, one of many sites where edibles are tested, reported that the testing is mostly for the local farmers because no one outside the area would really buy their rice. Similarly, a restaurant owner in a nearby area confessed that they use the tested Fukushima products for the restaurant at times but buy non-Fukushima products for their home. She is afraid of consuming radioactive foods: "I am ok but for my kids I want to be careful," she commented.

This concern for one's self and family not only speaks to an elevated sensitivity among the residents in Fukushima about the foods and radiation in general but also underlines the fact that rice, vegetables, fruits, or anything made in Fukushima would not sell outside Fukushima or even, perhaps, *inside* Fukushima. This fear and hesitance are bolstered by some testing results that have come back positive with a becquerel (Bq) level that surpasses the acceptable amount (note, though, that in 2011 Japan lowered its standard from 500 to 100 Bq/kg for many foodstuffs).[6]

At an agricultural cooperative I visited, every 30-kg bag of rice processed goes through a machine that measures the amount of becquerel per kilogram.[7] After one bag has passed through the machine, the computer attached to the machine read 9 Bq/kg, which is much lower than the standard of 100 Bq/kg. The farmer told me "20 Bq/kg is the highest amount we allowed for a 30 kg bag of rice," while attaching a sticker that guarantees the safety of the rice in the bag. A barcode is printed on each of these stickers, promising the transparency of the production process. This practice, started only in 2013, allows anyone to trace the source of an individual bag of rice down to the location of production and testing by scanning the barcode.[8]

Setting a significantly lower standard than the country's previous requirement has been a desperate attempt at countering the negative image of Fukushima circulating in the mind of consumers. Despite this effort, Fukushima foodstuffs do not in fact sell on the market. "In reality," said the farmer, "Fukushima products are the safest of all from the perspective of monitoring and testing. No other place in Japan actually has been testing as rigorously as we do here. They (non-Fukushima farmers) might be afraid to test and find out that their product is worse than Fukushima's, but the name of Fukushima has been contaminated, perhaps more so than the land itself." Such is the extent of the damage done by the harmful rumors spread throughout and about Fukushima, which parallels people's fear of the invisible threats emanating from all things Fukushima; the fear coincides with the projected

---

6. Note that becquerel is the amount of radioactivity released upon the decay of radioactive material per second. The higher the quantity of becquerel is, the higher the radioactivity of a given item, although becquerel does not correlate with the degree of the effect of radiation on a person's body since the strength and longevity of radiation vary depending on the type of radioactive substance (e.g., cesium vs. plutonium). Thus, during the screening, what is mostly being measured is the level of cesium-134 and -137, whose half-life are 2 and 30 years, respectively; the measurement does not express, explicitly, the physiological damages one risks upon consuming potentially contaminated food.
7. Detailed information about this process is described at the home page of Fukushima prefecture at http://wwwcms.pref.fukushima.jp/pcp_portal/PortalServlet;jsessionid=ADB46D6478109DA67DC63E7351B90110?DISPLAY_ID=DIRECT&NEXT_DISPLAY_ID=U000004&CONTENTS_ID=35674. A video of the process can be seen here: http://wwwcms.pref.fukushima.jp/pcp_portal/PortalServlet;jsessionid=E7CABF11EA457FF4247279BE0C13E4EF?DISPLAY_ID=DIRECT&NEXT_DISPLAY_ID=U000004&CONTENTS_ID=35675. For the EU standard see "Commission Implementing Regulation (EU) No 351/2011 of April 11, 2011, at http://eur-lex.europa.eu/LexUriServ/LexUriServ.do?uri=OJ:L:2011:097:0020:0023:EN:PDF. For the US Standard, see US Food and Drug Administration CPG Sec. 560.750 Radionuclides in Imported Foods-Levels of Concern at http://www.fda.gov/ICECI/ComplianceManuals/CompliancePolicyGuidanceManual/ucm074576.
8. In 2013, Fukushima prefecture tested 8,383,076 bags of rice, and of those only 12 bags (0.0001%) did not pass the 100-Bq/kg standard. One can trace every bag of rice in Fukushima here at https://fukumegu.org/ok/kome/.

prefectural boundary line drawn on the map, within which, in the minds of the rest, the invisible threat is *imaginatively* contained.

The suppression and/or forgetting of another map address this deep association between Fukushima and radiation. The fourth map (Figure 19.7) is

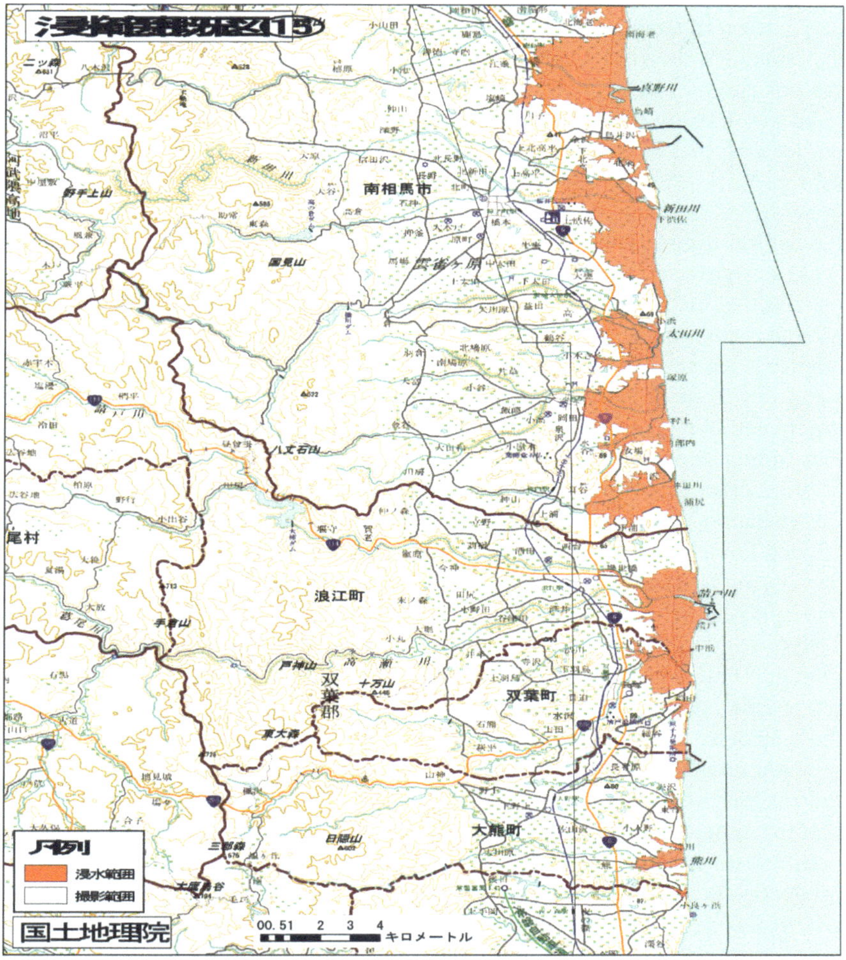

FIGURE 19.7 A Tsunami map of Fukushima costal side (Hamadōri) depicting Futaba town, Ōkuma, town, Namie town and a part of Minamisoma. The areas that are not shown in this map are Hirono town, Iwaki city, Naraha Town, Soma District, Tomioka Town (in alphabetical order). The red color indicates the areas where the tsunami reached.[9]

9. Maps of other areas of Fukushima prefecture can be found here at the website of Geospatial Information Authority of Japan at http://www.gsi.go.jp/common/000060134.pdf; http://www.gsi.go.jp/common/000060136.pdf; and http://www.gsi.go.jp/common/000060331.pdf.

an "unusual" map of Fukushima that does not map any radiation-related phenomenon. Instead, the map depicts a part of the physical areas where the tsunami reached inland in Fukushima (112 km^2 of 13,782.76 km^2 total). The map shows that severe damage was done to the coast of Fukushima, resulting in the deaths of around 1,600 people. However, *this* statistic has not been the major concern among those who hear about Fukushima in news. Ignorance of the tsunami's dreadful impact is partly due to the on-going fear surrounding the Fukushima Plant, which is still not under control. The first priority seems to be to somehow contain the continuing disaster. The lack of emphasis placed on the map that represents the initial damages from the natural disasters that subsequently caused the nuclear disaster exemplifies how Fukushima is imagined as the land of contamination, distinct from other areas damaged by the natural disasters in northeastern Japan.

One significant feature that distinguishes the rumor map from the other maps is that the prefectural boundary, as depicted by the rumor map, is marked as if the radiation from the disaster did not escape outside of Fukushima. The emphasis on the boundary line encircling the prefecture makes Fukushima stand out from its surrounding neighbors. In this sense, the map concretely articulates the binary opposition rendered by the Fukushima and non-Fukushima reterritorialization. Notice that this map is a "secondary elaboration" of other existing maps in that it selectively represents reality in terms of the radiation spread. That is, the scientifically ratified spread of radiation beyond Fukushima's boundaries illustrated by the SPEEDI analysis and aerial monitoring (Figure 19.8) made the prefectural containment urgent in order to underplay a major threat to immediately surrounding prefectures, including Tokyo, 250 km southwest of the plant. The rumor map is, therefore, a cartographic (object-mediated) materialization of the general public's cognitive reterritorialization that diagrams the on-going essentialization of Fukushima in the outsiders' understanding of radiation contamination and its all-too-wishful regional containment. In this particular sense, this specific map that projects the visible separation of Fukushima from the rest embodies the reality of harmful rumors with which the people of Fukushima live. Acting according to harmful rumors—for instance, by not buying "contaminated" Fukushima products—is a sociopolitical means to *actualize* and to *contain* the radiation within Fukushima as well as to make the map's projections *real*. The rumor map represents the already perceptible sociopolitical phenomena of reterritorialization of the country along the axis Fukushima or non-Fukushima.

By observing a temporal series of maps developed after the disaster, it becomes clear how the abstract containment of radiation contamination, via the representation of the disaster in map form, is concretized in the use of maps that acts almost as defense mechanisms that visualize distance, breadth, and prefectural divide. Although they are as invisible as the radiation is to the human eye, each boundary reified by the maps distinctively (and rather magically) assuages the possibility of invisible contamination. However, of the

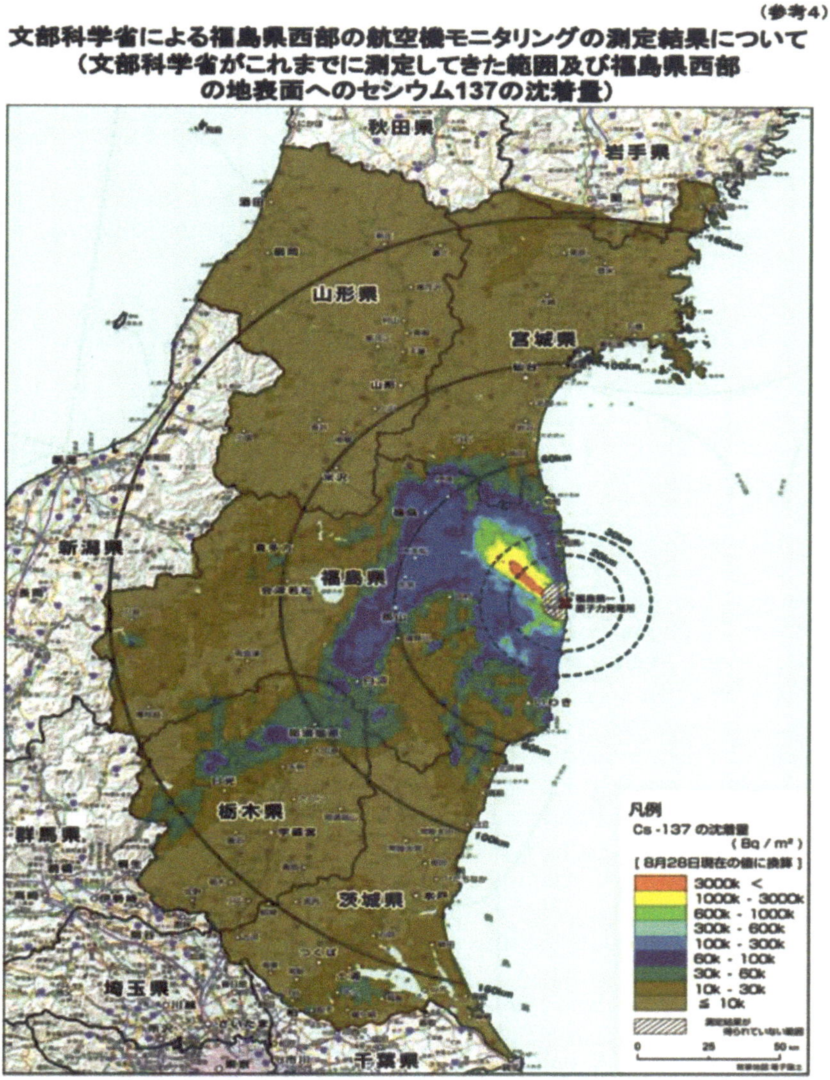

FIGURE 19.8 A wider view of the Cesium-137 precipitation from the MEXT aerial monitoring survey. This map indicates the spread of radiation beyond the prefectural boundary of Fukushima. *Taken from http://radioactivity.nsr.go.jp/ja/contents/5000/4894/24/1910_0912.pdf.*

four maps, the breadth, rumor, and tsunami maps are alterable: time is a major variable that can change the condition of the reality being depicted as in a weather map. For example, as time passes, the airborne radiation lowers in severity in many areas; likewise, knowledge about radiation, radioactivity, and the geography of Fukushima (an understanding that it extends far west from the plant) has somewhat helped to reduce the unfounded prejudice against people and products in Fukushima.

Nonetheless, if we change the scale of analysis, zooming in to the city of Minamisoma, it becomes noticeable that the distance map continues to thrive, despite its proven inaccuracy vis a vis the breadth map. In fact, since the disaster, it has been gaining even stronger presence and concreteness. The last section will reverse the vector of analysis by exploring how local people's lived experiences with the invisible may be visualized in a map.

19.6 TERRITORIAL PALIMPSEST: MINAMISOMA CLOSED-UP

In *Imagining Disaster*, Weinsenfeld (2012, p. 8) argues alongside Wood et al. (2010) that "maps are visual propositions that give form to social and political interests." While this statement seems to aptly characterize the rumor map discussed above, it is crucial to remember that maps also exist in a nonvisual and inchoate form. This last section attempts to explore the acculturation of a specific cognitive map of the disaster by the residents of Minamisoma city and how a particular sequence of cartographical and visual containment has been facilitating a localized naturalization of the disaster. In this section, I will use the concept of "cognitive map" (e.g., Lynch, 1960; Rengert and Pelfrey, 1997; Tolman, 1948) as a particular kind of map that is organized via the process of an apperception, that is, people's sense of place that is not materialized in a map. With this concept in mind, this last section traces back the mapping process and attempts to answer the question: "how can territories be mapped or materialized?" The preexisting knowledge of a territory makes "seeing" a map—a represented territory—analytically inseparable from the map "seen as" (see Galison, 2004, pp. 257–294). In this sense, the map's reversibility and susceptibility to mediations, as argued by Deleuze and Guattari (2004[1988], p. 13), are incomplete conceptualizations of the map, since they only consider the map as a single object and event with multiple frameworks.

A thought experiment I pose here is to assemble the four types of maps described in one single hypothetical map. This laying experiment illuminates some of the consequences of a series of attempts at containment by inquiring how the people inside the spatiotemporally, objectively, and sociopolitically marked territories of contamination imagine their own contaminated land. I argue that a map does not precede but rather *presupposes* a territory: a series of representations are partially assimilated into the preexisting (cognitive) map to create a new territory. If maps of nuclear disaster are used to contain the invisible radiation, a culture of containment may percolate throughout questions of how such attempts at containment are lived by the people who themselves live in the designated land of contamination.

We can begin to envision the local's cognitive map by examining a map that has been crafted by the city of Minamisoma, which surmises the 3.11 disasters from the perspective of its community members (Figure 19.9). As the messy lines coded with multiple colors indicates, Minamisoma has been severely reterritorialized due to the disaster. When all the maps thus far are merged, the sheer complexity of the geographic location of Minamisoma vis a vis the disasters becomes visible. Minamisoma has been damaged by the earthquake and tsunami (~4,366 houses and 1045 dead, of which 409 are disaster-related deaths). It is located within three different distance zones (10–20 km (107 km^2), 20–30 km (181 km^2), and outside 30 km (111 km^2)) and consists of three different evacuation/exclusion zones, which have gone through further rezoning. In addition, many sparse spots within the city have been designated as restricted areas recommended for evacuation. As of June 2013, 16,138 have evacuated from the city and 11,254 are evacuating within the city, and many have expressed that they are unsure if they will ever return (Reconstruction, 2013). In the two years and eight months that have passed since the disasters, this city, with its ever-shifting population (71,949 before the disasters, ~10,000 as of March 26, 2011, and a current population of around ~64,000 are registered and ~47,000 actually living in the city), is still in turmoil.

The blue dotted lines in the map (Figure 19.9) mark the preexisting territorial boundaries of three wards (from the top, Kashima, Haramachi, and the bottom is Odaka) within Minamisoma. Minamisoma was established in 2006 as a result of consolidating three municipalities. This union has been a major variable in the (problematic) decision to divide up the city along the lines drawn by the distance map. A close look at the map shows how the various lines of the disaster have been drawn without recourse to ward boundaries. Such a cartographic palimpsest of multiple containments has produced on the one hand, deterritorialization, dissolving the consolidation of the city, and on the other, reterritorialization, that is, the production of new spaces that belong to no ward. A dozen of Minamisoma residents voiced that the city should not have been consolidated. Because of the union, they criticized, people in the same ward had to be divided according to the various projective lines of containments after the disaster. In fact, the Nuclear Regulation Authority (formerly known as the Nuclear Safety Commission) suggested that the criteria for determining evacuation/exclusion zoning should not only be based on the level of radiation but also on other variables such as potential social and political tensions resulting from the total size of evacuees as well as the coherence of action within a local municipality (Minamisoma, 2012).

One major characteristic of Minamisoma residents is their heightened sense of the distance of their residences from and to the nuclear plant. "I am 31 km away and ...." "This place is about 12 km so...": such configurations of places vis a vis the distance from plant indicate that distance remains the principle reference in positioning themselves, as if the disaster is understood by and lived differently according to the distance. One resident who lives

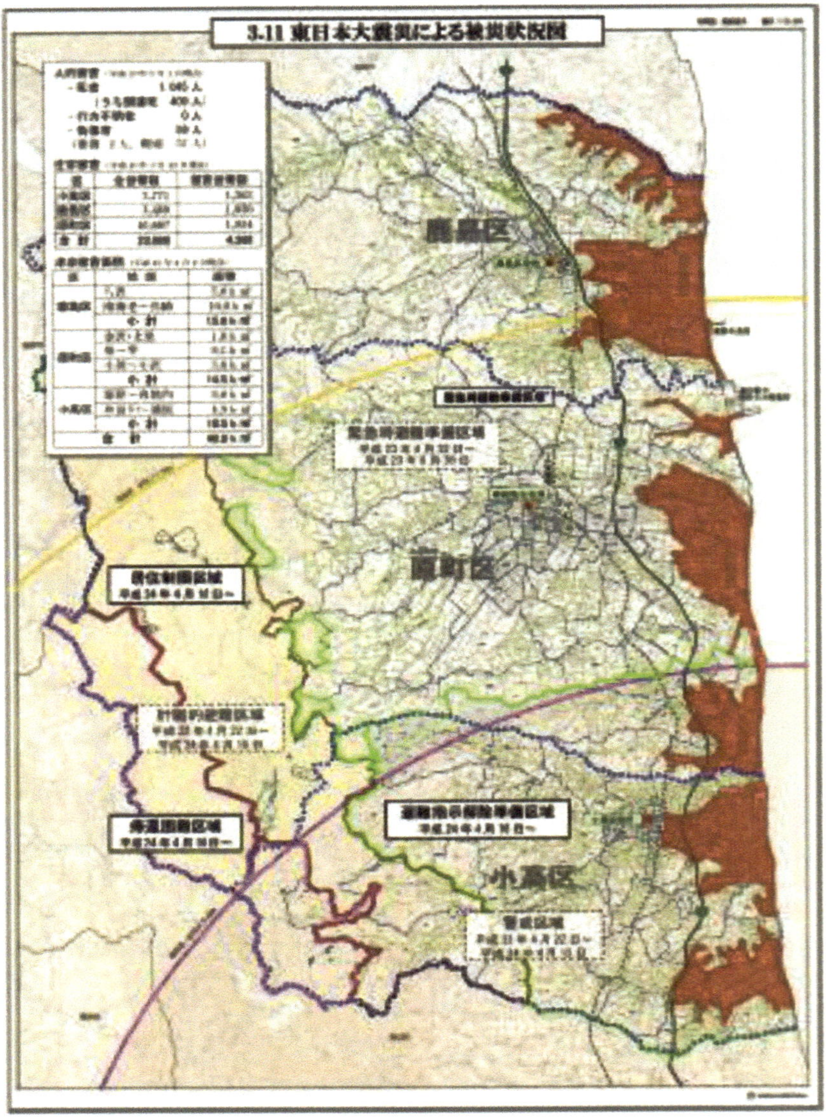

FIGURE 19.9 A close-up view of Minamisoma city, Fukushima. *Taken from http://www.mext.go.jp/b_menu/shingi/chousa/kaihatu/016/shiryo/_icsFiles/afieldfile/2013/06/24/1336650_1_1.pdf.*

outside of the 30-km zone but works inside the 20- to 30-km zone lamented that the governmentally regimented circles have created "invisible dividing walls" within the city. Interestingly, those who have evacuated from areas closer to the plant to further places, within Minamisoma, seem less likely to locate themselves on the axis of distance, or if they do they talk about the distance in the past tense, describing how far away the houses they previously lived in are from the plant. Now, they identify their places with reference to the name of the ward in which they used to live. Despite individual differences in reaction to the disaster, the nuclear plant made itself salient in the mind of people living in Minamisoma via its relative distance, which disoriented their sense of place. "I did not know that Odaka was this close," or "I live just outside of the 30 km line and am really victimized because my house was not close enough to receive compensation."[10] Encountering this sudden spatial disorientation, the local people seem to be using newly emerged disaster maps not to contain radiation but to reflect on their spaced-out sense of places and selves (cf. Galison, 2006) in relation to the closeness and farness to the plant, which was not included in their previous cognitive maps.

This is not to say that the empirically based radiation monitoring maps do not matter; they matter a great degree. Although many residents stopped carrying around radiation survey counters, which they used to rather obsessively, they have learned through experience that the reterritorialization of their place by the breadth maps does not accurately represent the actual mapping of airborne radiation in many areas, nor do radiation level readings determine which places are acceptable to live in according to the government. For instance, regions in the mid-Fukushima show much higher amounts of radiation than the coastal region, but it is places like Odaka, where large parts of it are, objectively, less contaminated, where people have not been allowed to live. The opaque logic used to determine the safety and evacuation/exclusion zones discredits the lines drawn on the breadth map, which seemed to express metaphorical divisions based on sociopolitical motivations rather than scientific facts. The evidence for such a skeptical attitude is visible on the distance map itself; though it supposedly represents mechanical circles, a zoomed-in view on the 20-km line reveals not-so-objective curvatures (Figure 19.10).

By responding to the mapped images held by many outsiders, it is not unusual for the locals to jokingly comment that they have gotten used to living in a contaminated land. In reality, however, they know very well, perhaps more so than the distanced view from the above or mechanically and sociopolitically drawn lines of containment, which places are *actually* more contaminated. Many locals avoid going to places with higher radiation. Those unsafe places are categorized as "high," and most places in the city are considered "low." Community members organized their places after the disaster according to the two major scales of "high/low" and "near/far."

10. The distance has been used to qualify the rights to apply for compensations from TEPCO. In fact, the differences in compensations between zones have been causing a huge problem in the coherency of Minamisoma city. This important subject matter deserves more careful attention than the chapter can do justice.

FIGURE 19.10 The close-up view of the 20-km radius line across Minamisoma city, showing how the supposedly circular line (in blue) is irregularly curved (in light blue) seen near. *Taken from Minamisoma city, Publication of 20- and 30-km maps. 20km_1. pdf (left) and 20 km_2. pdf (right) at http://www.city.minamisoma.lg.jp/index.cfm/10,3110,66,html.*

Layering these new signs onto the maps of the disaster, the local cognitive map appears to be much more evidence-based, comprehensive, and realistic (Figure 19.11). The local residents' comment on the habituation of living in this reterritorialized space, for good or bad, hints at the on-going reconfiguration of their preexisting sense of place with the new scales of distance to the source and of amounts of radiation. In this regard, the maps with different interpretative frameworks interact first by palimpsesting—that is, by spatially superimposing new lines and dots on an already lived territory—and second, by temporally producing the actual consequences on the territory mapped. In turn, by reading the same maps nonliterally they can instead be used as tools to reconfigure and acculturate one's self in order to locally naturalize the disaster. The fundamental difference between the other maps and the locals' cognitive map lies in materialization. Thus, what is materialized and thus visualized in a map seems to supplant the lived experience of the territories mapped by presenting itself not as a mere representation of a space but as a plausible argument about the place to be believed.

19.7 CONCLUSION: SIGNS OF MAPS AND SIGNS IN MAPS: SEMIOTIC RETERRITORIALIZATION

The chapter traced backward through time, identifying the processes by which the nuclear disaster has been interpreted by various interpretative frameworks in society. The shifts of maps from conventional to scientific, and to sociopolitical projections of the invisible were not diametrically opposed. Each layer of interpretations has rather been superimposed, materializing multiple abstract and defensive walls of containment in maps on the one hand, and on the other, stifling the unmapped, on-going cognitive acculturative process of containing the invisible threat of radiation in vivo. As the representations of the disaster, those maps—which are also a materialized tool of imaginative containment—did create a new configuration of cartographic territories vulnerable to essentialization. However, when the presupposing knowledge of those projected territories is taken into account, that is, when local knowledge about people's experiential sense of places is made ethnographically salient, another, more pragmatic view of the disaster emerged. In this sense, map as a mode of representation is a potent technology of signs that analogizes abstract ideas with concrete objects and vice versa. For disaster researchers, therefore, maps are "the object of a temporal apprenticeship, not an abstract knowledge" (Deleuze, 2000[1964], p. 4), indicating neither real nor not real, but, simultaneously, some contextual facts and technohistorical elaborations of a culture of containment.

Exploring how the residents of Minamisoma use the same set of maps apperceptively to negotiate the possibility of rehabituating their place illustrates the local strategies to pragmatically contain the disaster. This observation suggests the importance of historicizing the location of our study—predisaster of a

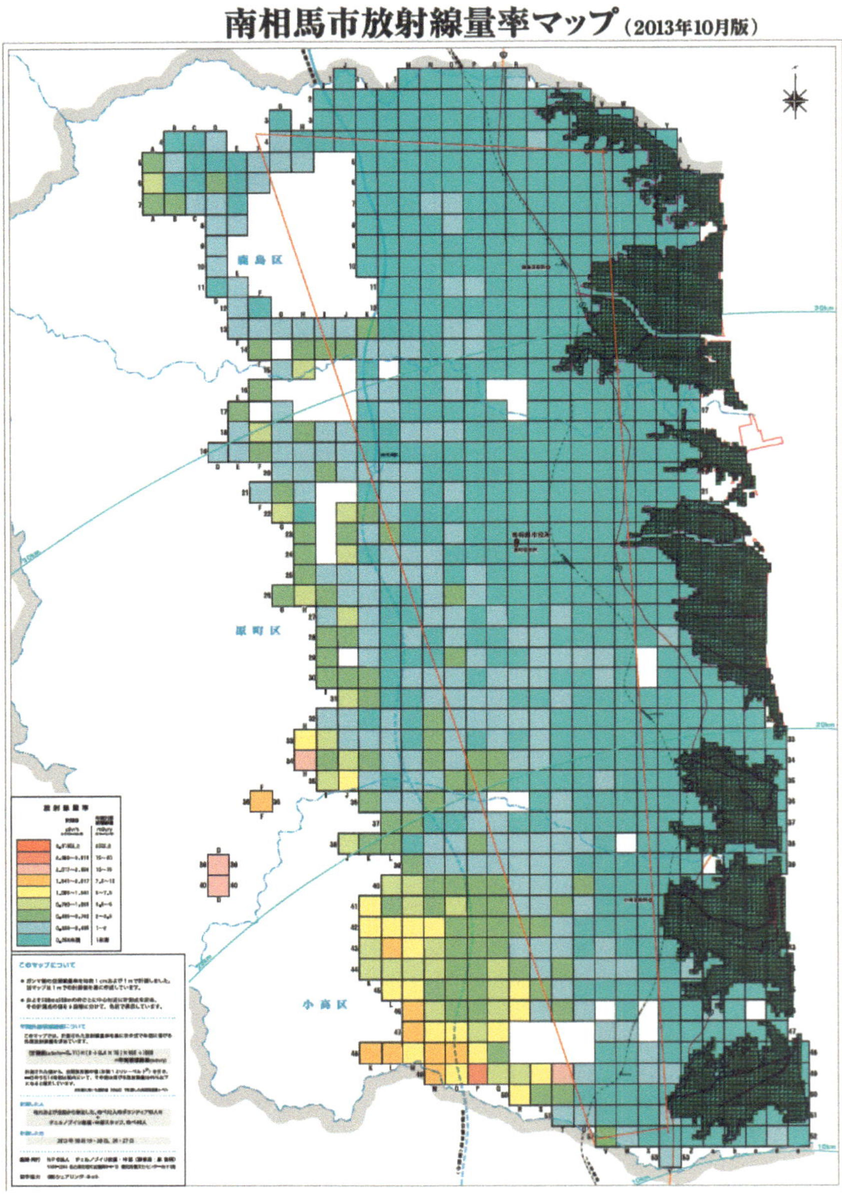

FIGURE 19.11 A hypothetical cognitive map of Minamisoma. The red lines in the middle of the map shows a culture of containment by the local residents. The radiation monitoring data were collected by NPO Chernobyl Support Group, Minamisoma. The Geographical Layers were added by Sharingnet. The original map has been distributed to the residents of Minamisoma by the NPO. The tsunami area layer was later added by Tsuyoshi Fujino.

disaster—in gaining a more "*vue d'ensemble*" of the disaster as well as locating culture materialized in a series of technical visions enabled by various maps. For disaster researchers, map(s), as a historically meaningful series of signs of society and reflectors of actions in society, are a useful analytical tool to use in identifying the work of *semiotic reterritorialization*—a materialization technology of a belief in which a propositional representation of an abstract layering of signs is presented and acted as if it is the only tangible, thus, plausible evidence of the belief. The harmful rumor map's parallelization of the outsiders' growing *imaginations* of contamination and their constitutive *actions* in reifying the regional containment map exemplifies a map's capacity as a technology of signs. As both ideas and objects, maps embody a struggle of interpretative frameworks among varying ideological bearings in society in presenting, as this case shows, the disaster as containable within a bounded space–time and associated risks as calculable by an appropriate parameterization. In light of the local knowledge in Minamisoma, a territory is subjected to the act of both literal and metaphorical palimpsest, yet layering does not erase what is underneath line by line.

There is, then, an imperative for disaster researchers to reject, yet again, the floating abstractions of society concretized into a single material framework, in favor of a close-up gaze at what does not fit into one neat representation and one interpretation, which takes each other recursively as alibi.

ACKNOWLEDGMENTS

My fieldwork was made possible by the generous support from the Japan Foundation 2013–14 Doctoral Fellowship. I wish to acknowledge the people of Minamisoma city for patiently talking to me about their perceptions of and experiences in their worlds. I would like to thank Professor Richard Parmentier at Brandeis University and Professor Peter Galison at Harvard University for reading earlier versions of this chapter and pointing out conceptual incoherencies. I would also like to thank Beth Semel for editing earlier versions of the chapter and for transforming them into something more intelligible each time. All errors in the paper are strictly mine.

REFERENCES

Baudrillard, J., 1994[1981]. Simulacra and Simulation. University of Michigan Press, Michigan.

Clark, L., Chess, C., 2008. Elites and panic: more to fear than fear itself. Soc. Forces 87 (2), 993–1014.

Daston, L., Galison, P., 2007. Objectivity. Zone Book, Cambridge.

Deleuze, G., 2000[1964]. Proust and Signs: The Complete Text (R. Howard, Trans.). University of Minnesota Press, Minneapolis.

Deleuze, G., Guttari, F., 2004[1988]. A Thousand Plateaus: Capitalism and Schizophrenia (B. Massumi, Trans.). Continuum, New York.

Fukushima Minpo, 2011. The Residents of Namie's Evacuation to Tsushima. http://www.minpo.jp/pub/topics/jishin2011/2011/12/post_2755.html (accessed on 08.11.13).

Galison, P., 2004. Image of Self. In: Datson, L. (Ed.), Things That Talk: Object Lessons from Art and Science. Zone Books, New york, pp. 257–294.

Galison, P., 2006. Breakdown. In: Selinger, E. (Ed.), Postphenomenology: A Critical Comparison to Ihde. State University of New York Press, New York, pp. 161–171.

Galison, P., 2012. Black-out spaces: Freud, censorship and the re-territorialization of mind. Br. J. Hist. Sci. 45 (2), 235–266.

Geertz, C., 2000[1983]. Local Knowledge: Further Essays in Interpretive Anthropology, third ed. Basic Books Classics, New York.

Gusterson, H., 2011. Social containment of disaster. Anthropol. News 52 (5), 4.

Haffner, J., 2013. The View from Above: The Science of Social Space. The MIT Press, Cambridge.

Japan Association of Science & Technology Journalists, 2013. A Thorough Validation of the Fukushima Nuclear Disaster: What Was the Problem [Tettei Kensho Fukushima Genpatsujiko: Naniga Mondai Dattanoka]. Kagaku Doujin, Kyoto.

Kahoku, S., 2011. Testimonies: The Lack of Knowledge Led Us to Escape to a Highly Contaminated Area—The 800 Residents of the Coast Side of Namie, Fukushima. Accessed on 18.11.13. at http://www.kahoku.co.jp/spe/spe_sys1071/20111109_01.htm.

Kurgan, L., 2013. Close up at a Distance: Mapping, Technology, and Politics. Zone Books, New York.

Lynch, K., 1960. The Image of the City. MIT Press, Cambridge.

Minamisoma City, 2012. Regarding the Setting of Evacuation Preparation Zone in Minamisoma City. http://www.city.minamisoma.lg.jp/index.cfm/10, 3190,c,html/3190/kuiki-settei.pdf.

Nagel, T., 1989. The View from Nowhere. Oxford University Press, Oxford.

Parmentier, R.J., 1997. The pragmatic semiotics of cultures. Semiotica 116 (1), 1–115 (special issue).

Philips, S., 2013. Fukushima is not Chernobyl? Don't be so sure. Accessed on 18.11.13. at http://somatosphere.net/2013/03/fukushima-is-not-chernobyl-dont-be-so-sure.html.

Prime Minister and His Cabinet, 2012. The Number 7th Nuclear Accident Taskforce Meeting. Accessed on 18.11.13. at. http://www.kantei.go.jp/jp/singi/genshiryoku/pdf/gensai_gaiyo_07.pdf.

Rengert, F.G., Pelfrey, V.W., 1997. Cognitive mapping of the city center: comparative perceptions of dangerous places. In: Weisburd, D., McEwen, T. (Eds.), From Crime Mapping and Crime Prevention. Criminal Justice Press, New York, pp. 193–217.

Sato, Y., 2012. Why Was SPEEDI not Used? [Houshasen Kakusan Yosoku Shisutemu SPEEDI, Naze Katsuyou Sarenakattanoa] Tōyō Shoten, Tokyo.

Silverstein, M., 2013. Discourse and the no-thing-ness of Culture. Signs Soc. 1 (2), 327–366.

Solnit, R., 2009. A Paradise Built in Hell: The Extraordinary Communities that Arise in Disaster. The Viking Press, New York.

Staring, R., 1998. Rethinking the power of maps: some reflections on paper landscapes. Ecumene 5 (1), 105–108.

Sugenoya, A., 2012. In Order to Get along with Radioactivity for 100 years from Now [Korekara Hyakunen Houshanouto Tsukiau Tameni]. Aki Shobou, Tokyo.

The Reconstruction Agency, 2013. Preliminary Report of the Minamisoma City Population Survey Concerning the Attitudes and Aims of Residents. http://www.reconstruction.go.jp/topics/main-cat1/sub-cat1-4/ikoucyousa/20131101_ikouchousa_sokuhouminamisouma.pdf.

Tierney, K., 2008. Hurricane Katrina: catastrophic impacts and alarming lessons. In: Quigley, J.M., Rosenthal, L.A. (Eds.), Risking House and Home: Disasters, Cities, Public Policy. Public Policy Press, Berkeley.

Tolman, E.C., 1948. Cognitive maps of rats and men. Psychol. Rev. 55, 189–208.

Watanabe, 2012. Kayahara Number 5. The Special Edition on the Great East Earthquake and Tsunami. Kashima Cultural History Club, Kashima.

Weisenfeld, G., 2012. Imagining Disaster: Tokyo and the Visual Culture of Japan's Great Earthquake of 1923. University of California Press, Berkeley.

Wood, D., Fels, J., Krygier, J., 2010. Rethinking the Power of Maps. The Guilford Press, New York.

Yomiuri, S., 2013. Nuclear Regulation Authority Indicates 20 mSv Is Safe [20 miri shiberutoikade anzen-kisei iinkaiga shishin].

Yoshida, K., Harada, S., 2012. An outline of evacuation processes among surrounding municipalities of the Fukushima Daiichi Plant [Genpatsusyuuhenjichitai no Hinan no Keii]. In: Yamashita, Y., Kainuma, H. (Eds.), A theory of Nuclear Evacuation [Genpatsu hinan Ron]. Akashi Shoten, Tokyo. pp. 365–389.

Chapter 20

Therapeutic Communities in the Context of Disaster

Brenda D. Phillips
Ohio University—Chillicothe, OH, USA

ABSTRACT
People traumatized by disaster also report that healing effects come to them as well. Known as either the therapeutic or altruistic community, survivors describe beneficial effects from interacting with those who come to help. This chapter describes the historical context of the therapeutic community, conditions that give rise to therapeutic benefits, and the range of the community's effects. The chapter also critiques the concept of a therapeutic community for empirical investigation. Relying on the extant, albeit limited, literature, the chapter also explains the ways in which practitioners might generate and use a therapeutic community.

Research suggests that postdisaster communities can be *corrosive*, such as when conflict erupts, or *therapeutic*, which occurs when an outpouring of aid arrives to address loss. Presumably, a therapeutic community can at least temporarily and positively transform disaster-stricken communities. To better understand a therapeutic community, this chapter defines the concept along with the conditions under which such a community arises and its consequences. The concept originally arose from the literature and practice within social work, mental health, and addiction treatment professionals.

Most disaster literature places the therapeutic effect within a limited time frame postimpact. Relevant examples include response time volunteerism, which tends to be fairly short lived but impactful. Such volunteerism is assumed to arise out of socially altruistic behaviors motivated by personal and collective belief systems. Other conditions that push people forward include influences by media accounts and by agents of socialization particularly peers, family, and faith connections.

Compelled by societal values and significant others that influence behavioral reactions, volunteerism reflects and appears to generate beneficial effects. Primarily, qualitative studies have described those consequences. Those effects seem to include a sense of relief by people facing the daunting tasks of

Hazards, Risks, and Disasters in Society. http://dx.doi.org/10.1016/B978-0-12-396451-9.00020-2

recovery, the restoration of hope by those dealing with despair, and a renewal of belief in the kindness of the human race.

These general accounts, though, do not move beyond the response time period. Less is known about the impact of local participation or the value for high-risk populations. Certainly, disasters reveal deeply entrenched social problems. Socially stratified people face more significant impacts and extenuated recoveries. However, research has yet to consider if therapeutic effects also reflect stratification effects. Nor has the disaster community examined the value of intentionally building a therapeutic community for targeted use as has been done within the addiction treatment community.

Accordingly, this chapter reviews the extant literature and generates recommendations in order to (1) elucidate the conceptual properties of a therapeutic community; (2) discuss the practical application of a therapeutic community; and (3) spur longitudinal research, especially for those at the highest risk for a lingering recovery.

20.1 DEFINING THE "THERAPEUTIC COMMUNITY"

Two distinct and rarely overlapping bodies of literature exist on the concept of the therapeutic community. Originally, the term came out of addiction treatment as an intentional, targeted strategy to help those at risk. Although addiction treatment professionals originated the concept in the 1940s, disaster scholars did not pick up the term until the 1960s and since then only in fairly limited instances. Nonetheless, the notion of the therapeutic community has been embraced as a reflection of typical altruistic response to disasters.

20.1.1 Historical Context of the "Therapeutic Community" Concept

The idea of the therapeutic community originated in the 1940s (Main, 1946) as a way to treat people facing various medical and psychiatric conditions including addiction (Jones, 1953; Manning, 1989; Bunt et al., 2008). Since then, practical, targeted use of residential settings as a means to create a therapeutic community has been both lauded and critiqued. Efforts to establish the concept have included development of practitioner organizations and a cader of academic researchers. For example, The World Federation of Therapeutic Communities became established in 1975, reflecting the creation of therapeutic communities in over 60 nations (Bunt et al., 2008). The *International Journal of Therapeutic Communities* followed with scholarly and applied research. The notion of the therapeutic community now includes not only residential settings (from small groups to total institutions) but also as a way of viewing patients or people within the setting (Clark, 1965a,b as noted in Filstead and Rossi, 1973). In time, the concept came to encompass "a method of organizing the social structure of a treatment setting to cultivate and take advantage of natural social relationships" (Filstead and Rossi, 1973, p. 10).

According to the U.S. National Institute on Drug Abuse (2002, p. 1), therapeutic communities are "drug-free residential settings that use a hierarchical model with treatment stages that reflect increased levels of personal and social responsibility" and encapsulate a range of residential settings. Peer interaction within the residential setting, coupled with structured treatment, is intended to move the resident toward "the behavioral skills, attitudes, and values associated with socialized living" (National Institute on Drug Abuse, 1976, p. 2). From one study of >65,000 individuals, therapeutic communities were found to have positive outcomes, with "lower levels of cocaine, heroin, and alcohol use; criminal behavior; unemployment; and indicators of depression than they had before treatment" (U.S. National Institute on Drug Abuse, 2002, p. 2; National Institute on Drug Abuse, 1976). Positive outcomes were "strongly related to treatment duration" of at least 90 days (p. 3). Not only do addicts appear to benefit from a therapeutic community but evidence also suggests that prisoners and gang members who experience a therapeutic community are less likely to repeat crimes or return to addictions (Yablonsky, 1997).

Research has supported the idea that, over time, a therapeutic community becomes a change agent, partly because it is residential (Yablonsky, 1997). Living in a therapeutic community is required, because a residential setting offers frequent, peer-connected interaction with people who have already recovered from addiction. Further, an intentionally constructed and highly structured environment seems to provide therapeutic benefits in terms of reducing risky behaviors. As one researcher noted, in contrast to the pull of addiction, a therapeutic community meets similar needs but "in a humanistic positive way" (Yablonsky, 1997, p. 150). This positive setting then integrates those at risk out of the therapeutic community and into the larger society.

Several aspects of traditional therapeutic communities appear similar to as well as different from disaster therapeutic communities. For instance, people in therapeutic communities remain in a relatively closed society that possesses structure and stability. Disaster-stricken communities are more open and fluid, with less than predictable environments. Furthermore, members of a therapeutic community are expected to participate fully while survivors in a disaster setting serve as more passive beneficiaries from altruistic efforts. Traditional therapeutic communities also have boundaries, tend to be relatively small (40–80) and can be located in neighborhoods, jails, prisons, shelters, and similar settings (U.S. National Institute on Drug Abuse, 2002).

Disaster therapeutic communities are likely to have irregular and unclear boundaries, cover significant geographic areas, and occur in a wide range of previously unspecified locations. Finally, entry into the therapeutic community occurs in stages: induction and early treatment, primary treatment, and reentry into the broader society (U.S. National Institute on Drug Abuse, 2002).

One might argue that disaster recovery also occurs in phases: short-term recovery, long-term recovery, and a return to a new normal (NHRAIC, 2005; Phillips, 2009).

20.1.2 The Therapeutic Community in Disaster Studies

Classic works in disaster studies describe a set of collective actions that seem to deliver psychologically beneficial, communitywide effects (e.g., Fritz, 1961; Barton, 1970; Quarantelli and Dynes, 1976). Termed alternatively as the "therapeutic community," or the "altruistic community," the concept encapsulates an *effect*. Implicit within discussions of therapeutic communities is the notion of social actors engaged in collective work toward some common, public good. Thus, the concept covers a range of action from individual concerns over those affected by disaster, to the arrival of volunteers who take action, to a collective feeling of goodwill or hope within the community they serve.

A therapeutic community "represents informal mass social and physical support; victims are rescued, sheltered and reassured by fellow community members" (Perry and Lindell, 2003, p. 110). The pain of the disaster, in Barton's altruistic community, is overcome by people who turn out to help. Barton's work (1970) represents the most thorough initial discussion of this phenomenon to date, which he named the altruistic community. In an extensive examination, Barton discussed 71 propositions about the therapeutic community. Of these 71 propositions, 39 focus on the individual, 23 on "contextual relationships" and 9 on "collective-level relationships" from "the simple and fairly obvious to the complex" (Quarantelli and Dynes, 1977, p. 26). Yet, as Barton (1969) states and Quarantelli and Dynes (1977, p. 26) note, "almost none of the propositions rests on solid empirical data." Barton's proposed deductive approach, interestingly, stands in contrast to the majority of therapeutic community studies that have been qualitative and inductive. Still, the number of studies remains quite limited. Early researchers generally seemed to assume that the altruistic or therapeutic community occurred in the location where the disaster happened. Some writers have contrasted a therapeutic with a corrosive community. Others have situated the debate within the type of disaster. All studies involve volunteerism of some kind.

In contrast to the assumption that therapeutic communities arise within the disaster zone, Miller (2007) examined what happened when hurricane Katrina displaced nearly one million people to new sites across the United States. Despite a distance of 365 miles (587 km) from the impact site, a therapeutic community seems to have emerged. Proximity, then, does not seem to be necessarily related to a therapeutic community (Miller, 2007). Nonetheless, normative mechanisms in this instance did influence altruistic responses, which were consistent with locally held values (Miller, 2007).

Similarly, Defede (2002, see also Scanlon, 2003) observed what happened when planes landed in Gander, Newfoundland (Canada), on September 11,

2001, after being denied entry into the United States. Though arriving passengers were not directly harmed by the events of September 11, they did need food, shelter, and medical care. The host community responded, assisting displaced passengers in need. In a different situation, the explosion of Swissair Flight #111 in 1998 compelled an impacted community to step up and support arriving responders, investigators, family, and media. Their experience as a host community and impact site resulted in unrecognized and untreated trauma from exposure to human remains and fatigue from dealing with outsiders. In a small sample study (N = 13), investigators determined that 46 percent of those interviewed met criteria for posttraumatic stress disorder, which was attributed to a lack of resources for processing a traumatic experience (Stewart et al., 2004). Clearly, efforts to create a therapeutic community turned out to have some detrimental effects.

Some writers have also contrasted a therapeutic community with a corrosive community. Essentially, the two types of postdisaster communities are treated like polarized opposites. In contrast to the therapeutic community, a corrosive community is "characterized by a loss of trust...a perceived loss of charity, concern, empathy and recovery resources; a fragmentation of community groups; and a breakdown of social relationships, both personal and institutional" (Jenkins, 2013, p. 405). As Miller (2006, p. 71) observed in his own community, "a corrosive community is characterized by social disruption, a lack of consensus about environmental degradation, and general uncertainty." Corrosiveness seems to appear when victims suffer mentally and physically, when government or organizations fail, and when litigation develops as documented after the *Exxon Valdez* oil spill in Alaska (Picou et al., 2004).

The United States witnessed such corrosiveness after hurricane Katrina in 2005 when residents and advocates raised accusations of racism and lack of concern for those affected. Allegations resulted in inquiries before the United States. Congress and changes within emergency management agencies (U.S. House of Representatives, 2006). Prior to Katrina, blaming erupted after September 11, 2001, in multiple nations. Corrosive relationships resulted in targeted hostilities toward Muslims who experienced increases in hate crimes including murder (Peek, 2011; Poynting and Noble, 2004).

For some researchers, technological disasters seem far more likely to generate a corrosive community (Freudenberg, 1997, 2000). Technological disasters seem to incorporate an element of blaming, such as when an oil spill occurs, a levee fails, or a corporation is perceived to be negligent (Blocker and Sherkat, 1992; Picou et al., 2004). One study did not find a therapeutic community when a technological disaster involving asbestos and pesticide occurred. Instead, conflict erupted into an "emotional climate" of "anger, frustration, resentment, bitterness, and anxiety" (Cuthbertson and Nigg, 1987, p. 480). Victim clusters emerged, generating an atmosphere of nonsupport among the divided groups.

Why does a therapeutic community not appear as commonly in technological disasters? Those who might otherwise participate in a therapeutic community may hold the view that the agent blamed for the disaster should make reparations. Indeed, "in the aftermath of a catastrophic technological failure or toxic contamination...the defining characteristic of the postdisaster phase is the emergence of a corrosive community" (Picou et al., 2004, p. 1496).

What seems consistent is that corrosiveness occurs when systems fail people at risk and injured parties demand accountability. Yet even within corrosive communities, therapeutic communities can still emerge. An appropriate example is hurricane Katrina in the United States, which generated significant conflict over levee failures and the deaths of 1,800 people and thousands of pets. Despite and perhaps because of the corrosive context, a massive, historic outpouring of volunteer effort created a therapeutic community with both tangible (repairs and rebuilt homes) and intangible (hope over despair) outcomes (Miller, 2007; Phillips, 2013; Jenkins, 2013). It may be that these two contrasting communities fall along a continuum (Gill, 2007) that stretches from conflict to consensus, with the two types of communities able to exist simultaneously or to follow on each other. Research remains to be done to sort out these significant and important distinctions between types of communities and the conditions under which they occur.

20.2 CONDITIONS UNDER WHICH THE THERAPEUTIC COMMUNITY ARISES IN DISASTERS

Studies that have examined therapeutic community effects typically do so within the fairly limited time frame of "response." Usually, response is defined as "actions taken at the time a disaster strikes that are intended to reduce threats to life safety, to care for victims, and to contain secondary hazards and community losses" (Tierney et al., 2001, p. 81). A number of social actions can be anticipated within the response time framework, which includes the arrival of spontaneous, unaffiliated volunteers (Orloff, 2011; Neal, 1993) as well as long-term, affiliated volunteers (Britton, 1991; Neal, 1993, 1992). It is the arrival of these volunteers that seems to precipitate the emergence of a therapeutic community.

In addition, certain conditions are believed to influence the appearance of a therapeutic community. First, prospective volunteers must hear about a need such as those generated by a disaster (Barton, 1970). They do so through both formal and informal means of communications. Although Barton wrote about more traditional means of communication (television, radio, and newspapers) today's more rapid dissemination through social media needs to be taken into account (Palen, 2008; Sutton et al., 2008; Palen and Hughes, 2009; Sutton, 2010). But just hearing about the event through formal means may not be enough to compel people to act. Hearing it from unofficial sources, particularly friends and family, seems to reinforce a belief that one must act.

Informal information sources, particularly family and friends, come from one's primary agents of socialization. Socialization is defined as the way that one is raised, including the views instilled in children by parents, teachers, clergy, or peers over a lifetime (Clausen, 1968). Socialization is informed by one's culture and especially by culturally influenced values. Value systems and related ideologies provide a meaningful framework in which to justify one's actions to help. Values define what is desirable or undesirable, such as helping people in distress. Values influence the content of norms, which are defined as behavioral guidelines, or how to act (Scott and Marshall, 2009). In discussing therapeutic communities, Barton (1970) described a *normative mechanism* that compels people toward prosocial behavior in disasters. The combination of these normative mechanisms and sociocultural imperatives produce altruistic responses to those in distress. Value-based, normatively driven behaviors seem to produce something more than the sum of the actions when volunteers help disaster volunteers. People describe general feelings that social and psychological benefits have resulted, which disaster researchers have named the therapeutic or altruistic community.

Such a community seems to exist experientially, socially, and psychologically. In one study, respondents reported feeling a sense of hope being restored after sometimes year of postdisaster despair (Phillips, 2013). Their comments juxtaposed heartfelt appreciation of volunteers coming to help total strangers with stories of fraudulent contractors and lack of government aid. In this post—hurricane Katrina study, respondents also situated their positive feelings vis-a-vis the corrosive community experienced right after the disaster. Charges of racism, homophobia, ableism, and injustice marked those days and historically marginalized populations felt abandoned again (Stukes, 2013). To make things worse, hurricane Rita followed two weeks after Katrina. Rita survivors felt abandoned by government and organizations focused on Katrina. Then, hurricane Ike happened in 2008 and struck parts of Texas and Louisiana. Respondents affected by Rita and Ike felt remembered by volunteers and described their prosocial actions as caring and uplifting in contrast to the neglect they felt from government.

Religious organizations and faith traditions also foster resilience in the face of disaster (Phillips and Thompson, 2013). Pastoral staff and those who provide spiritual care in particular promote healthy responses to grief through helping others to process their experiences (Chinnici, 1985; Jordan, 1976; Roberts and Ashley, 2008). All faith traditions offer some kinds of healing philosophies or rituals that can help people. Buddhists, for example, rely on the belief of karma (De Silva, 2006). Christians use prayer as a form of a "hazard adjustment" in the face of disaster impacts (Mitchell, 2003). Mechanisms within such faith traditions represent a form of cultural capital that can be relied on to prompt therapeutic benefits (Lawson and Thomas, 2007).

20.2.1 Why Do People Volunteer?

The development of a therapeutic community, then, seems to inherently rest upon socially altruistic responses, particularly postdisaster volunteering. Personal contact through volunteerism seems to enhance a feeling of "sympathetic identification" with disaster victims (Barton, 1970). Sympathetic identification prompts a sense of connection or responsibility for those who are harmed by disasters. Therapeutic community studies suggest that in response to socialization agents, cultural imperatives, normative mechanisms, and sympathetic identification, people turn out to help. Why else do people volunteer?

Studies identify several patterns associated with volunteerism. For example, parental and peer socialization is associated with volunteerism that is consistent with what researchers propose as a condition influencing the generation of therapeutic communities (Park and Jerry, 2000). Parents, peers, and other agents of socialization raise children to be of service to others and to cooperate toward common, public needs. Civic involvement norms characterize most societies, with people supporting civic associations, nongovernmental organizations, and voluntary efforts. Australian researchers, for example, found that volunteers felt personally obligated to help others (Amato et al., 1984). It would seem then that most societies have the potential to produce a therapeutic community.

Additional demographic variables influence people who turn out to volunteer. One pattern associated with volunteerism is gender. Women tend to volunteer more than men do, though gendered patterns diminish somewhat in Europe (Ruiter and DeGraaf, 2006). In disasters, as in more general opportunities for service work, women seem to be more likely to engage in emotional tasks and less likely to be placed into leadership positions. Despite a gendered division of labor, then, women seem well placed to help generate a therapeutic community. One's place in the life cycle may influence their volunteerism, with mothers of young children less able to volunteer (Wilson, 2000). Age influences volunteerism too, with younger and older people more likely to be able to volunteer. Service drops off during the middle years of life when people become busily engaged in raising families and working.

Race also may impact volunteerism, with possible consequences for what kinds of populations receive benefits from the therapeutic community (see upcoming section). Studies demonstrate that "much volunteer recruitment is intraracial" (Musick et al., 2000, p. 1560). What overcomes this pattern is asking people to volunteer. Such requests may be influenced by social networks, values and ideologies as "blacks are more influenced by their church than are whites" (Wilson, 2000, p. 228; see also Musick et al., 2000).

Faith also is associated with volunteerism, in part because of the influential socialization that occurs within a religious context (see Figure 20.1; Becker and Dhingra, 2001; Wilson and Musick, 1997). Faith provides a framework in

FIGURE 20.1 Mennonite Disaster Service volunteers rebuild a home destroyed along the U.S. Gulf Coast. *Photo courtesy: Mennonite Disaster Service.*

which people hear messages encouraging them to go forth and serve. However, studies also show that messages alone remain insufficient to compel people to act. Asking people to go on mission trips, disaster response teams, and similar efforts matters. Those who go forth to serve are more likely to do so with fellow believers (Ruiter and DeGraaf, 2006). Phillips (2013) found this to be true among the 852 survey respondents who volunteered with Mennonite Disaster Service after hurricanes Katrina, Rita, and Ike. Many went with mission teams from within their own congregations.

In addition to being motivated to serve others, people derive benefits from their service. Volunteers after September 11th, for example, reported feeling a stronger solidarity with others (Lowe and Fothergill, 2003). These largely intangible benefits plant a seed for future service. McAdam (1988) uncovered this when assessing volunteers for the civil rights movement in the United States. Women who volunteered in 1965 reported higher levels of later service work decades later. The consequences of volunteerism, then, seem to exist for both the giver as well as for the receiver within the therapeutic community.

20.3 CONSEQUENCES OF THE THERAPEUTIC COMMUNITY

What seems consistent in depictions and descriptions of a therapeutic community is how people see their lives transformed. They move from chaos into order, from isolation into interaction, and from exploitation into trust (Phillips, 2013). Landscapes change, from debris-filled neighborhoods where residents cannot even find their own homes to locations where reconstruction can now commence. People's outlooks change, from despair to hope (Phillips, 2013) and from the edge of suicide to stepping back from the metaphorical ledge (Kako and Ikeda, 2009). Indeed, the therapeutic community has been described as influential in reducing potentially negative psychological impacts (Perry and Lindell, 1978; see also Fritz, 1961; Barton, 1969).

The core of the therapeutic community seems to be caring. What recipients of volunteerism report is being uplifted and, in their own words, "healed" from a traumatic, disorienting experience (Phillips, 2013). The effects seem similar to the essence of caregiving, which "is about acknowledgment, concern, affirmation, assistance, responsibility, solidarity, and all the emotional and practical acts that enable life" (Kleinman and van der Geest, 2009, p. 161). Caregiving "happens when hope and consolation are abandoned" and "when all there is to do is to be present with the sufferer, sharing his/her suffering by simply and usually silently just being there" (Kleiinman and van der Geest, 2009, p. 161). The notion of caregiving is "concern and compassion, and, in a larger sense, love" (Kleinman and van der Geest, 2009, p. 161). Simply put, people impacted by disasters feel embraced.

Volunteers seem to benefit too. After hurricane Katrina, volunteers reported feeling a deeper connection to their faith, a greater understanding of disaster impacts, and a closer connection across cultural divides (Phillips, 2013). Non-disaster studies suggest that volunteers likely experience physical and psychological benefits (Wilson and Musick, 1997). Some occupational benefits may be experienced too, when volunteers learn new skills, develop leadership capacities, and transfer those abilities to work settings (Wilson and Musick, 1997).

Generally speaking, the consequences of the therapeutic community are described in rather positive and perhaps idealistic terms. A more critical examination, however, raises important questions. For example, negative consequences can also appear with the onset of a therapeutic community. In most disasters, significant numbers of volunteers arrive unannounced. Such *personnel convergence* (Fritz and Marks, 1954) has great potential to remove debris, muck out homes, and move people from response into recovery (see Figure 20.2).

FIGURE 20.2 After a disaster, many volunteers arrive to help with debris. *Photo courtesy: Mennonite Disaster Service.*

Researchers, though, know well that spontaneous, unplanned volunteers or "SUVs" can also cause problems when beleaguered communities cannot meet the logistical needs of unexpected help (Orloff, 2011; Neal, 1993).

Their appearance also raises the question of *whose* therapeutic community do the volunteers create? Certainly, the overwhelming number of volunteers (and even donations) creates challenges for emergency managers, volunteer center coordinators, and hosting organizations. Those tasked with managing disasters may experience convergence as a problem rather than a psychologically beneficial relief to be experienced. To date, then, studies of disaster-time therapeutic communities emanate from the perspectives of aid recipients rather than from the viewpoint of organizations, agencies, and governments.

But even with a focus on recipients, researchers need to ask *which* recipients. For example, must one be a direct beneficiary of aid to experience the therapeutic community? Do people outside of the direct beneficiaries, such as nondisaster affected residents, also vicariously experience a therapeutic effect? Further, who is the *community*? Disaster scholars have raised this question previously, and noted that diverse communities actually exist within any given geographic location including those impacted by disasters (Marsh and Buckle, 2001).

Conversely, extant studies seem to imply that a homogeneity exists within a therapeutic community. Given the reality that any community is actually composed of diverse communities within it—how uniformly is the therapeutic community experienced? Are there consequences because of the characteristics of the arriving volunteers?

Following hurricane Hugo in the United States in 1989, one study found that higher than typical levels of aid were rendered. However, they also discovered that "mutual help is not distributed equally or randomly" (Kaniasty and Norris, 1995, p. 474). Examining a therapeutic, altruistic or "emergent helping community" (p. 466, terms used interchangeably in the study), researchers found that personal characteristics influenced the amount and kind of aid received. Simply put, whites received more tangible support than blacks and women received more than men when researchers controlled for disaster losses. Similar to disaster studies, concern exists within the therapeutic community setting for groups that may have population-specific needs. Interestingly, women-only programs have reported more success than mixed-gender programs (U.S. National Institute on Drug Abuse, 2002). Why? Because women-only programs tend to have more services specific to women, something that stands in contrast to critiques of disaster response and recovery (e.g., Enarson and Morrow, 1998).

Thus, a more critical examination of the therapeutic community concept suggests that while useful consequences develop, the same inequities that plague disasters may reappear in postdisaster volunteerism. In addition, the therapeutic community should not be viewed as entirely beneficial although it can have considerable practical qualities should it appear.

20.4 PRACTICAL IMPLICATIONS

From a practical perspective, can an intentional therapeutic community be built for targeted use? Certainly this is the case within the practitioner community treating people with medical, psychiatric, or addictive health issues. As Jones (1973, p. 335) noted, the nondisaster therapeutic community practitioner sector "represented an early model for change" that reflected the "direction of a more value-conscious society." Can something similar be created within disaster-stricken locations?

The implications of building an intentional therapeutic community, similar to what is seen in residential treatment facilities, are certainly provocative. For example, in a corrosive community might the intentional construction of a therapeutic community reduce conflict? Ethically, would it be appropriate to do so? Conflict might in reality be an emergent activity pointing out critical but unmet needs. Creating a therapeutic community might reduce conflict and also thwart emergence essential for underserved populations and areas. Conversely, creating an intentional therapeutic community could benefit those with unmet needs if a manager or official works to connect arriving volunteers with those hit hard by disaster.

Regardless of the likely challenges associated with intentional development and targeted use of a therapeutic community, doing so seems wise within certain caveats. First, the therapeutic community often synonymously develops with the arrival of volunteers who are eager to help. Advance planning for the arrival of spontaneous unplanned volunteers is critical in order to leverage the social capital they bring (Orloff, 2011; Neal, 1993). In a catastrophic event, volunteers would likely need some degree of training and certainly would benefit from integration into organized efforts that match with local needs (Clizbe, 2004; Paton, 1994, 1996).

However, volunteers tend to leave and disaster volunteerism declines when media coverage wanes. Sustaining a therapeutic community would demand a collaborative, coordinated effort with established disaster volunteer organizations. One means to do this in the United States has been through encouraging long-term recovery groups or committees (LTRCs). Traditionally, the Federal Emergency Management Agency tasks a Voluntary Agency Liaison with helping voluntary and faith-based organizations to connect and coordinate. One outcome may be the LTRC or a similar entity. When the LTRC meets, local leaders convene and lead the meeting, where outside organizations (e.g., Presbyterian Disaster Assistance, Adventist Disaster Response, the Buddhist Tzu Chi Foundation, or the Humane Society) discuss a local family's needs. These organizations then pool and/or leverage their resources and volunteer labor to tackle repairs or reconstruction. In one five-year study of a faith-based organization, these types of outsider–insider relationships as enacted through the LTRC proved critical. Both aid recipients and LTRC members remarked on experiences consistent with those described as part of a therapeutic community (Phillips, 2013). Practitioners interested in fostering a therapeutic community would be well advised to work with established disaster volunteer organizations that bring in supervised volunteers over a long period of time (see Figure 20.3).

Internationally, additional admonitions should be offered. When disasters strike nations with meager resources, outsiders tend to assume what type of aid should be offered, how, and when. However, making such assumptions can actually cause significant logistical problems for locals and may be culturally, socially, politically, and economically insensitive (Phillips and Thompson, 2013; Kumaran and Torris, 2011; Sugimoto et al., 2011). Not surprisingly, intentional efforts aimed at producing a therapeutic community are likely to fail. Recommendations to overcome outsiders' misguided efforts to help should focus on supporting local organizations (Real, 2010). The diversity of faith traditions, reflected also in cultural diversity, suggests that a range of frameworks would likely be needed in generating a therapeutic community (Rowell et al., 2011; Schreurs, 2011; Gillard and Paton, 1999).

FIGURE 20.3 Volunteers associated with experienced disaster service organizations provide the most effective kinds of assistance. *Photo courtesy: Mennonite Disaster Service.*

The element of time matters also in disasters (Neal, 2013). Even though the arrival of response time volunteers seems to produce a therapeutic community, the majority of postdisaster needs occur in the recovery phase (Quarantelli, 1999; Mileti, 1999; Phillips, 2009). Thus, sustaining a labor force well-suited to aiding people appropriately over time emerges as a critical challenge for practitioners. What might be needed to resource a therapeutic community? For how long? The arrival of outsiders can be problematic, when locals do not know whom to trust or when some outsiders appear to arrive for a photograph opportunity and then leave (Phillips, 2013). It takes time for trust to develop, especially so if a corrosive community has been in place prior to and especially after a disaster. What seemed to work after hurricanes, Katrina, Rita, and Ike, was the arrival of a well-organized set of volunteers operating consistently out of an established belief system focused on service. Oversight from the local community mattered (Phillips, 2013).

20.5 RESEARCH RECOMMENDATIONS

Practitioners benefit from well-done, empirical research that provides trustworthy guidance. Unfortunately, despite positive depictions of the therapeutic community, research remains considerably underdeveloped. Such research is needed. Sound, empirical research begins with a well-considered conceptual definition of the phenomenon under study. Yet, even within the practitioner sector that treats addictions and medical problems, concern remains over the conceptualization of "therapeutic community." As Filstead and Rossi (1973, p. 6) explain, "The concept in terms of a precise definition and method of implementation has never been quite clear." The same is true in disasters including the use of the term: therapeutic or altruistic community? Given the lack of psychological measurement for therapeutic effects, it would seem that the term altruistic community might be more appropriate. Disaster researchers have had more success in studying altruistic responses, although even disaster volunteerism requires further study. And, in studying the phenomenon of a therapeutic or altruistic community, which conditions or variables should be pursued first? Disaster volunteerism, while perhaps the most easily measured, may only be part of why the phenomenon appears.

In studying the phenomenon further, researchers must decide if the therapeutic (or altruistic) community is a truly distinct phenomenon? Researchers have clearly documented that many communities experience a decrease in community conflict and an increase in helping behavior (Dynes and Quarantelli, 1976; Quarantelli and Dynes, 1977; Perry and Lindell, 2003; Rodríguez et al., 2006). Is this decrease in predisaster conflict associated with the emergence of a therapeutic community? Could they be synonymous? Given that decreases in community conflict are usually short lived and that studies of therapeutic communities tend to focus on short time frames (e.g., the response phase), how can we know for sure? Does the appearance of a therapeutic or altruistic community counter corrosiveness? If so, in what ways, to what extent, and for which populations? Past studies suggest that predisaster patterns (e.g., community conflict) will disappear after disaster but then reappear in time (Tierney et al., 2001). Does the appearance of a therapeutic community influence the time and trajectory of that return to predisaster patterns—or does it have the potential to transform a community completely to a new, postdisaster pattern?

Additional questions remain, suggesting that much research remains to be undertaken. For example, under what conditions does the therapeutic or altruistic community develop, endure, and survive? How can the presumed social and psychological benefits be measured to capture both tangible (e.g., debris removal) and intangible consequences (e.g., the restoration of hope). Clearly, crossdisciplinary research is needed to address these kinds of questions.

Given the concerns raised here over equity, how widespread is the therapeutic or altruistic community? Is the "feeling" of a therapeutic effect widely disseminated or experienced only by direct recipients? Are people who hear about good works like postdisaster volunteerism indirectly impacted by what they see and hear? How do other actors involved in the therapeutic community experience its effects, including emergency managers and volunteer coordinators? What is the role of local committees who manage long-term recovery in the evolution and maintenance of a therapeutic or altruistic community over time, particularly those that coordinate with outsiders?

Because of these lingering questions and the fact that most studies focus on response time, longitudinal research should be conducted. As Perry and Lindell (2003, p. 53) note, "research on the persistence of the therapeutic community response has been insufficient to permit confident acceptance of Barton's hypothesis of long-term persistence." Perry and Lindell (1978, p. 106) suggest that what is needed is "an explicit, formal, theoretical framework which could guide the directions of scientific inquiry and against which empirical findings may be evaluated." In addition, research must incorporate the concerns of practitioners who seek to generate and apply the positive consequences of a therapeutic or altruistic community for the greater good of all affected by disasters.

REFERENCES

Amato, P.R., Ho, R., Partridge, S., 1984. Responsibility attribution and helping behaviour in the ash wednesday bushfires. Aust. J. Psychol. 36 (2), 191–203.

Barton, Alan, 1969. Communities in Disaster. Anchor Books, NY.

Barton, A., 1970. Communities in Disaster: A Sociological Analysis. Anchor Books, Garden City, NY.

Becker, P.E., Dhingra, P.H., 2001. Religious involvement and volunteering: implications for civil society. Sociol. Relig. 62 (3), 315–335.

Blocker, T.J., Sherkat, D.E., 1992. In the eyes of the beholder: technological and naturalistic interpretations of a disaster. Organ. Environ. 6 (2), 153–166.
Britton, N.R., 1991. Permanent Disaster Volunteers: Where Do They Fit? Nonprofit and Voluntary Sector Quarterly 20 (4), 395–414.
Bunt, G.C., Muehlbach, B., Moed, C.O., 2008. The therapeutic community: an international perspective. Subst. Abuse 29 (3), 81–87.
Chinnici, R., 1985. Pastoral care following a natural disaster. Pastoral Psychol. 33 (4), 245–254.
Clark, D.H., 1965a. The developing concept of the therapeutic community. Psychother. Psychosom. 13, 238–245.
Clark, D.H., 1965b. The therapeutic community—concept, practice, and future. Br. J. Psychiat. 111, 947–954.
Clausen, J.A. (Ed.), 1968. Socialization and Society. Little, Brown, and Company, Boston.
Clizbe, J.B., 2004. Challenges in managing volunteers during bioterrorism response. Biosecur. Bioterrorism: Biodef. Strategy, Pract. Sci. 2 (4), 294–300.
Cuthbertson, B., Nigg, J.M., 1987. Technological disaster and the nontherapeutic community: a question of true victimization. Environ. Behav. 19 (4), 462–482.
De Silva, P., 2006. The tsunami and its aftermath in Sri Lanka: explorations of a buddhist perspective. Int. Rev. Psychiat. 18 (3), 281–297.
Defede, J., 2002. The Day the World Came to Town: 9/11 in Gender. Harper Colllins, Newfoundland, NY.
Dynes, R., Quarantelli, E.L., 1976. The family and community context of individual reactions to disaster. In: Parad, H., Resnik, L., Parad, L. (Eds.), Emergency and Disaster Management. The Charles Press, Bowie, MD, pp. 231–244.
Enarson, E., Morrow, B., 1998. The Gendered Terrain of Disaster: Through Women's Eyes. Praeger, Santa Barbara, CA.
Filstead, Rossi, J.J., 1973. Therapeutic milieu, therapeutic community and milieu therapy: some conceptual and definitional distinctions. In: Rossi, J.J., Filstead, W.J. (Eds.), The Therapeutic Community: A Sourcebook of Readings. Behavioral Publications, NY, pp. 3–16.
Freudenburg, W.R., 1997. Contamination, corrosion and the social order: an overview. Curr. Sociol. 45 (3), 19–39.
Freudenburg, W.R., 2000. The "risk society" reconsidered: recreancy, the division of labor, and risk to the social fabric. In: Cohen, M.J. (Ed.), Risk in the Modern Age: Social Theory, Science and Environmental Decision-Making. St. Martin's Press, New York, pp. 107–120.
Fritz, C., 1961. Disaster. In: Merton, R., Nisbet, R. (Eds.), Contemporary Social Problems. Harcourt, Brace and World, NY, pp. 651–694. Chapter 14.
Fritz, C., Marks, E., 1954. The NORC studies of human behavior in disaster. J. Soc. Issues 10 (3), 26–41.
Gill, D.A., 2007. Secondary trauma or secondary disaster? Insights from hurricane Katrina. Sociol. Spectrum 27, 613–632.
Gillard, M., Paton, D., 1999. Disaster stress following a hurricane: the role of religious differences in the Fijian Islands. Aust. J. Disaster 2, 2–9.
Jenkins, P., 2013. The nature of human communities. In: Thomas, D.S.K., Phillips, B.D., Fothergill, A., Lovekamp, W. (Eds.), Social Vulnerability to Disasters. CRC Press, Boca Raton, FL, pp. 397–414.
Jones, M., 1953. The Therapeutic Community: A New Treatment Method in Psychiatry. Basic Books, Inc., NY.
Jones, M., 1973. Therapeutic community concepts and the future. In: Rossi, J.J., Filstead, W.J. (Eds.), The Therapeutic Community: A Sourcebook of Readings. Behavioral Publications, NY, pp. 325–336.

Jordan, C., 1976. Pastoral care and chronic disaster victims: the buffalo creek experience. J. Pastoral Care 30 (3), 159–170.

Kako, M., Ikeda, S., 2009. Volunteer experiences in community housing during the great Hanshin-Awaji earthquake, Japan. Nurs. Health Sci. 11 (4), 357–359.

Kaniasty, K., Norris, F.H., 1995. In search of altruistic community: patterns of social support mobilization following hurricane Hugo. Am. J. Community Psychol. 23 (4), 447–477.

Kleinman, A., van der Geest, S., 2009. "Care" in health care: remaking the moral world of medicine. Med. Antrcpol. 221 (1), 159–179.

Kumaran, M., Torris, T., 2011. The role of NGOs in tsunami relief and reconstruction in Cuddalore district, South India. In: Karan, P.P., Subbiah, S.P. (Eds.), The Indian Ocean Tsunami. University of Kentucky Press, Lexington, pp. 183–212.

Lawson, E., Thomas, C., 2007. Wading in the waters: spirituality and older black katrina survivors. J. Health Care Poor Underserved 18, 341–354.

Lowe, S., Fothergill, A., 2003. A need to help: emergent volunteer behavior after September 11th. In: Monday, J. (Ed.), Beyond September 11th: An Account of Post-Disaster Research. Natural Hazards Research and Applications Information Center, Boulder, CO, pp. 293–314.

McAdam, D., 1988. Freedom Summer. Oxford University Press, NY.

Main, T.F., 1946. The hospital as a therapeutic institution. Bull. Menninger Clin. 10, 66–70.

Manning, N., 1989. The Therapeutic Community Movement: Charisma and Routinization. Routledge, London.

Marsh, G., Buckle, P., 2001. Community: the concept of community in the risk and emergency management context. Aust. J. Emerg. Manage. 16 (1), 1324–1540.

Mileti, D., 1999. Disasters by Design. Joseph Henry Press, Washington DC.

Miller, D.S., 2006. Visualizing the corrosive community: looting in the aftermath of hurricane Katrina. Space Cult. 9 (1), 71–73.

Miller, L., 2007. Collective disaster responses to Katrina and Rita: exploring therapeutic community, social capital and social control. Southern Rural Sociol. 22 (2), 45–63.

Mitchell, J., 2003. Prayer in disaster: case study of christian clergy. Nat. Hazards Rev. 4 (1), 20–26.

Musick, M., Wilson, J., Bynum Jr., W.B., 2000. Race and formal volunteering: the differential effects of class and religion. Soc. Forces 78 (4), 1539–1571.

National Institute on Drug Abuse, 1976. The Therapeutic Community: Proceedings of Therapeutic Communities of America Planning Conference. Department of Health, Education and Welfare, Washington DC, US.

NHRAIC (Natural Hazards Reduction and Applications Information Center), 2005. Holistic Disaster Recovery. NHRAIC, Boulder, CO.

Neal, D.M., 1992. The local Red Cross in time of disaster: characteristics and conditions of organizational effectiveness during the Loma Prieta earthquake and central Texas floods. J. Volunteer Adminis 11 (2), 6–16.

Neal, D.M., 2013. Social time and disaster. Int. J. Mass Emerg. Disaster 31 (2), 247–269.

Neal, D.M., 1993. Volunteer organizations in disaster: a comparison of the Red Cross following the Loma Prieta earthquake and central Texas floods (Winter). J. Volunteer Admin. 6–17.

Orloff, L., 2011. Managing Spontaneous Community Volunteers in Disasters: A Field Manual. CRC Press, Boca Raton, FL.

Palen, L., 2008. Online social media in crisis events. Educause Quart. 3, 76–78. Available at. http://net.educause.edu/ir/library/pdf/eqm08313.pdf.

Palen, L., Hughes, A., 2009. Twitter adoption and use in mass convergence and emergency events. In: Paper Presented at the 6th International ISCRAM Conference, Gothenburg, Sweden.

Park, Jerry Z., Smith, Christian, 2000. " 'To Whom Much Has Been Given...': Religious Capital and Community Voluntarism among Churchgoing Protestants." J. Scientific Study Religion 39, 272–286.

Paton, D., 1994. Disaster relief work: an example of training effectiveness. J. Trauma. Stress 7 (2), 275–288.

Paton, D., 1996. Training disaster workers: promoting well-being and operational effectiveness. Disaster Prev. Manage. 5 (5), 11–18.

Peek, L., 2011. Behind the Backlash: Muslim Americans after 9/11. Temple University Press, Philadelphia.

Perry, R.W., Lindell, M.K., 1978. The psychological consequences of natural disaster: a review of research on american communities. Mass Emerg. 3, 105–115.

Perry, R.W., Lindell, M.K., 2003. Understanding citizen response to disasters with implications for terrorism. J. Conting. Crisis Manage. 11 (2), 49–60.

Phillips, B.D., 2009. Disaster Recovery. CRC Press, Boca Raton, FL.

Phillips, B.D., 2013. Mennonite Disaster Service: Building a Therapeutic Community after the Gulf Coast Storms. Lexington Books, Lanham, MD.

Phillips, B.D., Thompson, M., 2013. Religion, faith, and faith-based organizations. In: Thomas, D.S.K., Phillips, B.D., Fothergill, A., Lovekamp, W. (Eds.), Social Vulnerability to Disasters, second ed. CRC Press, Boca Raton, FL, pp. 341–363.

Phillips, Brenda, 2014. Mennonite Disaster Service: building a therapeutic community after the Gulf Coast Storms. Lexington Books, Lanham, MD.

Picou, J.S., Marshall, B.K., Gill, D.A., 2004. Disaster, litigation, and the corrosive community. Soc. Forces 82 (4), 1493–1522.

Poynting, S., Noble, G., 2004. Living with Racism: The Experience and Reporting by Arab and Muslim Australians of Discrimination, Abuse and Violence since 11 September 2001. Report to the Human Rights and Equal Opportunity Commission, Sidney, Australia.

Quarantelli, E.L., 1999. The Disaster Recovery Process: What We Know and Do Now Know from Research. University of Delaware. Disaster Research Center Preliminary Paper #286.

Quarantelli, E.L., Dynes, R.R., 1976. Community conflict: its absence and its presence in natural disasters. Mass Emerg. 1, 139–152.

Quarantelli, E.L., Dynes, R.R., 1977. Response to social crisis and disaster. Annu. Rev. Sociol. 3, 23–49.

Real, A., 2010. Acts of god: the role of religion in disaster risk reduction. Humanitarian Exch. Mag. Available at http://www.odihpn.org/humanitarian-exchange-magazine/issue-48/acts-of-gods-the-role-of-religion-in-disaster-risk-reduction (accessed 9.12.13.).

Roberts, S.B., Ashley, W., 2008. Disaster Spiritual Care: Practical Clergy Responses to Community, Regional, and National Tragedy. Sky Lights Path Publishing, Woodstock, VT.

Rodríguez, H., Trainor, J., Quarantelli, E.L., 2006. Rising to the challenges of a catastrophe: the emergent and prosocial behavior following hurricane Katrina. Ann. Am. Acad. Polit. Soc. Sci. 604, 82–101.

Rowell, R.L., Mercer, L.A., Gichomo, G., 2011. Role of pastors in disasters curriculum development project: preparing faith-based leaders to be agents of safety. J. Homeland Secur. Emerg. Manage. 8 (1), 1–32.

Ruiter, S., De Graaf, N.D., 2006. National context, religiosity, and volunteering: results from 53 countries. Am. Sociol. Rev. 71 (2), 191–210.

Scanlon, T.J., 2003. Helping the other victims of September 11: gander uses multiple EOCs to deal with 38 diverted flights. Aust. J. Emerg. Manage. 18 (2), 108–121.

Schreurs, M., 2011. Improving governance structures for natural disaster response: lessons from the Indian ocean tsunami. In: Karan, P., Subbiah, S. (Eds.), The Indian Ocean Tsunami. University of Kentucky Press, Lexington, pp. 261–282.

Scott, J., Marshall, G., 2009. A Dictionary of Sociology. Oxford University Press, NY.

Stewart, Sherry H., Terry, L. Mitchell, Kristi, D. Wright, Pamela, Loba, 2004. The Relations of PTSD Symptoms to Alcohol Use and Coping Drinking in Volunteers who Responded to the Swissair Flight 111 Airline Disaster. J. Anxiety Disord 18 (1), 51–68.

Stukes, P., 2013. Gay Christian Service: Exploring Social Vulnerability and Capacity Building of Lesbian, Gay, Bisexual, Transgender and Intersex Identified Individuals and Organizational Advocacy in Two Post-Katrina Disaster Environments. Department of Sociology, Denton, Texas. Doctoral Dissertation.

Sugimoto, S., Sagayaraj, A., Sugimoto, Y., 2011. Sociocultural frame, religions networks, miracles: experiences from tsunami disaster management in South India. In: Karan, P.P., Subbiah, S.P. (Eds.), The Indian Ocean Tsunami. University of Kentucky Press, Lexington, pp. 213–235.

Sutton, J., Palen, L., Shklovski, I., 2008. Backchannels on the front lines: emergent uses of social Media in the 2007 Southern California wildfires. In: Paper Presented at the 5th International ISCRAM Conference, Washington DC.

Sutton, J.L., 2010. Twittering Tennessee: distributed networks and collaboration following a technological disaster. In: Proceedings of the 7th International ISCRAM Conference, Seattle, Washington.

Tierney, K., Lindell, M., Perry, R., 2001. Facing the Unexpected. Joseph Henry Press, Washington DC.

U.S. House of Representatives, 2006. A Failure of Initiative: Final Report of the Select Bipartisan Committee to Investigate the Preparation for and Response to Hurricane Katrina. U.S. House of Representatives, Washington DC.

U.S. National Institute on Drug Abuse, 2002. Therapeutic Community. Research Report Series, NIH Publication #02–4877. U.S Department of Health and Human Services, Washington DC.

Wilson, J., 2000. Volunteering. Annu. Rev. Sociol. 26, 215–240.

Wilson, John, Musick, Marc, 1997. Who Cares: Toward and Integrated Theory of Volunteer Work. Amer. Sociol. Rev. 62 (5), 694–713.

Yablonsky, L., 1997. Gangsters: Fifty Years of Madness, Drugs and Death on the Streets of America. New York University Press, NY.

Chapter 21

View of Abrahamic Religions on Natural Disaster Risk Reduction

Mohsen Ghafory-Ashtiany
International Institute of Earthquake Engineering and Seismology (IIEES), Tehran, Islamic Republic of Iran

ABSTRACT

The purpose of this chapter is to show the benefit of using the richness of Abrahamic religions toward developing an effective risk communication and education approach, especially among a religious community. The chapter approach is to present the guiding principle of disaster risk reduction (DRR); the views of Abrahamic religions and specially Islam on disaster and earthquake in particular God's bounty, earth, good deeds, human behavior, safety, and vitality to clear the existing misconception of assuming that disasters are due to God's anger, etc.; and finally the issues safety and human life are discussed in terms of religious teaching. Findings shows that the dissemination of this type of knowledge has helped to clear the misconceptions and increase people's understanding of and knowledge about disaster-related issues; as a necessary step in the process of DRR and improving safety and development, all of which can be viewed as demonstrations of God's love for humankind. It is the first time that this noble correlation between religion and risk reduction has been explained in a paper, and it is expected to open the road for research and discussion on this topic.

21.1 INTRODUCTION

In ancient times when science was in its infancy people believed that disasters, especially earthquakes, represented the power of Mother Nature. A knowledge of thunderstorms, floods, droughts, earthquakes, eclipses, and other natural events was too scarce to enable them to be interpreted correctly. Humans were vulnerable to natural extremes and were dominated by the need to survive amid the ferocity of nature, which was the main goal of their entire lives. Their limited knowledge suggested that powerful but arbitrary forces were responsible for their miseries. The Druids told them that disasters were the will of the

Hazards, Risks, and Disasters in Society. http://dx.doi.org/10.1016/B978-0-12-396451-9.00021-4

gods and they had no choice but to trust them, unless they assured them that such disasters would not happen again. The ancient Maya people, for example, had a god of thunderstorms called *Hurakan* (the root of the word "hurricane"), a god of earthquakes (called *Pillan*), a god of thunder, and a chief of all the gods. Aided by cohorts of evil spirits, Pillan caused earthquakes and volcanic eruptions. Greek, Roman, and Egyptian civilizations all had gods that killed people who disobeyed them. Also, they explained natural disasters in terms of the anger of the gods. In fact, history is full of evidence that priests, religious leaders, and rulers took advantage of people's fears and consequently oppressed them in the name of a god or gods. The lesson of history is that the accumulation of misconceptions on the cause of natural disasters among traditional peoples and societies needs to be clarified using the same method with which it was created.

Today, a great amount of knowledge exists about planet earth, the governing physical laws, and how and why natural events happen. Despite this, natural phenomena cannot be controlled by human beings, and most of the time environmental and social forces do not act as though they have been modeled, programed, or predicted by experts. Moreover, failure to utilize existing knowledge, research, guidelines, codes, and recommendations designed to promote safety stems from failure to transform these instruments into something that people can understand, believe in, and accept. As a result of inherited misconceptions and superstition, many people still believe that little can be done to stop disasters. In both ancient and modern times, people have tended to consider natural events in terms of myth, religion, and psychology. A persistent belief exists that science has failed to protect people adequately and as a result people search for alternative interpretations of natural events. Even today citizens, rulers, and religious leaders try to take advantage of earthquakes and other natural events in the name of religion. This unfortunate tendency was evident in the United States after Hurricane Katrina in 2005; in Indonesia, Sri Lanka, and India after the 2004 tsunami, and in Pakistan after the 2005 Muzaffarabad earthquake, and in Iran after the 1986 Manjil earthquakes. Moreover, when speaking of "God's wrath" and the arbitrariness of "why this is happening to us," we totally neglect the scientific evidence, which suggests it is criminal negligence that causes the terrible losses of lives and property and points out that human awareness can save lives.

Today, the constant threat of natural disasters and loss of human resources requires new initiatives in order to communicate and mitigate risk to traditional societies. A new approach is needed in order to change the mentality of traditional communities regarding the concepts that disasters are God's will or expressions of His wrath. It is necessary to clarify misconceptions such as those relating to fatalism and God's will about whether individuals will be saved or sacrificed to disaster, and that mosques and holy places are immune to damage, as was believed during the Izmit earthquake in Turkey in 1999 and the Indian Ocean tsunami of 2004. In seeking guidance from the Holy Qur'an we find more references to God's kindness than to His wrath. It should be also noted that no statement by the Prophet Mohammad (Peace Be Upon Him-PBUH) ever mentioned that earthquakes or other disasters are expressions of the wrath of God or the result of disobedience or infidelity. Instead, there are many statements that show the need to prepare for disasters and prevent them from happening.

The traditional fatalistic view of disaster (Act of God, God's punishment, etc.) has to change, and today, it is time to realize that by understanding the importance of science, research, technology, and common sense, we can save precious human lives and resources, but only if we stop attributing our negligence and bad deeds to God's wrath. We cannot survive if we refuse to use the knowledge and wisdom that God has given us.

It should be noted that the author of this paper is not an expert in theology, but with the consideration of multidimensional meanings of the holy books, have tried to show that religious teaching can be adapted to the principle of DRR and one can benefit from the religious teaching in order to create a stronger incentive and will for risk reduction. Also, based on the author's understanding of Islam, the introduced concepts can be related to other religions that believe in one God. The discussion can also be used in other aspects of our lives and is not limited to disasters, as there are important objectives of human safety, vitality, and development to be considered. In the future, this process will need to be strengthened significantly with the input of more expert knowledge.

21.2 KEY ELEMENTS OF DISASTER RISK REDUCTION

In order to appreciate the religious view of natural disasters and human safety, it is necessary to outline the key concepts of Disaster Risk Management (DRM), with special reference to earthquakes, as these are a very important source of hazard in some of the world's most populous countries. Disaster risk is defined as a complex combination of hazard, vulnerability of the built environment, and human socioeconomic and cultural impacts, as well as in terms of a country's level of governance, capacity, preparedness, and response to catastrophic events when they occur (Ghafory-Ashtiany, 2009).

$$\text{Disaster RISK} = \frac{\text{Hazard} \times \text{Physical Vulnerability} \times \text{Socio} - \text{Economic Value}}{\text{Governance, Capacity, Resilience}}.$$

Natural events, such as earthquake, are unavoidable in their nature and cannot be controlled by humans; and can be considered as God's bounty in terms of the earth stability. The secondary effects of earthquake including tsunamis and geotechnical hazards, such as landslides and liquefaction phenomena, and rain including floods should also be accepted as ineluctable facts of earth and nature. The risk potential that people are exposed to is determined by vulnerability in terms of the probability of failure of the built environment due to poor or inadequate urban development, planning, design, construction, and structural weaknesses. Disaster risk should also be understood in the context of socioeconomic systems, which determine people's health, income and access to information at the time of disaster. Finally, the effectiveness of emergency response is an important variable. This last category has increased in importance both as a result of the greater value accorded to human life today in relation to the past and the higher levels of investment in each individual in terms of training and education that have resulted from improved economic development.

Based on the above explanation, the "guiding principles" of Disaster Risk Reduction (DRR) can be outlined as (Ghafory-Ashtiany, 2009):

1. Belief in the facts of nature and existence of natural hazard of the earth, as well as in the fact that we can only survive natural events if we are prepared and structures are built to be safe and compatible with natural forces and hazards.
2. Comprehensiveness: effective and successful DRR requires a comprehensive or all-embracing approach to disaster that extends from their sources of occurrence to their effects and impacts.
3. Expert leadership and good governance are essential to the success of DRR programs. Risk cannot be reduced without knowledge and scientific management. Thus, the role of people is to understand, believe, and implement knowledge and follow the guidance and recommendations of experts.
4. Risk reduction should have the objective of sustainable development, but this can only be achieved effectively through collaborative efforts devoted to promoting good management.
5. Risk reduction requires long-term actions based on consistent policies and backed by tolerance of the slow pace of implementation. One must also ensure full accountability for actions taken. In other words, managing earthquake risk requires collaboration, patience, and transparency of process.

The following section considers how religious teaching can be used to strengthen the implementation of these principles and how, by increasing awareness of disaster risks, a collective prevention and safety culture can be promoted, and the public can be alerted to their role in the reduction of vulnerability.

21.3 KEY CONCEPTS IN ABRAHAMIC BELIEF

All the Abrahamic religions (Islam, Christianity, and Judaism) believe in one God and the prophets as the god messenger to the human beings. The Quran states that: "We believe in Allah and that which is revealed to us and that which was revealed to Abraham, and Ishmael, and Isaac, and Jacob, and the tribes, and that which Moses and Jesus received, and that which the prophets received from their Lord. We make no distinction between any of them, and to Him we must surrender" (Qur'an-Baqare; 2:136). This verse of Quran shows that the basic concepts of these religions are same.

Considering that the author has a better knowledge of Islam than other religions, the Quran has not changed through time, and in the view of the author Islam has a more sound approach to disaster in comparison with other religions, the chapter has more emphasis on Islamic ideas on safety and vitality than on other Abrahamic religions.

21.3.1 Islam

Islam (Shirazi), which is based on a perfect conception of the universe, is a comprehensive and realistic school whose principal objective is to address or answer all questions regarding human needs, whether they are spiritual, intellectual, sentimental, individualistic, or social and whether they pertain to this world or the next. Some of the key concepts in Islamic teaching that relate to the principles of risk reduction listed above and that can be used to benefit DRR can be outlined as follows:

1. "Islam" (Mottahari) means submission to the will and guidance of Allah (God), as the creator of the Universe, which can be traced according to the signs of knowledge and power through the different aspects of anything that exists in the Universe. In today's meaning, Islam relates to the guidance that has been revealed through the teachings of the prophet Mohammad (PBUH) about 1430 years ago, which in principle are valid and can be used all the time, even on the issues to safety and disasters.
2. In Islam, submission to God should be based on cognition, wisdom, belief, faith, and revelation, which should result in "good deeds," vitality, and happiness. Human deeds and behavior should be based on belief, and belief should be complemented with good deeds. Belief in God and good deeds is not sufficient, for "Whoever submits himself wholly to Allah, he is the doer of good..." (Qur'an-Luqman; 31:22). Thus, believing in and accepting that earthquakes occur should be accompanied by appropriate good deeds.
3. Holy Qur'an (al-Islam organization): Allah's Book of Guidance is comprehensive, multifunctional, and capable of multidimensional interpretations such that it is valid for all time and can be applied to all matters relating to the life of human beings: "We have revealed to you the book which clarifies every matter" (Qur'an-Nahl; 16:89). In other words, the simple and literal text of the verses of the Qur'an, which are written at the level of common human cognizance, can unfold multiple meanings and can exemplify features that can be read by human beings with

different levels of knowledge and used for the perfection of human life and behavior, both spiritually and physically.

4. Leadership of wise, elite, intellect, and experts (Olel-amr and Olel-albab): "Are those who know equal to those who do not?" (Qur'an-Zumar; 30:9). "Oh you who believe! Obey Allah, and obey the Messenger, and those charged with authority among you" (Olel-amr) (Qur'an-Nisa; 4:59). Thus, humans should live with knowledge and cognition. One of the objectives of Islam is to establish the rule of the wise and convert the domination of ignorance into wisdom and knowledge.
5. High value is attributed to group work, social activities, social responsibilities, cooperation, and consultation in various aspects of life, including giving full consideration to others. "And their business is [conducted] through consultation among themselves." (Qur'an-Shora; 42:38). God's help can be found within group work (Prophet Mohammad-PBUH). Strong emphasis is given to group prayers as a symbolic act of group activity.
6. Human beings are born innocent, pure with full of ability to grow and achieve success, well-being, and completeness (Saadat). Islam encourages humans to maintain their purity and improve their abilities.
7. God's will (Taghdir and Sonnat Elahi) is based on human deeds and behavior. Without struggle, good, hard work, and constant belief, one cannot expect to achieve perfection and earn Gods goodwill.
8. In all aspects of life, God has given human beings free will and the right to distinguish good deeds from bad. He has also given them options about what to do and how to avoid determinism and being forced to do things. Taking the path of safety is a human choice and requires appropriate actions.
9. Human beings do not have any right to harm themselves or others. Causing harm or vulnerability to others is an unforgivable sin, unless the person who has been harmed forgives the action.
10. The highest importance is given to respecting and observing "Human Rights."
11. Regarding human activities, people must respect the rules of law and conduct, whether they pertain to this world or the next, and whether they are individualistic or collective.
12. Through the concept of "Amr-bil-Maruf," people are supervised and invited and encouraged to do good deeds. Conversely, "Nahye-anil-Munkar"—forbids them to do wrong: "And from among you there should be a party who invites to good and enjoins what is right and forbids the wrong, and these it is that shall be successful" (Qur'an-Al-e-Imran, 3:104).
13. Sin means violating God's guidance by ignoring facts and knowledge, being negligent or failing to use knowledge.

14. Qur'an and the Hadith (the Prophet Muhammad's sayings and doings) contain three dimensions: Tohid (unity of god), Iman (the belief), and Amal (the practice).

Thus, it can easily be seen that many of the key concepts in Islam support the guiding principles of DRR and DRM. In the next section, after looking at earthquake-related issues (as the most common disaster) in the Qur'an, I will attempt to clarify misconceptions about disaster and its relation to sin and God's wrath.

21.3.2 Christianity

The foundations of Christian theology are expressed in ecumenical creeds that are accepted by followers of the Christian faith. From the 10 Commandments that concern matters of fundamental importance in Christianity are the greatest obligation is to worship only God, the greatest injury to a person is to kill a human, the greatest injury to family bonds is adultery, the greatest injury to commerce and law is bearing false witness, the greatest intergenerational obligation is honor and respect to parents, the greatest obligation to community is being truthful and honest, and the greatest injury to moveable property is to steal (Khanlari). The 10 Commandments were written with room for varying interpretation because they are fundamental. They are not as explicit or detailed as rules and regulations or many other biblical laws and commandments, because they provide guiding principles that apply universally, across changing circumstances. They do not specify punishments for their violation. Their precise import must be worked out in each separate situation. Thus, we can draw almost same principle and guidance that we have taken from Islam.

Some Doctrines about Earthquakes from the Holy Bible are

- "Moreover the law entered, that the offence might abound. But where sin abounded, grace did much more abound" (Romans, Chapters 5, 20).
- "For nation shall rise against nation, and kingdom against kingdom. And there shall be famines, and pestilences, and earthquakes, in diverse places" (St. Matthew, Chapters 24, 7).
- "And great earthquakes shall be in diverse places" (St. Luke, Chapters 21, 11).
- "When these things begin to happen, we should start looking up, because our redemption draws near" (Luke 21:25–28).
- "And behold, there was a great earthquake, for an angel of the Lord descended from heaven and came and rolled back the stone and sat on it" (St. Matthew, Chapters 28, 2)."And behold, the curtain of the temple was torn in two, from top to bottom. And the earth shook, and the rocks were split" (St. Matthew, Chapters 27, 51).
- "And I beheld when he had opened the sixth seal, and, lo, there was a great earthquake; and the sun became black as sackcloth of hair, and the moon became as blood" (Revelation, Chapters 6, 12).

- "And the same hour was there a great earthquake, and the tenth part of the city fell, and in the earthquake were slain of men seven thousand: and the remnant were affrighted, and gave glory to the God of heaven" (Revelation, Chapters 11, 13).
- "And every island fled away, and no mountains were to be found" (Revelation, Chapters 16, 20).
- "And the temple of God was opened in heaven, and there was seen in his temple the ark of his testament: and there were lightning, and voices, and thundering, and an earthquake, and great hail" (Revelation, Chapters 11, 19).
- "And there were voices, and thunders, and lightning's; and there was a great earthquake, such as was not since men were upon the earth, so mighty an earthquake, and so great" (Revelation, Chapters 16, 18).
- "And suddenly there was a great earthquake, so that the foundations of the prison were shaken: and immediately all the doors were opened, and every one's bands were loosed" (Acts of the Apostles, Chapters 16, 26).
- "You have made the land to quake; you have torn it open; repair its breaches, for it totters" (Psalm, Chapters 60, 2).
- "Therefore we will not fear though the earth gives way, though the mountains be moved into the heart of the sea" (Psalm, Chapters 46, 2).
- "You have loved righteousness and hated wickedness. Therefore God, your God, has anointed you with the oil of gladness beyond your companions" (Psalm, Chapters 45, 7).
- "Then the earth reeled and rocked; the foundations also of the mountains trembled and quaked, because he was angry" (Psalm, Chapters 18, 7).
- "The earth is utterly broken, the earth is split apart, and the earth is violently shaken" (Isaiah, Chapters 24, 19).
- "Therefore I will make the heavens tremble, and the earth will be shaken out of its place, at the wrath of the Lord of hosts in the day of his fierce anger." (Isaiah, Chapters 13, 13).
- "You will be visited by the Lord of hosts with thunder and with earthquake and great noise, with whirlwind and tempest, and the flame of a devouring fire" (Isaiah, Chapters 29, 6).
- "He who Removes Mountains, and they know it not, when he overturns them in his anger, who shakes the earth out of its place, and its pillars tremble" (Job, Chapter 9, 5–6).
- "And there was a panic in the camp, in the field, and among all the people. The garrison and even the raiders trembled, the earth quaked, and it became a very great panic" (1 Samuel, Chapters 14, 15).

From the above, one can draw different points such as: Occurrence of natural disasters is a sign of judgment day; God works through "natural laws" and brings good out of what appears to be evil; We have to accept a natural event as a fact of life and we have to learn to live with it; One should not lose hope after a disaster and learn how to move forward; Disaster as a divine punishment of god for our sin; etc.

The statement by many clergy as stated in Gary Stern's book on "Can God Intervene?" indicates that many Christian clergy somehow see the natural disaster due to sin and human bad deeds; and godlessness and sinfulness were responsible for natural disasters. Also, the church believes in original sin and the fall of humankind. No way exists for a Christian to look at natural disasters without taking into account the sinful, fallen nature of the world. As far as I see, this view is common among all religious people with different beliefs.

A group of Christian scholars also believe in our free human actions through which we do evil deeds and have brought on the brunt of natural disasters; or how God's enduring love can cause bad for the human. Some others believe that when we think that God is in control of everything that goes on in the world and every single thing that happens is planned by God, then you have to hold God responsible for a disaster. I believe that this type of thought discredits the God that Abrahamic religion believes in.

Stern in his book states that "Vinson Synan, professor of divinity at Regent university, focuses on Revelation 11:15, which says 'And the seventh angel sounded; and there were great voices in heaven, saying, the kingdoms of this world (have) become the kingdoms of our Lord, and of his Christ, and he shall reign for ever and ever" states that "When the whole created order is redeemed, there will be no more natural disasters," and "Until then, there will be random disasters that God has permitted. I don't know that God likes it to happen, but it says in Scripture that God was in the whirlwind." He also quotes Isaiah 11:9, which says "in all my holy mountain, nothing shall hurt nor shall destroy." This is a promise that after the final redemption of all things, the curse on the land will be lifted and the physical world will no longer cause pain. "In the meantime, from the fall of Adam until the final redemption of all things, there will be natural disasters," Synan explained casually. "Man is subject to this. God allows it. Maybe the Lord allows Satan, who is the destroyer, to do this. In the end, when the natural disaster hits, people are mute. It is an act of God and shows us there are powers far beyond anything man can do." (Stern, 2007)

21.3.3 Judaism

Jews believe that Torah was revealed to Moses at Sinai in oral form, and handed down from teacher to pupil as a book of guidance that also provides the 10 Commandments that are the sacred act of central importance and the 603 commandments relating to various duties to be performed by a Jewish person. Jews also read their holy book through generations of interpretations and acknowledge the evolution of human understanding of God. The Talmudic image of God is vastly different from the image of God presented in Islam or Christianity. The God described in Talmud is not responsible for what we call "Acts of God." Thus, God does not interfere with the natural course of the world. Earthquakes happen because of laws of nature. People suffer as a result, but not because God has willed this specific tragedy to occur. We cannot depend on miracles. But we are responsible for the rickety ladders in our lives. When building a nuclear plant so close to a fault line, we will be responsible for its possible disasters as well as the extent to which global warming is created by human beings, the dangers posed by depending on energy sources that are dangerous, and the connection between our consumption and the planet's inability to sustain all of us. We cannot depend on miracles, only on our resolve to take responsibility for what we can change to make the world safer.

21.4 THE QUR'ANIC VIEW OF EARTHQUAKES

Here, only the view of Quran is presented with consideration of the author's knowledge and understanding of Islam. The Holy Qur'an is the principal source of religious thought in Islam; with its multidimensional meanings, it is also the principal source of guidance on all aspects of life and can comprehensively be adapted to all times and matters. In order to better understand this view and avoid misunderstandings and misconceptions in Islamic society, this section of the chapter looks in depth at the verses of the Qur'an that relate to earthquakes and the principles described above and relates them to actions needed for DRR.

The term "Earthquake = Zelzal" is mentioned directly in Chapter 99 of the Qur'an and indirectly in various verses, such as 7:78, 7:91, 7:155, 7:171, 16:26, 17:37, 17:68, 29:37, 34:9, 67:16, and 69:5.

21.4.1 Direct Reference to Earthquakes

The verses with the direct reference to earth and earthquakes are as follows (Holy Qur'an):

1. "And in the Earth are neighboring pieces" (Qur'an-Raad; 13:4). This verse shows that the concepts of earth's plate tectonics and the evolution of earth that has been identified by earth scientists in recent centuries were mentioned in the Qur'an more than 1400 years ago.
2. "So the earthquake took them unawares, and they lay prostrate in their homes in [before] the morning! So they became motionless bodies in their abode" (Qur'an-A'araf; 7:78 and 91). This verse can be related to the concept of the sudden occurrence of disasters and in particular earthquakes.
3. Qur'an-Earthquake; Chapter 99:

In the name of Allah, the Beneficent, the Merciful. When Earth is shaken with her (final) earthquake...So, anyone who has done an atom's weight of good, see it! And anyone who has done an atom's weight of badness, shall see it.

Most Islamic scholars relate this chapter to the Day of Judgment, or the earthquake that will mark the end of the world. However, based on the simple and straightforward meaning of the first verse of this chapter, it can be considered as the description of an earthquake event. With this view, Verses 8 and 9 can refer to the time when humans see the results of their deeds on what they have done. In terms of construction, those who have constructed well shall see its safe results and those who have violated the principles of safe building will in sorrow see collapse and other unhappy results. It is signifying that even a very small good deed in respect of earthquake safety and safe construction and development would lead to positive results. On the other hand, even a very small violation or attempt to ignore safety will produce manifest results. The construction of even one safe building will lead to a lower dependency of the community on the emergency response system. Similarly, any seismically unsafe construction or wrong acts will increase dependency on the emergency system and consequently require greater post-earthquake recovery efforts.

21.4.2 Belief and Good Deeds in the Qur'an

A key concept in risk reduction is safe and compatible design of structures with respect to the expected level of hazard, along with construction of good quality. This concept can be achieved by believing and accepting the existence of earthquakes and by respecting and implemented codes and doing good deeds, as an action required for safe design and construction. This issue has been addressed in the Qur'an through the two complementary concepts of "belief" and "doing good" as one of the key principles of Islam and, in fact, of virtually all religions; and as key guiding principle of DRR. Some relevant chapters and verses in the Qur'an are as follows:

1. In Chapter 101, "Asr" (The Age or The Time), multilayered meanings can be applied to the reduction of losses from natural events such as earthquakes. The verses of this chapter read as follows:

In the name of Allah, the Beneficent, the Merciful. I swear by the time. Most surely humans are in a state of loss. Except those who "believe" or have "faith" and do "good deeds," and join together in the mutual teaching of "truth" or "justice" and of "patience and constancy."

The key teachings in this chapter are

- In terms of faith, it is essential to recognize and believe in God and his wisdom and guidance in all aspects of life and for all things that exist, including the facts of nature and the existence of earthquakes as natural phenomena, and to trust in God's mercy as a fundamental motivation for carrying out acts for safety and vitality.

- Good Deeds or acts that are based on knowledge, wisdom, and recognition are encouraged and are prerequisites for success and for avoiding losses.
- With its positive meaning in the Qur'an, patience causes motivation and incentive for doing acts that prevent losses. Patience and hope are key factors in risk mitigation and loss reduction.

2. And those who believe and do good deeds, shall be rewarded by entering into heaven... (Qur'an-Nisa; 4:122).
3. Surely those who believe and do good deeds, will be rewarded by heaven (gardens of bliss) (Qur'an-Luqman; 31:8).
4. Surely the Qur'an guides one to that which is most upright and gives good news to the believers who do good so that they shall have a great reward (Qur'an-Israa; 17:9).
5. Surely those who believe and do good deeds, we do not waste the reward of him who does good work (Qur'an-Kahf; 18:30).
6. And spend of your substance in the cause of Allah, and make not your own hands contribute to destruction; but do good; for Allah loves those who do good deeds and are beneficent (Qur'an-Baqare; 2:195).

It can be seen that the Qur'an considers "belief" to be a requirement for doing a good deed if people desire to reach success and be rewarded by being safe during natural events. Thus, their actions should consist of good deeds and high quality work. It can therefore be concluded that

- Believing that our creator's guidance is toward the best human performance and better living. Humans are encouraged with their free will, awareness, and knowledge to follow that guidance, which is comprehensive and covers all aspects of life.
- Believing in wisdom, facts, and expertise; as well as accepting, respecting, and following the spiritual, individual, social, and technical laws; rules; and regulations.
- Doing our best acts when circumstances demand it: acts that are based on appropriate belief and knowledge.
- Ensuring the correct use of God's bounty and nature through applying knowledge and wisdom.
- Good deeds in relation to earthquakes means following expert leadership, ensuring that planning and development are compatible with hazards, constructing according to building codes and regulations; and constructing buildings in a seismically safe manner.

If human beings as individuals or societies follow these principles in all aspects of their daily life, natural events will not turn into disasters and safety will become an integral part of the development process of each nation.

21.4.3 Public Supervision (Amr-bil-Maruf)

Good management and governance are necessary for the effective implementation of risk reduction and the implementation of "zero tolerance" in respect of building code violations. This requires comprehensive supervision, both by the authorities and especially by the people, who can play an important role in developing countries without imposing a high cost. To encourage public and community participation in the implementation of supervision, the concept of "Amr-bil-Maruf" and "Nahye-anil-Munkar" can be used. These mean "inviting people and encouraging them to do good deeds and forbidding harm." They are key concepts in Islam and can be used toward the process of effective implementation of risk reduction programs and to ensure that codes and regulations are not violated. As an example, verses 104 and 110 of Chapter "Al-e-Imran" of the Qur'an, which address this concept are as follows:

And from among you there should be a party which invites one to do to good and enjoins what is right and forbids the wrong, and this it is that shall be successful (Al-e-Imran; 3:104).

Ye are the best of peoples, evolved for mankind, enjoining what is right, forbidding what is wrong, and believing in Allah. If only the People of the Book had faith, it would be best for them: among them are some who have faith, but most of them are perverted transgressors (Al-e-Imran; 3:110).

Considering that the violation of building codes and standards, and improper use of land are common issues in developing countries are key obstacles to the success of earthquake risk reduction programs, the implementation of this key concept in Islamic societies can enhance community participation and supervision, which would be very useful as a means of stopping violations, especially by illegal developers.

21.4.4 Sin

Another key concept that has been addressed in all religions and beliefs is the concept of sin. In the past, in many traditional societies, disasters were viewed as the result of people's sins. This view can be further clarified by a detailed look at the meaning and description of sin in Islam, and by attempting to see how sin can result from disaster. In Islam, sin is a violation of God's guidance due to ignorance, lack of belief in the facts, and performing acts against self-consciousness. In more detail, the concept of Sin that can apply to DRR can be outlined as follows:

- Misusing the most important of God's bounties, which are human ability, wisdom, and free will; or using human ability in an incorrect way toward an action that causes damage or disaster.
- Refusing the use of knowledge and wisdom that God has given us along with our free will and, instead, doing a bad deed or one that causes damage and loss to oneself or others.

- Misusing Nature, for example, by promoting development and construction, which cause damage to the environment.
- Causing damage, harm, or vulnerability to oneself. Harm can be related to human health, safety, and assets, as well as to the environment.
- Violation of human rights and social laws.
- Carrying out acts that cause mistrust and corruption among people and in society.

The Qur'an, in relation to human actions and responses, sin, and bad deeds emphasize that "good and evil, blessings and punishment, are from Allah, but no distress afflicts a man unless he earns it by his own mistakes or blunders." In other words, if we assume that disasters are caused by sin, the sins are from the doer of bad acts and God never does harm or wants bad things to happen to human beings. With this meaning, the process of sinning can result in disaster. Most disasters and human losses in earthquakes are due to failure to use available engineering principles and bad construction. These verses read as follows:

1. *So We took each one in his sin*; of them was he on whom We sent a strong storm, and of them was he who was overtaken by the Cry, and of them was he whom We caused the earth to swallow, and of them was he whom We drowned**.** *It was not for Allah to wrong them, but they wronged themselves* (Qur'an-Ankaboot; 29:40).
2. ... If some good befalls them, they say, "This is from Allah"; but if evil occurs, they say, "This is from yourself" ... (Qur'an-Nisa; 4:78).
3. O'man: Whatever benefit comes to you, it is from Allah, and whatever misfortune or distress befalls on you, it is from yourself, ...(Qur'an-Nisa; 4:79).

In a similar manner to sin, corruption or mischief in the development and building process will result in unsafe construction and will cause damage and disasters. In this case, in the Qur'an, God says: And do not seek to make mischief or corrupt in the land, surely Allah does not love the mischief-makers (Qur'an-Qesas; 28:77).

Thus, when humans as developers, engineers, builders, owners, rulers, or policy makers lack faith in the common laws of nature and belief in scientific facts, and when they misuse existing know-how on safety and engineering principles, or violate building regulations with free will, this can be considered as committing a sin, which in turn results in vulnerability, disaster, and destruction.

21.4.5 Related Statements from the Prophet (Hadith)

The statements and lectures (Hadith) of the Prophet Mohammad (PBUH) and his companion are a valuable source of information on Islamic teaching and are complementary to the Qur'an. Some of these statements can be related to the need to do good deeds and also to issues related to earthquakes, as follows:

- "God loves those who do a deed or job correctly, perfectly and completely" (perfectionism).
- "People should have a strong desire to do good deeds."
- "People's deeds or actions that have good intentions and produce good results are a means of worshipping God."
- "A person is not a believer when neighbors are not safe from his misconduct and when he thus behaves in a manner that will increase vulnerability." Our building or neighborhood developments should not be constructed in a way that creates problems for neighbors or cause damage to a neighbor's property.
- "Those who cheat and violate the law and people's rights are not true Muslims." This statement should be a moral and religious incentive toward the implementation of a "zero tolerance" policy for any type of violation of codes and regulations.
- "Ignorance is the root of all problems and faults." This shows the responsibility of Muslims to learn about earthquakes and to be prepared and safe; as well as the need to use existing know-how in all aspects of their lives.
- "Short-term actions do not last." Short-term solutions to DRR are neither implementable nor effective. Sustainable DRR requires long-term efforts that would be implemented with consistency and patience.

These statements are among the best guidance for Muslims toward good quality and safe construction and developments and as a means of stopping negligence in order to avoid seismic losses and earthquake disasters.

Based on the principle of Abrahamic religions and what I have read and heard from experts, I can say that most of the above-mentioned teaching of Islam can also be found in Christian and Jewish belief.

21.5 CORRELATION BETWEEN GOD'S GUIDANCE AND RISK-REDUCTION PRINCIPLES

Considering the multidimensional meanings of the holy books, I have tried to show that religious teaching can be adapted to the principle of DRR and that it can be used as a basis or incentive for enhancing risk reduction among public, especially among religious societies. Given that the most important part of God's bounty to human beings is His guidance, ability, wisdom, and knowledge, given in order for people to do good deeds, and based on knowledge and cognition, the effects of "good deeds" and "bad deeds" on construction, community, and environment can be summarized in general form in the diagram shown in Figure 21.1.

According to this view, the rewards for "doing good deeds," which are based on God's guidance and have to be accompanied by "belief," results in

FIGURE 21.1 The relationship between our acts and protection against disaster.

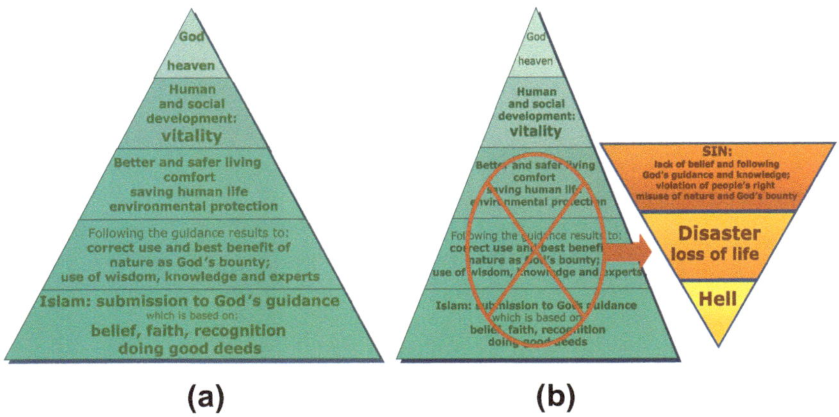

FIGURE 21.2 (a) The process (bottom to top) of achieving safety, human and social development, vitality, and finally "heaven" by following God's guidance versus (b) The process of causing disaster, loss of life, and "hell" by not following the guidance and by doing sin and "bad deeds."

sustainable development, safety, and vitality in human societies. Figure 21.2 shows how, by following God's guidance, people can realize themselves and in religious terms, as an ultimate goal, reach heaven. In this world "heaven" means a productive, safe, healthy, happy, and peaceful life. According to this, "belief" and "doing good deeds" can be interpreted in most as follows:

- *Belief*: believing that our Creator's guidance is for the best of human performance and better living. People are encouraged by being given free will, awareness, and knowledge to follow God's guidance, which is comprehensive and covers all aspects of life. Believing in wisdom, facts, and expertise; as well as accepting, respecting, and following spiritual, individual, social, and technical laws; rules and regulations. Thus, Abrahamic religions encourages people to believe in earthquakes as natural phenomena and facts of the Earth and, based on this belief, encourages them to do good deeds as follows:
- *Doing good deeds*: doing the best acts possible, based on the most correct beliefs and best knowledge. In relation to earthquakes and safety, this means planning and development that are compatible with hazards; obedience to building codes and regulations; following the leadership of experts; using seismically safe construction; and, finally, using knowledge and wisdom to make the most correct use of God's bounty and nature.

In the case that people perform bad deeds and do not follow the path shown in Figure 21.2(a), their punishment is destruction and loss of life, which will result in disaster, as is schematically shown in Figure 21.2(b). In other words, bad deeds that are carried out on the basis of ignorance or negligence and without using appropriate logic in theological language are called "Sin." Thus, losses and disasters mainly result from people's bad, incorrect, and inappropriate deeds. This is the simple explanation on the concepts and statements that disasters are resulted from sin. Hell, which is the result of sin, refers to a life with misery, destruction, lack of community development, and so on. Thus, disaster is not God's wrath or his anger with regard to humanity; it is simply the result of people's bad deeds and failure to follow God's guidance in all aspects of our lives.

21.6 NOAH'S ARK: A CLEAR EXAMPLE OF HOW TO BE SAFE IN DISASTER

The story of Prophet Noah (PBUH) and his experience with disaster is one of the best examples for teaching reported in Abrahamic religions. In this respect, the Qur'an says: So We revealed to him, saying: Construct the Ark within Our sight and under Our guidance; then when comes Our command, and the fountains of the earth gush forth...; and address Me not in favor of the wrongdoers; for they shall be drowned (Qur'an-Al-Mumenoon: 23,27). The lessons that should be learned from the story of "Noah's Ark" are (1) God instructed his prophet Noah to build the Ark in order to be safe in a disastrous typhoon and storm; (2) safety can only be achieved when one is prepared, even in the case of most dangerous disaster; and (3) God's protection of his prophets is also achieved through respecting the laws of nature, which are based on wisdom, knowledge, and human ability. There are many reported examples in Islam that offer similar teaching on dealing with disasters.

21.7 CONCLUSION

In this paper, I have attempted to show the view of Abrahamic religion and especially Islam on disaster and how it can create a religious incentive among people to take disaster risks seriously and, by understanding and using existing know-how, do good deeds and move more effectively toward earthquake risk reduction policies and actions. The discussion can be summarized as (1) Risk communication and knowledge dissemination should be compatible with people's beliefs and cultures. I believe that the use of religious teaching and knowledge of disaster and development can be an effective way of promoting safety and (2) Sustainable development and "Vitality" can be achieved through faith, knowledge, and the conduct of good deeds. If we reject or ignore wisdom and refuse to use the know-how that God has given us as his bounty, we will do sin and face "disaster."

REFERENCES

Allameh, A., Al-Mizan, T. Interpretation of holy Qur'an. www.almizan.org/tafseer.

Ghafory-Ashtiany, M., 2009. View of Islam on earthquakes, human vitality and disaster. Disaster Prevent. Manage. 18 (3), 218–232.

Holy Qur'an: English translation by Yusuf-Ali, A., Pickthal, M., Shakir, M.H. www.Qur'an.al-Shia.com.

Khanlari, K. Personal communication and discussion on Christianity Doctrine.

Makarem Shirazi, N. Our Belief: A Brief Description of Islam. www.makaremshirazi.org.

Mottahari, M. Islam: A Comprehensive School. www.al-shia.com.

Stern, G., 2007. Can God Intervene? Greenwood Publishing Group, Inc.

The value of Qur'an in the eyes of the Muslims. www.al-Islam.org.

Chapter 22

Conclusion: Hazards, Risks, and Disasters in Society

Andrew E. Collins [1], Janaka Jayawickrama [2,4], Samantha Jones [1] and Bernard Manyena [3,4]

[1] *Department of Geography/Disaster and Development Network (DDN), Northumbria University, Newcastle, UK,* [2] *Post-war Reconstruction and Development Unit, University of York, York, UK,* [3] *Humanitarian and Conflict Response Institute, Manchester University, Manchester, UK,* [4] *Disaster and Development Network (DDN), Northumbria University, Newcastle, UK*

ABSTRACT

This closing chapter draws together overall emerging points from the volume, having resonance with societal aspects of the wider series of which this is a part. It offers reflection and guidance for an expanding field of study built from the rationale outlined in the Introduction with summative comments influenced by the three chapter sections. It is concluded that dealing with hazards, risks, and disasters in society requires an understanding of cognition and behavior and would require a cultural shift for much of humankind. Although a definitive checklist of actions to be taken is beyond the realms of any one volume, previous learning and the insights that are on the whole empirically presented within this volume indicate processes that are required to progress for the better.

22.1 MORE ON THE APPROACH

In the introductory chapter, an outline of core approaches to linking environmental hazards and risks with society had suggested forward-looking analyses for this subject matter to be through an integrated and multihazard approach. This is evidenced by many of the chapters that followed. However, it was acknowledged that much of the learning for this field of study and practice also takes place around individual hazard categories and events. Referring in the Introduction to flood-risk management as one of many examples, it was suggested that the relationships and reactions to hazards were complex since they involved environmental change, people's exposure to risk, and the peculiarity of prevention and response. By way of extending that case example further, exceptional incidents of UK flooding of 2014 present further observations.

Hazards, Risks, and Disasters in Society. http://dx.doi.org/10.1016/B978-0-12-396451-9.00022-6

Faced with hitherto unexperienced challenges, elements of the emergency management sector at the Thames Barrier, which protects London from combinations of high tides and excessive runoff, called on Londoners to "know their flood drill." This operations-oriented call was understandable given the circumstances, though inevitably raised concerns for those with "no drill." A distinction between a personal drill and that which might be more widely dictated from the authorities may sum up one of the key challenges in disaster prevention and response of our times—how best to combine understanding, management, and governance of risk. Good risk assessment whether in relation to personal or wider institutional authority should lead to improved risk management, but ultimately this process becomes subject to questions of whose responsibility and varied reactions to risks there may be.

The questions posed by complex hazards and risks are partly so because early warning and early action (IFRC, 2009) on environmental events, including floods, are subject to no less than the evolving physical hazards, human behavior, and politics. The example here, nonetheless, is where minimum standards for flood-drill guidance could be revised over and again for effectiveness—dealing with flooding disasters as both consequence and cause of the development Londoners would want to see. In the meantime, the more practical and short-term advice had to be to prepare to move belongings to upper floors, disconnect household mains supplies, and evacuate away from lower-lying areas.

This example of a response to hazards, risks, and disasters can be reinforced by more in-depth analytical considerations that would facilitate a revision of behavior for all involved over the longer time frame, addressing the underlying influences on what people end up experiencing. A conclusion here is that opportunity exists in relation to hazard and risk to engage people, research and evaluation around multiple perceptions, variations in human reactions, and uncertainty. For the public, practitioners, and politicians, this requires addressing both the experiential and nonexperiential knowledge accumulated (Collins, 2014), weighing hazards, risks and disasters in the light of values, and assumptions concerning society. This remains one of the main conundrums exposed through the information provided by contributions to this volume.

A further conclusion from the more people-centered or health-centered approaches to disaster resilience and sustainable development is to recognize the need for "integrated health-risk reduction" through proactive, human–ecological perspectives. This is proactive, radical, and effective because it is everyday people who interact with, and adapt to hazards, manage risks, demand rights, develop resilience, and secure livelihood niches. Supportive research and governance contexts are needed to harness engaged and innovative aspects of society in more hopeful and aspirational desires for the future. The argument underpinning practice and theory in this instance is simply that to invest in sustainable health, safety, and well-being is to mitigate the impacts of environmental degradation and destructive global changes.

22.2 NEED FOR A NEW DISCOURSE

The volume has by necessity used some of the mainstream terminology of this field that is interpretable in regards to development, vulnerability, resilience, capacity, exposure, prevention, and response. These aspects of hazards, risks, and disasters have also been incorporated more institutionally into versions of Disaster Risk Reduction (DRR), Climate Change Adaptation (CCA), and sustainable development, formally reflected in initiatives of the United Nations, national development programs, development organizations, and the plans and accounts of various other organizations and institutions (see Chapter 1). However, considering why greater progress in building safer and more secure societies is not forthcoming suggests a need to extend to less easily categorized or more aspirational discourses. For example, an apparent fixation with the identification and application of resilience may be inappropriate to future generations aspiring for development and well-being. Alternatively, the fixation might be appropriate, but the discourse around resilience needs reshaping to clarify its varied meanings in relation to different operational contexts, so as to better support a postresilience framework. It is suggested that addressing hazards, risks, and disasters in society is about understanding and acting upon both the totality and specificity of critical processes and outcomes of societal change. This includes getting development out of disaster, recognizing and instilling the hope and engagement that comes from within society itself.

22.3 FURTHER SUMMATIVE OUTFLOW OF THIS VOLUME

By way of providing opportunity for more in-depth analysis, case-study material, new or hitherto less well-documented ideas, chapters were solicited across a broad swathe of the subject represented by this book. Systematic focus was maintained through grouping chapters into underlying issues of (1) prevention and response, (2) hazards of change, and (3) less frequently represented ways of reducing hazards, risks, and disasters into the future. Empirical detail, exploration, consolidation, and innovation should be recognizable within what has emerged.

Overall emerging points from the first section of chapters included the finding that much of the human side of disaster analysis and management has remained focused on hazard and risk quantification, attempts to mitigate impact, delivery of response, and development of frameworks, or conceptual development to assist more unified agreements on strategy. The recent drive for a more polycentric and people-centric view has comprised a substantive part of this, being about how individuals, groups, or formalized institutions engage with disaster management. However, this is yet to fully take hold as a driver of change and human security within most societies. What we learn from this volume, particularly six chapters referencing to the theme of prevention and response in the first section is that current practice (what we are doing) is variously supporting and critiquing the attempts being made to confront disaster. Humanity is engaged in varied versions of cost-effective avoidance, preventing and responding to disaster through contrasting forms of social-economic organization and administration that is contested, requiring of deeper analysis, more representation, and ongoing revision.

Second, and as highlighted in the chapters of the second section, it is evident that the field of hazards, risks, and disasters has necessarily expanded to include a much wider range of issues than had been anticipated by earlier natural hazards-focused initiatives. Although tectonic, fluvial, climatic, and other aspects of the physical environment remain important to this field, and support a well-established research agenda, the overall subject has adapted to include greater attention to human exposure to hazards and risks composed of multiple influences. What is increasingly better understood, but will be faced into the future, is the need to consider a spectrum of ways in which people experience loss and damage through varying exposures and interpretations of hazards and risks. Human systems have become all but dependent on cyber-space, presenting new hazards of operability and the means to prevent and respond to fundamental breakdowns of the services upon which society depends. Although, much of the developing world had been less exposed to this, the future is ever more technologically globalized.

Moreover, the notion of a hazard, risk, or disaster as being somehow randomly generated by nature has become dwarfed by disruptions to life and livelihood that are triggered by development activity and violent conflict. In mass and localized emergencies involving environmental hazards or conflicts, those needing to respond to either are often the same agencies and skilled personnel. Societies affected by the threat of environmental disaster often also experience conflict. The means to response will need to develop in ways that are capable of adjusting to the multiple circumstances of instability within which new and secondary hazards and risks are generated. Humanitarian assistance already takes place through philanthropy, the military, as part of trade agreements, or by the survivors themselves.

A further option for many facing the onset of human-induced hazard and risk is to resist change or to fight back with new ways of surviving. As such impoverished subsistence level communities displaced by large-scale agricultural concerns in Brazil are highly active in taking back the land, though as a consequence at added risk of losing their lives at the hands of lawless landowners. These cases of infringement of a right to survival are ironically rarely considered an aspect of hazard and risk reduction, though they are grounded in issues of environmental, social, and economic sustainability.

Resistance, whether or not involving migration, can also be considered another form of adaptation. The challenges of what we face for the future is overall one in which response to emergency, whether immediate or in the long-term is a political as well as a moral concern. The process of adaptation, or of building resilience, is where heterogeneous people experience hazards, risks, and disasters varyingly on the basis of age, gender, location, wealth, culture, perception, and other potential predictors. These ongoing aspects of prevention and response for the hazards of social, technological, and political-economic change remain accompanied in varying measure by intersecting environmental hazards that are often unpredictable.

Third, as highlighted in the chapters of the third section, the findings of the volume point to a field that will be challenged to provide innovative, adaptive, radical, and hitherto underdeveloped approaches. Evidence exists from the writings here that opportunities for this drive will typically draw from multiple disciplinary areas of research and practice, though much that is already known is not acted upon. In particular, legal frameworks can be set up to protect humanity against the excesses of inappropriate development that heighten the risk of calamity. Establishing people's rights to be protected through legislation will need balancing with the development of responsibility and capacity within society. In that much of the principles of this have been experienced or known throughout the millennia, further learning is through observation and awareness of what had already worked in terms of human self-governance that could enable self-help supported by social care as human entitlement. This requires the building of (or return to) a sustainability based on values of social participation, cooperation, trust, and collective support from caring authorities to name some characteristics. Conversely, it can be argued that the hazard, risk, and disaster scenarios experienced are where human cooperation has become replaced by overly competitive behavior.

22.4 IMPROVED DEALING WITH HAZARDS, RISKS, AND DISASTERS IN SOCIETY

Dealing with disaster becomes one of understanding cognition and behavior in the face of hazards, risks, and disasters. It has long been realized that knowledge alone is insufficient to prevent catastrophic decision making by sentient people. More cooperative forms of society in relation to hazards, risks, and disasters are clearly an imperative. However, current world systems and blind avenues of unsustainable development, selfishness, human pride, and insecurity threaten the scope of opportunity to act now to prevent deterioration later.

As in the past, many (perhaps a growing number) will continue to find resolution in the interpretations of religion, personal faith, and the knowledge that can deliver. All will seek to survive better. The tools for doing so include the values of caring and support one for the other already alluded to in this volume, technical skills and resources and the means to better communicate, learn, interpret, empathize, and come up with smart solutions for future sustainability.

The above synopsis of hazards, risks, and disasters in society solicits a vast cultural shift for most of humankind, advances in science and technology, and better political prioritization. It is not possible to provide a definitive checklist of all items to be attended to, rather at least a knowing of the right means. Some of the knowing has been illustrated by the contributors to this volume. This is part of a process that will continue to make progress subject to the following means being upheld:

- Working from the "bottom," that is, through the community level, particularly young people and those more exposed to hazards and risks
- Action research whereby learning is by doing, and in relation to resilience of sustainable values
- Crossing of disciplinary boundaries to induce the full balance of hazard and risk influences
- Linking practice with reinterpretations of theory—practitioners alongside analysts
- Speaking truth to power … driven by evidence
- Working with the "top," that is, the gatekeepers
- Heightening awareness
- Behavioral change
- Capacity building
- Communication
- Regulation
- Innovation
- Financing

Facilitation of a process for improved outcomes in living with hazards and risks to uphold disaster as "not inevitable" requires evidence-based practices and policies to be going on across noncompetitive and sentient public, private, and civil–societal sectors. This requires a new form of professionalization hitherto more possible in the realms of speculation. However, it would engage options for a future of being more resilient to hazards, risks, and disasters by understanding the options for longer-term sustainability that raises quality of life. Predicting how the new ways of thinking and living will be more fully expressed personally, institutionally, and operationally is however beyond the speculative limitations of this volume. The contents of this book are on the whole empirically derived from the actual and current world of hazards, risks, and disasters in society.

REFERENCES

Collins, A.E., 2014. Beyond experiential learning in disaster and development communication. In: Egner, H., Schorch, M., Voss, M. (Eds.), Learning and Calamities: Practices, Interpretations, Patterns. Routledge, London.

IFRC, 2009. International Federation of the Red Cross and Red Crescent Societies (IFRC). Focus on Early Warning, Early Action, Geneva. World Disaster Report.

Index

Note: Page numbers followed by "b", "f" and "t" indicate boxes, figures and tables respectively

D

I

J

K

L

U

V

W

FSC
www.fsc.org
MIX
Papier | Fördert gute Waldnutzung
FSC® C083411

Kolibri 360 GmbH
Ferdinand-Jühlke-Straße 7
99095 Erfurt, Deutschland
produktsicherheit@kolibri360.de